Neutron Scattering–1981
(Argonne National Laboratory)

AIP Conference Proceedings
Series Editor: Hugh C. Wolfe
Number 89

Neutron Scattering–1981
(Argonne National Laboratory)

Editor
John Faber, Jr.
Argonne National Laboratory

American Institute of Physics
New York   1982

L.C. Catalog Card No. 82-073094
ISBN 0-88318-188-6
DOE CONF- 810840

PREFACE

The Symposium on Neutron Scattering was held at Argonne
National Laboratory, Argonne, Illinois, U.S.A., August 12-14, 1981
under the auspices of the commission on Neutron Diffraction of the
Inter-national Union of Crystallography in association with the
XIIth Congress and General Assembly of the Union held at Ottawa,
Canada, August 1981. Sponsorship with financial assistance was
provided by Argonne National Laboratory, Argonne Universities
Association, International Union for Crystallography, the Materials
Sciences Division of the National Science Foundation, U.S.A.
National Committee for Crystallography, U.S. Department of Energy
and the University of Chicago.

There were 148 registered participants at the conference from
20 countries, 66 from outside the U.S.A., and 82 from the U.S.A. of
which 29 were from Argonne National Laboratory. The program
included 22 invited papers, and 61 contributed papers presented in a
poster session. The program consisted of invited plenary lectures
in five broad areas of neutron scattering: pulsed neutron research,
instrumentation and techniques, magnetism, disordered solids, and
small angle scattering and biology. In addition, two informal discussion sessions were organized to discuss Rietveld analysis of
powder diffraction data, and recent developments in techniques and
instrumentation. Seventy-six of the papers are contained in these
Proceedings.

From all accounts, the conference served well the purposes for
which it was organized, namely the interchange of ideas between
active researchers in the many fields of science in which neutron
scattering contributes. As an important probe of condensed matter,
the ubiquitous nature of neutrons is well-demonstrated by noting
that important contributions are presented here spanning biology,
chemistry, materials science and solid state physics. This
diversity was particularly evident from the European contributions. The higher level of support for neutrons there, and the
much wider university involvement compared to North America has led
to a number of new initiatives that were ably demonstrated in the
papers presented at the Conference.

The recognition that neutron scattering experiments are essentially flux-limited led to an emphasis in two major areas: pulsed
neutron sources and new innovative instrumentation at existing high-
flux reactors in the world. That pulsed sources offer great expectation for advanced ultra-high flux neutron sources is amply
demonstrated by noting the international level of activity in this
area. Pulsed neutron sources are operational in Japan, USSR, United
Kingdom and the United States. A high intensity pulsed source is
also being considered in Germany. The timing and location of the
conference were particularly important since Argonne's Intense

Pulsed Neutron Source, IPNS-1, had just become operational. Symposium participants were able to tour the facility and observe the first results obtained.

We are grateful to all whose contributed efforts helped to ensure the success of this conference, including the Program Committee chaired by David Cox and the Local Committee chaired by Melvin Mueller. The members of these committees are listed below. We appreciate the cooperation of Hugh C. Wolfe, Series Editor for the AIP Conference Proceedings and of all the authors who have responded promptly to our call for timely submission of their papers and to our editorial requests for modifications. We thank Miriam Holden for her helpful expertise in conference planning and Helen Mirenic for her excellent secretarial skills during conference organization and preparation of this volume.

John Faber

Program Committee

D. E. Cox, Chairman, Brookhaven, U.S.A.
A. F. Andresen, Kjeller, Norway
G. Dolling, Chalk River, Canada
S. Hoshino, Tokyo, Japan
G. H. Lander, Argonne, U.S.A.
M. S. Lehmann, Grenoble, France
W. Press, Jülich, Germany
B. T. M. Willis, Harwell, U.K.

Local Committee

M. H. Mueller, Chairman, Materials Science Division
M. L. Holden, Conference Planning
J. E. Epperson, Materials Science Division
J. D. Jorgensen, Solid State Division
T. I. Morrison, Solid State Division
D. L. Price, Pulsed Source
F. J. Rotella, Solid State Division
M. Schiffer, Biology and Medicine
A. J. Schultz, Chemistry Division
S. K. Sinha, Solid State Division
R. G. Teller, Chemistry Division
J. M. Williams, Chemistry Division

Publications Committee

J. Faber, Chairman, Materials Science Division

TABLE OF CONTENTS

PREFACE

Section 1. PULSED NEUTRON RESEARCH

Section 4   MAGNETISM

Section 5   SMALL ANGLE SCATTERING AND BIOLOGY

NEUTRON DIFFRACTION PERFORMANCE
ON PULSED AND STEADY SOURCES

C. G. Windsor

Materials Physics Division, B.418, A.E.R.E. Harwell, OX11 0RA, U.K.

ABSTRACT

A comparison is made between diffraction experiment counting
rates at specified resolution on a pulsed source and on a reactor
using crystal monochromators. We show that performance is nearly
two orders of magnitude higher than would be predicted from the
equivalent peak thermal flux, even at thermal energies.

INTRODUCTION

A recent report[1] attempted to assess pulsed source performance
through the yardstick of peak thermal flux stating that: "The
numbers are approximately equivalent to steady state source fluxes
assuming an optimised repetition rate." I can only repeat what I
said in 1977[2]: "Comparisons based on peak fluxes could be misleading,
by beyond two orders of magnitude!" This had long been clear to ex-
perimenters.[3] This paper seeks to put the record straight for the
special case of diffraction experiments. More general comparisons
have been treated at length elsewhere.[4]

DIFFRACTION ON PULSED SOURCES AND REACTORS

Figure 1 illustrates diffraction for powder, liquid or amorphous
samples on the two types of source. Pulsed source diffraction is
universally performed by analyzing the time-of-flight t, over a
flight path L of several metres using one or several counters at
fixed angles $2\Theta_s$. Reactor diffraction is nearly universally performed by monochromating the neutrons by Bragg refection at some
fixed angle $2\Theta_M$ and measuring the scattered neutrons as a function
of scattering angle by a single moveable counter, by a moveable
bank of counters, or by a position sensitive detector. One fundamental difference is in the nature and range of the scan in scattering vector $Q = \frac{4\pi}{\lambda} \sin\Theta_s$:

$$Q = \frac{2m\, L\, \sin\Theta_s}{ht} \quad \text{or} \quad Q = \frac{4}{d_M}\, \frac{\sin\Theta_s}{\sin\Theta_M} . \tag{1}$$

In the pulsed case the upper limit in Q is essentially unbounded
within the range of interest to diffraction, however, each counter
angle has its own lower limit $Q_{min} = 2mL\sin\Theta_s/h\tau$ when the flight
time equals the pulse source period τ and frame overlap occurs. In
the reactor case there is a maximum Q value $Q_{max} = 4d_M\sin\Theta_M$, but
essentially no minimum Q value.

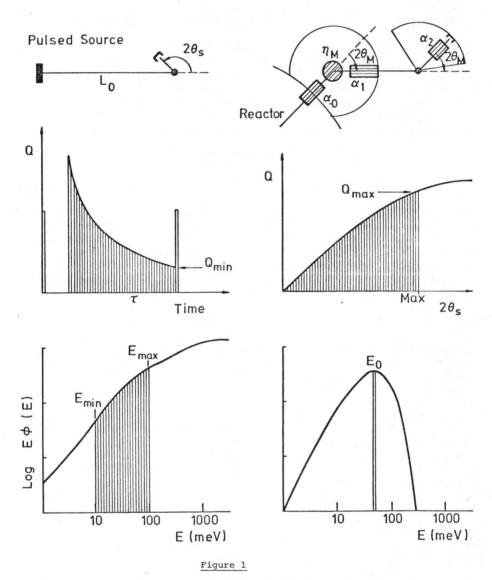

Figure 1

Powder diffraction on pulsed and reactor sources. The pulsed source
moderator shines directly on the sample with the Q scan being performed
on a polychromatic beam by time-of-flight. The reactor beam shines on
a crystal monochromator and must be collimated to produce the desired
wavelength resolution.

The resolution in Q value $R = \Delta Q/Q$ is given in either case by differentiating equations (1).

$$R = \frac{\delta_m(E)}{L} + \cot\Theta_s \, \Delta\Theta_s \qquad ; \qquad R = \cot\Theta_M \, \Delta\Theta_M + \cot\Theta_s \, \Delta\Theta_s \quad (2)$$

Here we have assumed in the pulsed case that all timing errors have been converted into an equivalent distance $\delta_m(E)$. If we now assume matched spectrometers and neglect focussing we may express the major parameters of the diffractometers in terms of the desired resolution R

$$L = \sqrt{2}\,\delta_m/R; \qquad \Delta\Theta_M = R/\sqrt{2}\cot\Theta_M;$$

$$\Delta\Theta_s = R/\sqrt{2}\cot\Theta_s; \qquad \Delta\Theta_s = R/\sqrt{2}\cot\Theta_s. \qquad (3)$$

Our performance comparison is now to evaluate the count-rate for pulsed and reactor diffractometers of the same resolution. We see immediately the crucial defect of comparing moderator peak fluxes. The pulsed moderator shines directly onto the sample and allows monochromatization by time-of-flight. The reactor moderator must shine first on to the crystal monochromator and it is the flux after monochromatization by Bragg reflection which should be compared with the pulsed source flux on the sample. In the pulsed case the sum of diffracted neutrons are shared during the total run time over all the various flight times corresponding to the different Q values. In the reactor case the sum of diffracted neutrons are shared during the total run time over the various counter angles corresponding to the scan in Q.

MODERATOR FLUX DEFINITIONS

The flux distribution, even the peak flux distribution, tells only half the story on pulsed neutron performance. Pulse duration is equally important since it defines the sample flux at a given resolution. The conventional measures of pulsed source performance are $\phi(E)$ the time averaged flux $cm^{-2}\ eV^{-1}$, and $\Delta t(E)$ the pulse duration. Both of these measures have the defect of changing rather rapidly with energy. We shall therefore use two related functions, the lethargy efficiency $\eta_m(E)$ and the effective moderator thickness δ_m. We define these as

$$E\phi(E) = n_F \, \eta_m(E) \qquad ; \qquad \delta_m(E) = \sqrt{\frac{2E}{m}}\,\Delta t(E) \qquad (4)$$

Here n_F is the time averaged fast neutron production rate, say 3.4×10^{15} fast neutron s^{-1} for IPNS I. $\eta_m(E)$ is a measure of moderator efficiency, and its value at 1 eV is often quoted where it is typically a few times 10^{-4} neutrons per steradian per incident

4

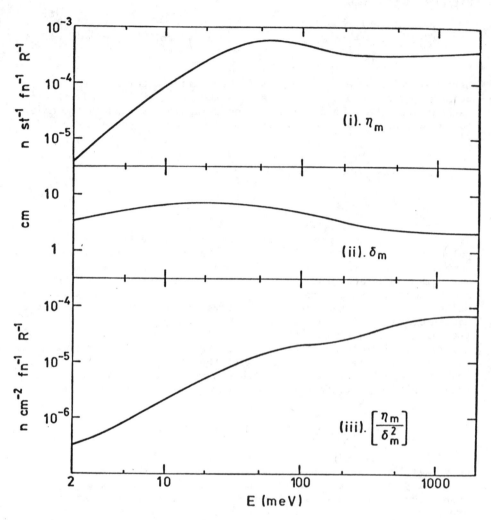

<u>Figure 2</u>

Pulsed source performance depends on two measurements, one showing the flux
distribution, one the pulse widths. These curves show the lethargy flux
η_m, per unit $\Delta E/E$, as an efficiency per incident fast neutron, and the
effective moderator width δ_m calculated for the IPNS moderator C^5.
Pulsed source performance is proportional to $\eta_m/\delta_m{}^2$.

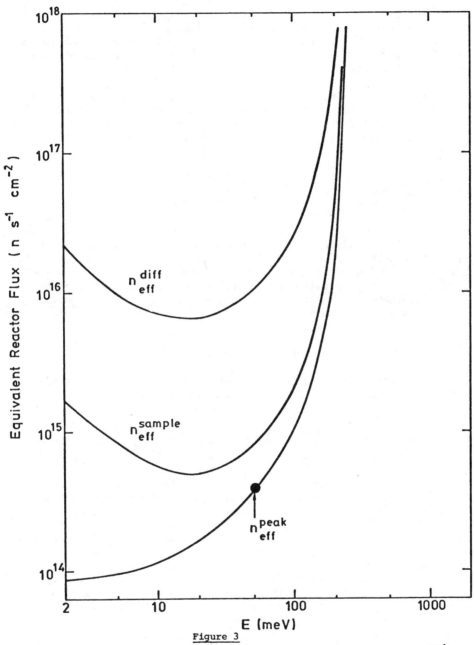

Figure 3

Effective total fluxes for the IPNS I source. The lower line shows n_{eff}^{peak} - the peak moderator flux; the middle line that reactor flux which gives the same useful flux at given resolution at the sample n_{eff}^{sample}; the upper line that reactor flux which gives the same count rate at the detector in diffraction n_{eff}^{diff}.

fast neutron. Figure 2(i) shows the values of $\eta_m(E)$ and Figure 2(ii) $\delta_m(E)$ calculated for the IPNS moderator C[5]. It is seen that neither η_m or δ_m have much dependence on energy except for a fall-off below thermal energies.

The effective reactor flux n_{eff}, defined in the IPNS report[5], is that reactor flux which equals the peak flux of the pulsed source at a nominal energy of 50 meV. The peak flux is the time averaged value, multiplied by the ratio of the period to the pulse width, so that

$$\frac{n_{eff} \, A_m}{4\pi} \left[\left(\frac{E}{kT} \right)^2 e^{-E/kT} \right] = n_F \, \eta_m \cdot \frac{\tau}{\Delta t(E)} \tag{5}$$

Here A_m is the moderator area, 100 cm^2 in the IPNS case, and τ the period of 16,666 μs (60 Hz). This formula gives IPNS I an equivalent flux of 4.10^{14} n cm^{-2}s^{-1} and IPNS II an equivalent flux of 10^{16} n cm^{-2}s^{-1}. Of course this flux depends on the reference energy of 50 meV. The lower curve in Figure 3 shows the peak flux of equation 5 evaluated as a function of energy. The rapid energy dependence shows just one danger in its use.

FLUX AT THE SAMPLE FOR GIVEN RESOLUTION

We first consider pulsed sources. The time averaged sample flux in neutrons cm^{-2}s^{-1} per fractional energy $\Delta E/E$ is given by

$$n_s(E) \cdot \frac{\Delta E}{E} = \frac{n_F \, \eta_m}{L_o^2} \cdot \frac{\Delta E}{E} \simeq \frac{n_F \, R^2}{2} \left[\frac{\eta_m(E)}{\delta_m^2(E)} \right] \cdot \frac{\Delta E}{E} \tag{6}$$

Here we have substituted for the incident flight path L_o, the value for the total flight path L given by equation (3a) assuming the usual case of $L_1 \ll L_o$. The "figure of merit" function in square brackets is plotted in Figure 2(iii), and represents a fair measure of performance for pulsed sources. We see its remarkable constancy from very high energy down to thermal energies, although it does drop off sharply for cold neutrons. To calculate a useful flux on the sample we must include all neutron energies giving rise to useful parts of the diffraction pattern. Let us suppose this is from energies E_{min} to E_{max} with corresponding Q values Q_{min} and Q_{max}. The total useful sample flux is then

$$n_s = \frac{n_F R^2}{2} \int_{E_{min}}^{E_{max}} \frac{\eta_m(E)}{\delta_m^2(E)} \cdot \frac{dE}{E} \simeq n_F R^2 \left[\frac{\bar{\eta}_m}{\bar{\delta}_m^2} \right] \cdot \log_e \left(\frac{Q_{max}}{Q_{min}} \right). \tag{7}$$

For this analysis we shall put $Q_{max}/Q_{min} = \sqrt{10}$ corresponding to a factor 10 change in energy, over which the figure of merit can reasonably be given a mean value. Figure 4(i) plots this neutron sample flux as a function of mean energy for a resolution $R = \Delta Q/Q = 0.01$ and the IPNS I source parameters.

We can calculate the sample flux from a crystal monochromator on a reactor of total flux n_r of say 10^{15} n cm^{-2}s^{-1}. In terms of the conventional Cooper – Nathans notation this is

$$n_s = n_r \left(\frac{E}{4\pi\,(kT)^2} e^{-E/kT} \right) \cdot 2E \cot\theta_M \frac{\alpha_o\,\alpha_1\,\eta_M}{(\alpha_o^2 + \alpha_1^2 + 4\eta_M^2)^{\frac{1}{2}}} \cdot$$

$$\frac{\beta_o\,\beta_1 \cdot P_M}{(\beta_o^2 + \beta_1^2 + 4\eta_M^2 \sin^2\theta_M)^{\frac{1}{2}}} \qquad (8)$$

P_M is a term including reflectivity losses, air absorption losses and flux depression at the source block. This analysis leaves it a unity, but it is more typically around 0.1. We see that the flux depends on the collimations before and after the monochromator α_o and α_1 and on the monochromator mosaic η_M. At the same time the resolution is given using equation (3b) as

$$R = \sqrt{2} \cot\theta_M \left(\frac{\alpha_o^2\eta_M^2 + \alpha_1^2\eta_M^2 + \alpha_o^2\alpha_1^2}{\alpha_o^2 + \alpha_1^2 + 4\eta_M^2} \right)^{\frac{1}{2}} \qquad (9)$$

Optimum performance maximizes the flux for given resolution. As Kalus and Dorner[6] showed, this is given for $\alpha_o = \alpha_1 \ll \eta_M$. A common condition which we shall assume is that the vertical resolution after the monochromator β_1 is large, so that $\beta_1 \gg \beta_o, \eta_M$. The flux then depends only on the resolution before the monochromator. β_o is typically determined by a source block of some 10 cm at a distance of say 5 m so that $\beta_o \sim 0.02$ radians. Assuming these optimized conditions and perfect reflectivity, the sample flux and resolution are

$$n_s = \frac{n_r}{4\pi} \cdot \left[\left(\frac{E}{kT} \right)^2 e^{-E/kT} \right] 2\cot\theta_M \cdot \frac{\alpha_1^2}{2} P_M \beta_o \; ; \quad R = \cot\theta_M \alpha . \qquad (10)$$

We may eliminate α_1 from the equations and so give the sample flux for given resolution as

$$n_s = \frac{n_r}{4\pi} \left[\left(\frac{E}{kT} \right)^2 e^{-E/kT} \right] \cdot \frac{R^2}{\cot\theta_M} \cdot \beta_o \cdot P_M \qquad (11)$$

8

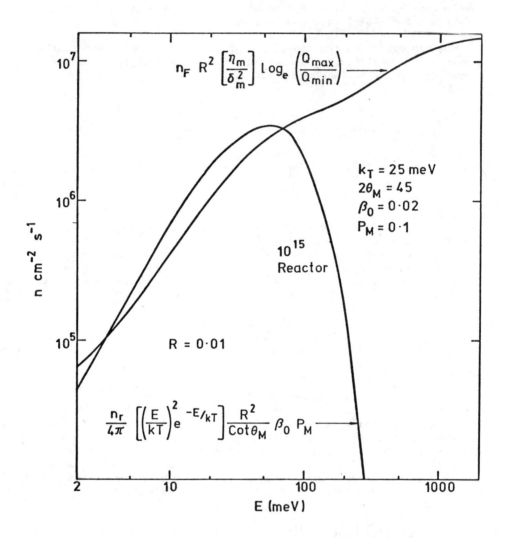

$$n_F \; R^2 \left[\frac{\eta_m}{\delta_m^2} \right] \log_e \left(\frac{Q_{max}}{Q_{min}} \right) \longrightarrow$$

$k_T = 25 \, meV$
$2\theta_M = 45$
$\beta_0 = 0 \cdot 02$
$P_M = 0 \cdot 1$

10^{15}
Reactor

$R = 0 \cdot 01$

$$\frac{n_r}{4\pi} \left[\left(\frac{E}{kT} \right)^2 e^{-E/kT} \right] \frac{R^2}{Cot\theta_M} \, \beta_0 \, P_M \longrightarrow$$

Figure 4

Calculated neutron flux at the sample in diffraction experiments to a specified resolution, R=0.01. The reactor curve reflects the lethargy flux, $\left[\left(\frac{E}{kT} \right)^2 e^{-E/kT} \right]$. The pulsed source curve reflects the figure of merit (η_m / δ_m^2).

This equation can break down when the calculated α_1 becomes greater than the natural geometric collimation allowed by the reactor source block. In practice monochromator angles are often fixed, and we shall assume $2\theta_M = 45°$. Figure 5(iii) plots out the sample flux for a 10^{15} reactor for this idealized crystal monochromator. We see that it reflects the reactor lethargy flux, and so is very different in energy dependence from the pulsed source reflecting η_m/δ_m^2. We also see that it can be orders of magnitude higher! We may express this fact by asking what reactor flux using equation 11 would give the same sample flux as equation 7. This is shown by the middle curve in figure 3.

COUNTING RATE IN DIFFRACTION AT GIVEN RESOLUTION

The detector counting rate n_d will be given for either source by a product of the macroscopic cross-section $d\sigma/d\Omega$, and the solid angle $\alpha_2\beta_2$. Assuming a perfect detector efficiency

$$n_d = n_s \cdot \frac{d\sigma}{d\Omega} \cdot \alpha_2\beta_2 \tag{12}$$

In our matched diffractometers $\Delta\theta_s$ will be given by equation (3c) or (3d). In pulsed experiments using near back scattering $\Delta\theta_s$ becomes large, and so the scattered collimation α_2 dominates over the incident collimation α_1, so that $\alpha_2 \simeq 2\Delta\theta_s = \sqrt{2} R/\cot\theta_s$. Thus for $2\theta_s = 150°$, and $R = 0.01$, $\alpha_2 \simeq 3°$. On the other hand for most reactor experiments and for low-angle pulsed experiments we need to match α_2 with α_1. Thus $\alpha_2 \simeq \sqrt{2}\Delta\theta_s \simeq R/\cot\theta_s$ and we have for $2\theta_s = 45°$ and $R = 0.01$, $\alpha_2 = 0.237°$. The detector count rates for the high-angle pulsed experiments and the low angle reactor experiments are therefore

$$n_d = n_F \frac{R^3}{\sqrt{2}} \left[\frac{\eta_m}{\delta_m^2}\right] \cdot \frac{d\sigma}{d\Omega} \cdot \frac{1}{\cot\theta_s} \cdot \log_e \frac{Q_{max}}{Q_{min}} \tag{13}$$

$$n_d = \frac{n_r}{4\pi} \cdot R^3 \left[\left(\frac{E}{kT}\right)^2 e^{-E/kT}\right] \cdot \frac{d\sigma}{d\Omega} \cdot \frac{\beta_o}{\cot\theta_s \cot\theta_M} \tag{14}$$

Thus high angle pulsed neutron diffraction gives an additional performance.

$$\sqrt{2}\cot\theta_s^{reactor}/\cot\theta_s^{pulsed} = 12.7 \text{ for } 2\theta_s^{reactor} = 45°$$

$$\text{and } 2\theta_s^{pulsed} = 150°.$$

The top curve in figure 3 shows the equivalent rector flux for this case. So much for peak thermal fluxes!

DISCUSSION AND CONCLUSION

A similar analysis leading to a performance comparison similar to figure 4 holds for several experiments where the wide incident energy range from a pulsed source can be used instead of a reactor monochromator scan. These include small angle scattering and vibrational spectroscopy using a beryllium filter or crystal analyzer. Significantly, the range of useful energies available with a pulsed source is less for single crystal diffraction or for coherent inelastic studies since useful data is concentrated into a few discrete peaks.

In conclusion, for many experiments done with crystal monochromators on reactors the IPNS I source is roughly equivalent to a 10^{15} reactor except above 100 meV where its performance becomes vastly superior. For diffraction experiments the possibility of back scattering allows a further order of magnitude in performance.

REFERENCES

1. W. F. Brinkman, J. McTague, J. J. Rush, R. Birgeneau, B. Batterman and F. Vook, Ames Laboratory Report 15-4761 (1980).
2. C. G. Windsor, "Neutron Inelastic Scattering" Vol. 1, p. 3, IAEA, Vienna (1978).
3. J. C. Dore and J. H. Clarke, Nucl. Inst. Meth. **136**, 79 (1976).
4. Argonne National Laboratory report ANL-78-88 (1978).
5. J. Kalus and B. Dorner, Acta, Cryst. A29, 526 (1973).

APPLICATION OF POSITION SENSITIVE DETECTORS TO STRUCTURE ANALYSIS USING PULSED NEUTRON SOURCES

N. Niimura
Laboratory of Nuclear Science, Tohoku University, Sendai 982, Japan

Y. Ishikawa, M. Arai and M. Furusaka
Physics Department, Tohoku University, Sendai 980, Japan

ABSTRACT

Techniques have been developed to use position sensitive detectors (PSD) at pulsed neutron sources. Two different types of TOF spectrometer equipped with PSD have been installed at two different pulsed neutron sources. The first is a single crystal diffractometer with a one-dimensional (1D) PSD and the second is a small-angle scattering machine with a two dimensional (2D) PSD. The former has been used to detect the diffuse scattering from a single crystal of Cu_3Mn. The latter has been used to observe the long-period helical structure of a single crystal of MnSi as well as for studies of purple membrane and collagen.

INTRODUCTION

Position sensitive detectors have been widely used for neutron scattering experiments with a monochromatic beam. The PSD allows one to record a large area of data at the same time. However, except in the special case of neutron small angle scattering, it has little advantage over a conventional detector for structure analyst of single crystals. On the other hand, when a PSD is used in a time of flight (TOF) single crystal diffractometer, a wide area of reciprocal lattice space can be observed simultaneously. When a one-dimensional PSD is used, a two-dimensional area of the reciprocal lattice space can be accessed. The portion of reciprocal space accessible by use of $\lambda_{min} < \lambda < \lambda_{max}$ and the angles $\phi_{min} < \phi < \phi_{max}$ is illustrated in Fig. 1(a).
The circles with radii k_{min} and k_{max} are drawn through the origin, and the direction of the radii represents the direction of the incident beam. All reciprocal lattice points within the shaded area will simultaneously diffract and be sampled by the PSD. This distinctive feature makes such an instrument promising for the measurement of intensity distributions in the reciprocal lattice space. Note that, in case of monochromatic beam, the number of the Bragg points on a Ewald sphere spanned by the PSD is quite limited, resulting in loosing the advantage of the PSD. In the present work, this has been applied to the measurement of the thermal diffuse scattering of a Cu_3Mn single crystal and to the collection of many Bragg reflections from a NaCl single crystal for the determination of the temperature parameters.
In case of the two dimensional (2D)-PSD coupled with a TOF

ISSN:0094-243X/82/890011-12$3.00 Copyright 1982 American Institute of Physics

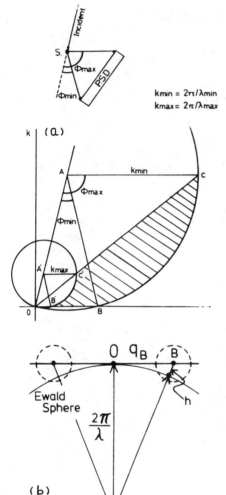

kmin = 2π/λmin
kmax = 2π/λmax

Fig. 1(a) The portion of reciprocal space accessible by use of a 1D-PSD.

(b) The deviation from the Bragg point to the Ewald sphere in the case of a 2D-PSD coupled with a small angle scattering experiment.

small angle scattering spectrometer (SANS), the situation is somewhat different from the high angle scattering as mentioned previously. This is because the deviation of the Ewald sphere from the reciprocal lattice plane perpendicular to the incident beam is quite small and the Bragg points in this plane always nearly satisfy the Bragg condition even if the monochromatic neutron beam is used. The situation is illustrated in Fig. 1(b) [or in an inset to Fig. 7]. The deviation distance h from the Bragg point to the Ewald sphere is given by $h = q_B^2 \lambda / 4\pi$ with q_B the wave vector of the Bragg point. If we assume that $q_B = 0.035$ A^{-1} (in case of MnSi) and $\lambda = 5$ A, h is calculated to be 4.88×10^{-4} A^{-1} which can be smaller than the intrinsic (mosaic) spread of the Bragg point. This is the reason why the satellite Bragg reflections have been observed for a single crystal of MnSi with the conventional SAND D11 (ILL). Note that $\Delta h/h$ by the instrumental resolution is usually very good, because $\Delta h/h$ is nearly equal to $\Delta\lambda/\lambda$ which is far less than 1.

If the small angle scattering is observed for a single crystal with TOF spectrometer equipped with 2D-PSD, however, Bragg reflections by different wavelengths neutrons are detected simultaneously at different positions in 2D-PSD, much detailed information can be obtained on the Bragg reflections by giving the wavelength dependence. A good example is provided by small angle scattering from a single crystal of MnSi which is described later. The results from the biological materials as purple membrane and deer collagen are also presented in this paper, for comparison.

The most troublesome problem of the TOF-PSD system is that (i)

it records a huge amount of data, all of which should be accepted
and displayed effectively and quickly, and (ii) the accumulated data
should be corrected for the incident neutron intensity spectrum $i(\lambda)$
to get a meaningful result.

1D-PSD

A single crystal diffractometer with a one dimensional PSD has
been constructed and installed at the pulsed neutron source of the
Tohoku University 300 MeV electron linac.[1] The PSD , obtained from
AERE Harwell, is a 4 atm ^{3}He counter 2.5 cm in diameter and about 50
cm in active length.   In case of a resistive wire type of PSD, the
positional information may be obtained in one of three ways.   One
is to measure the rise-time of the pulse from one end only, and the
second is to measure the timing of the zero crossings of the doubly
differentiated pulses from each end of the proportional counter tube.
The third method takes the ratio of the heights of the two pulses
which has a linear relation with position of incident excitaion.
In order to avoid confusion of the time measurement between the time
of flight and the electonics timing, we have adopted the third method, that is, the pulse height division method.

THE MEASUREMENT OF DIFFUSE SCATTERING
ARISING FROM SHORT RANGE ATOMIC ORDER

Measurement of the diffuse scattering intensity distribution in
reciprocal space is one of the good application of the 1D-PSD, because the scans should be made over the whole Brillouin zones.
The diffuse scattering due to the atomic short range order in annealed Cu_3Mn crystal has then been studied with the spectrometer.
A typical results   is
shown in Fig. 2 which
is an on-line display
of the data.
The signal-to-back-
ground ratio is quite
good and the diffuse
scattering intensity
distribution is clearly
separated from the
background.   In com-
parison with a single
counter TOF spectro-
meter, the measure-
ment time is dramati-
cally reduced and,
further, normaliza-
tion of different
measurements is gen-
erally unnecessary
since a 2-dimensional

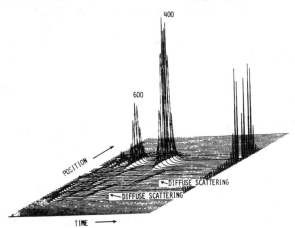

Fig. 2   Intensity distribution of the diffuse scattering of Cu_3Mn measured by use of
the 1D-PSD.

area is observed simlutaneously.   Note that the results in Fig. 2 were obtained in 6 hrs, using Tohoku University pulsed neutron source, the peak intensity of the thermal neutron flux of which is order of $\sim 10^{12}$ n/cm^2·sec.

The diffuse scattering distribution I_D at the position P and wavelength λ was calculated by

$$I_D(p,\lambda) = \frac{[\ I_{OBS}(p,\lambda) - I_B(p,\lambda)\ ]}{N \cdot i(\lambda) \cdot C(1-C) \cdot (b_{Cu} - b_{Mn})^2} A \cdot M \qquad (1)$$

where N is a scaling parameter, $i(\lambda)$ the incident neutron intensity spectrum as a function of wavelength λ, N the number of atoms in the specimen, C the content of manganese, and b the nuclear scattering amplitude.   A and M are absorption and multiple scattering corrections respectively. $N \cdot i(\lambda)$ is determined by measuring the incoherent scattering intensity from a known quantity of vanadium. Fig. 3 shows the diffuse scattering distribution, I_D, around (310) in q-space obtained by use of eq. (1). Diffuse maxima are visible around (3±1/2, 1, 0) and (3,1±1/2, 0). The most remarkable result which we found from Fig. 3 is that the diffuse scattering is not isotropic[2] but is elongated towards (310), which was overlooked in the previous works.[3,4] A recent X-ray study, however, has found a similar anisotropic diffuse scattering[5] in Ag$_3$Mn, suggesting a common feature of the short range order in both alloys.

By disregarding the anisotropy the Warren-Cowley SRO parameters α_{1mn} are calculated from

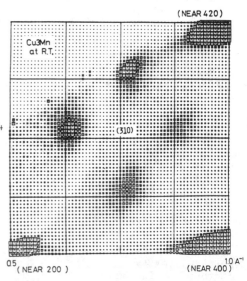

Fig. 3   The diffuse scattering intensity distribution of the Cu$_3$Mn crystal around (310) in reciprocal space.

$$\alpha_{1mn} = \underset{h_1 h_2 h_3}{\Sigma \Sigma \Sigma} I_D$$

$$\times \cos(h_1 l + h_2 m + h_3 n) \qquad (2)$$

The evaluated parameters α_{lmn} are plotted in Fig. 4. The observed values of $|\alpha_{lmn}|$ are consistent with the results of Wells and Smith. They are smaller than those of Hirabayashi et al[3].

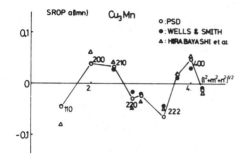

Fig. 4   The Warren-Cowley short range order parameters, α_{lmn}, of Cu_3Mn obtained by use of the 1D-PSD, compared with previously reported results.

THE TEMPERATURE PARAMETERS OF A SINGLE CRYSTAL OF NACL

The key question in the application of the PSD system to the structure analyses is whether the integrated intensity can be measured accurately. We have attempted to answer this question by determing temperature parameters from a NaCl single crystal, having monitored the integrated intensities of many Bragg reflections. On average in any one frame of measurement about 10 Bragg reflections, the number of observable reflections depending on the orientation of the crystal, are recorded. In total, about 100 Bragg reflections were collected. The maximum of $\sin^2\theta/\lambda^2$ was about 1.7 A^{-2}. A small computer program which automatically locates peaks and presents them intergrated areas, centroids, and calculates structure factors has been written.[6] Fig. 5 shows the observed integrated intensities correspond to measurements at different wavelengths. The differing intensities at the different wavelengths result from the extinction effect. As the extinction effect is large when the wavelength is long, the observed intensity decreases as $\sin^2\theta/\lambda^2$ decreases. The decrease in observed intensity at large

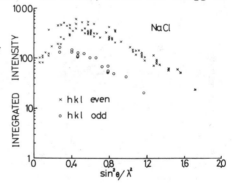

Fig. 5   The $\sin^2\theta/\lambda^2$ dependence of the observed integrated intensities of the Bragg reflections recorded from a single crystal of NaCl.

values of $\sin^2\theta/\lambda^2$ derives from the Debye-Waller factor. The magnitude of the extinction effect can be estimated from the wavelength dependence of the integrated intensity, and the temperature parameters of Na and Cl can therefore also be calculated.

The integrated intensity, I_{hkl}, for Bragg scattering in the case of TOF measurements is expressed by

16

$$I_{hkl} = i(\lambda)\,|F_{hkl}|^2\,d^2\lambda^2 A(\lambda) \cdot Y(\lambda, F) \tag{3}$$

where $i(\lambda)$ the intensity distribution of the incident neutron, F_{hkl} the structure factor, d the interplaner spacing, $A(\lambda)$ the absorption factor and $Y(\lambda, F)$ the extinction factor. For NaCl, F_{hkl} is written as

$$F_{hkl} = b_{Na}\exp(-B_{Na}\,\frac{\sin^2\theta}{\lambda^2}) + (-1)^{h+k+1}\,b_{Cl}\exp(-B_{Cl}\,\frac{\sin^2\theta}{\lambda^2})$$

the extinction factor $Y(\lambda, F)$ is calculated from the formula of Becker& Coppens[7] by adjusting the mosaic size and spread parameters by least-sequares techniques. The values for the temerature factors that were obtained are $B_{Na} = 1.31\ A^{-2}$ and $B_{Cl} = 1.41\ A^{-2}$ respectively.

2D-PSD

A 2D-PSD is installed in the small angle scattering spectrometer SAN at KENS[8]. The 2D-PSD is consists of an array of 43 1D-PSD's 1/2" in diameter and 24" in active length ^{3}He counter (6 atm)

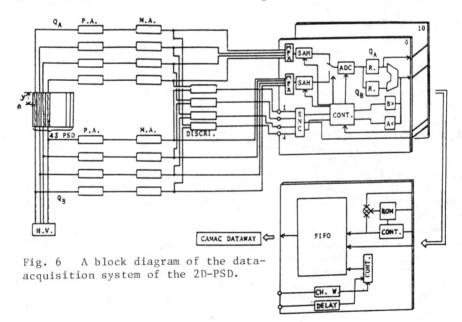

Fig. 6   A block diagram of the data-acquisition system of the 2D-PSD.

originally developed by Missouri University[13](RS-P4-0810-204).
Once the PSD with a bank of 90 pre-amplifiers is located inside a vacuum chamber with a typical operating pressure of 8×10^{-2} mmHg, a
voltage as high as 1900 V can safely be applied to the PSD, enabling
a distance resolution of about 10 mm to be achieved.

A block diagram of the data processing system for the 2D-PSD is
shown in Fig. 6. The position is determined by the following expression,

$$y \propto \frac{Q_B}{Q_A + Q_B} \tag{4}$$

where Q_A and Q_B are the charges detected at each end of a detector.
Q_A and Q_B are directed to the main amplifiers. The main amplifi-
er outputs are summed and fed to a single channel analyzer (Discrimi-
nator). When the sum of Q_A and Q_B is within the pre-set limits,
standard step singnals are generated, which are used as the gate
signals for the sample & holder (S&H), the analogue to digital con-
verter (ADC) and the time nalyzer. Since the conversion speed of
the ADC is fast ($\sim 5\mu sec$), a single ADC can handle the signals from
4 different detectors. The converted 8-bits data are then transferred to the registers and the summation of Q_A and Q_B is carried
out. The values of Q_B and $Q_A + Q_B$ are tarnsferred to the address
register of the P-ROM, in which positions calculated according to expression (4) are tabulated. The operation time for calculation of
expression (4) is equal to the access time of P-ROM, that is, 1 μsec.
A datum of the detector number is represented in 6-bits. The dis-
criminator outputs are used to determine the incident neutron time
of flight. Finally, the data of the position, the detector number,
and the time of flight are combined into a 18, 19 or 20 bits datum
point which is directed into the main computer memory through the
CAMAC dataway.

The quantity of data generated by the 2D-PSD using TOF scans is
extremely large (43 detectors × 64 position channels × 32 time channels = 88,064 discrete data points) and so direct correction and display of the results is very difficult with an on-line system. We
have therefore developed three simple on-line display systems[9].
The intensity in the ith position of the ith detector and in the
k^{th} time channel is given[10] by

$$I_{ijk} = \Phi_0(\lambda_k) \cdot \Delta\lambda_k \cdot \eta(\lambda_k) \; AS \; \frac{d\sigma}{d\Omega} (Q_{ijk}) \Delta\Omega_{ij} \tag{5}$$

where $\Phi_0(\lambda)$ is the time averaged flux at the sample per unit wavelength, A is the sample area, S is the sample thickness, $d\sigma(Q_{ijk})/d\sigma$

18

is the differential cross section, and

$$Q_{ijk} = 2\pi r_{ij}/L\lambda_k \quad \text{and} \quad \Delta\Omega_{ij} = \Delta x_i \cdot \Delta y_j / L^2 \quad ,$$

L is the distance from the sample to the detector, and r_{ij} is the distance from the i,jth element to the center of the detector, and $\Delta x_i \cdot \Delta y_j$ is the size of this i,jth element.

In order to obtain a quick display of $d\sigma(Q_{ijk})/d\Omega$, the expression (5) may be modified in the following three ways.

1) Integration over one parameter, either time or position.

$$d\sigma(Q_{ij})/d\Omega \rightarrow \sum_k I_{ijk} \quad \text{or} \quad d\sigma(Q_{ik})/d\Omega \rightarrow \sum_j I_{ijk} \tag{6}$$

2) Integration of the data from all directions and all wavelengths to provide a one-dimensional display of the data in Q space.

$$d\sigma(Q)/d\Omega \rightarrow (\sum_{ijk} \delta_{ijk} \cdot I_{ijk})/(\sum_{ijk} \delta_{ijk}) \tag{7}$$

where $\delta_{ijk} = 1$ if Q_{ijk} is within ΔQ around Q or $\delta_{ijk} = 0$ otherwise.

3) Integration over all wavelengths to provide a two dimensional display of the data in Q space obtained by

$$d\sigma(Q_{ij})/d\Omega \rightarrow (\sum_k \delta_{ijk} \cdot I_{ijk})/(\sum_k \delta_{ijk}) \tag{8}$$

Although modifications are used only for a quick on-line display, we have found that such displays are very useful to understand over all feature of scattering (see below).

TOF SMALL ANGLE SCATTERING EXPERIMENT BY USE OF A 2D-PSD

a) MnSi

MnSi is a typical itinerant helimagnet with the propagation wave vector $q_B = 0.035$ A^{-1} along the <111> direction. This material is particularly suitable to examine the feasibility of the TOF-SANS with 2D-PSD, for the single crystal scattering. Fig. 7(a) is an typical example of an on-line two dimensional display of the data. The satellites are elongated in the four <111> directions because the figure includes all the data obtained with the wavelengths between 4 and 9.5 A. Fig. 7(b) is, on the other hand, the two dimensional display of the data after the positions in the 2D-PSD are converted to the wave vector and the intensity is corrected for i(λ). Now four isotropic Bragg reflections are clearly detected at q = 0.035 A^{-1} along <111> from the center. Other spots would be due to multiple scattering. In Fig. 8(a), (b) and (c) are shown the wavelength dependences of the satellite wave vector, q_B, the satellite intensity, I(λ), and the spread of the satellite peak, $\Delta q/q$, obtained by the simultaneously measurement respectively. In Fig. 8(a) the q_B is found to be independent of wavelength, providing the experimental confirmation that the positions as well as the center in the

2D-PSD are accurately determined.   The value of q_B at 13 K is determined to be $q_B = 0.0345 \pm 0.005$ A^{-1} in good agreement with the previous measurement at ILL.[11]   In Fig. 8(b) the significant decrease

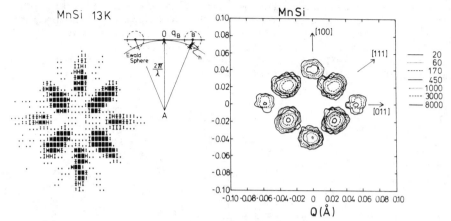

Fig 7(a)   A two dimensional on-line display of the scattered intensity distribution form MnSi.   (b)   A contour map of the scattered intensity distribution from MnSi obtained by use of equation (5).

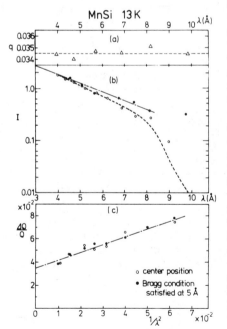

of the intensity with increasing the wavelength suggests that the deviation becomes the same magnitude as the intrinsic spread of the Bragg spot for $\lambda = 9A$. Actually the broken line in the figure was calculated with an assumption that the Bragg reflections have a Gaussian spread with a FWHM of 1.55×10^{-3} A^{-1}.   The same FWHM was also obtained by direct rotation of crystal.   Therefore the minimum size of the magnetic satellite is 4000 A.   Thus the TOF diffractometer gives us information on the Bragg spot size, without rotation of a crystal. By extrapolating I(q) to $\lambda = 0$, we can also get an extinction free intensity.   In Fig. 8(c) the spread of the Bragg reflections in the 2D-PSD $(\Delta q/q)^2$ is plotted against $1/\lambda^2$.   Since a simple consideration suggests   the $1/\lambda^2$

Fig. 8 (a)   The wavelength dependence of the satellite wave vector q_B.   (b) The wavelength dependence of the satellite intensity. (c) The $1/\lambda^2$ dependence of the $\Delta Q/Q$ of the satellite.

dependence is a result of instrumental resolution, the intersection
to the ordinate should give the intrinsic spread $(\Delta q/q)^2$ in the reciprocal plane, which is turned out to be much bigger than $\Delta q/q$ in
the direction perpendicular to the plane.   A careful consideration
suggests that such an anisotropical Δq is only apparent and seems to
reflect the special characteristic of our neutron guide system.
        These arguments   indicate that the measurement of Bragg re-
flections with 2D-PSD is quite effective, giving many information including that of the instrumental resolution.

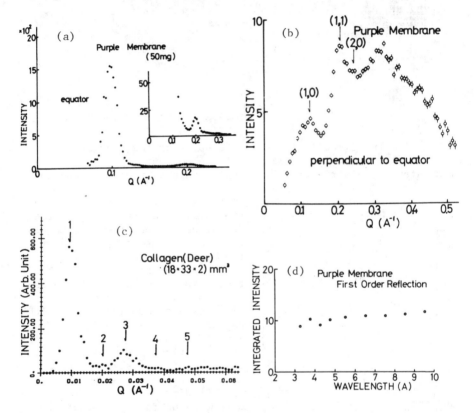

Fig. 9(a)   Equatorial neutron scattering data from purple membrane.
(b). Neutron scattering data from perple membrane perpendicular to the
equator.   (c) The $1/\lambda^2$ dependence of the $\Delta Q/Q$ of the satellite.
(d)   The wavelength dependence of the first order reflection of
purple membrane.

        b)   Purple Membrane

        Purple Membrane is one of the plasma membrane of Halobacterium
halobium, an organism that lives in environments containing high concentrations of salt.   This membrane containes only one protein,

bacteriorhodopsin, which exists in a two-dimensional crystalline lattice in the membrane.   Lipids occur in the gaps between the protein and form a double layer structure.   The thickness of the layer is about 60 A.

For the experiments in the SAN spectrometer purple membrane was oriented on three sheets of aluminium foil (8 × 15 mm in area). The distance between the foils was 1 mm.   Fig. 9(a)   and (b) show the neutron scatteirng data along the equator and perpenaucular to the equator respectively.   Fig. 9(a) was obtained from equation (5) by summing the data recorded at the equator over all wavelength. The thickness of the layer in this purple membrane was 60.4 A. In Fig. 9(b), (1,0) , (1,1) and (2,0) reflections were assigned. The first order Bragg reflection along the equator which has a peak at about $Q \sim 0.1 \ A^{-1}$ is intense suggesting the possibility of an extinction effect.   However, the intensity of this first order reflection, calculated according to equation (5) using a selected narrow wavelength band does not depend on the neutron wavelength. Fig. 9(d) indicates that such extinction effects are not significant. At the same time, however, these results demonstrate that the use of a 2D-PSD combined with the TOF is feasible even though essentially white incident neutron radiation is used.

c)   Collagen

Collagen is a typical biologival fibre which has the advantages both of ease of handling and of an occurrence of only a single type of rod-shaped molecule.   The collagen molecules are staggered by 670 A with respect to each other; this leads to a series of strong meridional reflections of this spacing in the neutron diffraction patterns.

The results of small angle scatteirng measurements on deer collagen are shown in Fig. 9(c).   The diffraction was observed up to 5th orders with a period of 690 A.   The observed intensity of the each reflection was explained with the assumption that the alignment of molecules is deduced by Doyle et al.[12]   The intensity of the first order reflection was found to be independent of wavelength.

CONCLUSION

(1)   In a 1D-PSD or a 2D-PSD combined with TOF techniques, huge amount of data should be accepted and displayed, the system of which has been established.
(2)   A time of flight single crystal diffractometer combined with 1D-PSD is very powerful and enables to observe the over all view of the reciprocal space like a photography, because the scans should be made over the whole Brillouin zones simultaneously, and
(3)   many Bragg reflecitons can be collected simultaneously and the wavelength dependence of the Bragg intensities provides the extinction correction.
(4)   The use of a 2D-PSD combined with the TOF small angle

scattering is feasible even though essentially white incident
neutron radiation is used, and
(5) moreover, in the case of the MnSi satellite, the wavelength dependences of the satellite wave vector, the satellite intessity
and the spread of the satellite peak gave many information including that of the instrumental resolution.

ACKNOWLEDGEMENT

The study of biological materials reported in this paper was
carried out in collaboration with members of KENS soft group. We
would especially acknowledge Prof. T. Mitsui, an organizer of the
group.

REFERENCES

1.   N. Niimura et al., Nucl. Instrum. Meth. 173, 517 (1980)
2.   M. Arai et al., Res. Rep. Lab. Sci. 13, 261 (1980)
3.   M. Hirabayashi et al., J. Phys. Soc. Jpn 45, 1591 (1978)
4.   P. Wells and J.H. Smith, J. Phys. F1, 763 (1971)
5.   H. Bouchiat et al., Phys. Rev. B23, 1375 (1981)
6.   C.G. Windsor, Private communication
7.   P.J. Becker and P. Coppens, Acta Cryst. A30, 129, 148 (1979)
8.   Y. Ishikawa et al., Proc. ICANS-IV 563 (1981)
9.   N. Niimura et al., KENS Report I. 163 (1980)
10. C.S. Borso et al., J. Appl. Crys. to be published
11. Y. Ishikawa et al., Solid State Commun. 19, 525 (1976)
12. B.B. Doyle et al., J. Proc. R. Soc. London. Ser. b 187, 37
     (1974)
13. R. Berliner et al., Final Report on a Small Angle Neutron scat-
     tering Spectrometer at MURR (1979) April

AN INSTRUMENT FOR LIQUIDS, AMORPHOUS AND POWDER DIFFRACTION.

Alan K. Soper
Los Alamos National Laboratory, Los Alamos, NM 87544

ABSTRACT

A time of flight diffractometer, which has been built at the Los Alamos pulsed neutron source, is described. The concept of resolution focussing is discussed and the application of the instrument to liquid structure over a broad range of momentum transfers is presented.

INTRODUCTION

The pulsed neutron source at Los Alamos (WNR) currently operates at a repetition rate of 120 Hz with the proton pulse width about 6 μsec, and an average proton current of about 4 μAmps. The protons impinge on a tungsten target, which produces approximately 15 neutrons per proton. The target is surrounded by a T-shaped water pre-moderator Fig. 1 (inset) and the neutron scattering flight paths view a 1.3 cm thick polyethylene moderator which is decoupled from the pre-moderator by a gadolinium foil, thickness 0.025 mm. There are 9 flight paths for conventional neutron scattering and 3 others are allocated to neutron physics experiments. Generally, the two types of experiments have different proton requirements so that the available beam time is shared between the condensed matter group, P-8, and the nuclear physics group, P-3. In practice, this means that P-8 has beam for at least 50 percent of the time that LAMPF is on, and lately much more than this. There is also a smaller amount of time set aside for target and moderator development in a second, low current target area.

The proton beam enters the main target area Fig. 1 above the ceiling of the experimental floor and is then bent sharply downwards onto the target below. The forward peaked high energy neutron distribution is thus dumped into the ground, which acts as a natural beam stop, whilst allowing the flight tubes to view the target from all horizontal directions. In fact the neutron scattering flight tubes do not view the target directly but are offset by a few inches, so that they only see moderated neutrons. This helps to reduce the number of high energy neutrons which enter the experimental area. Even so, gold foil activation measurements indicate that about 1 in 20 of the neutrons coming down the flight tubes have an energy greater than 1 MeV, so that shielding both experiments and people is a major headache.

There are five instruments currently in operation at WNR at various stages of development. All of these are to be regarded as

24

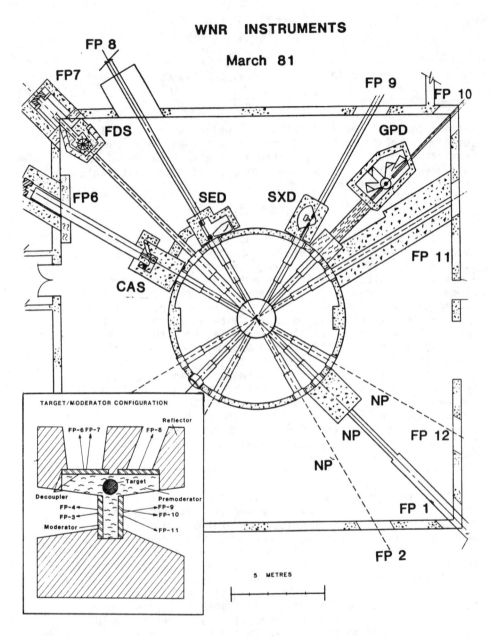

WNR INSTRUMENTS

March 81

FP 8
FP7
FDS
FP 9
FP 10
GPD
FP6
SED
SXD
FP 11
CAS
NP
NP
FP 12
NP
FP 1
FP 2

TARGET/MODERATOR CONFIGURATION

Reflector
FP-6 FP-7
FP-8
Decoupler
Target
Premoderator
FP-4
FP-9
FP-3
FP-10
Moderator
FP-11

5 METRES

Figure 1. Layout of the main experimental area at WNR. The five
instruments shown are the Crystal Analyser Spectrometer, (CAS),
Filter Detector Spectrometer (FDS), Special Environment
Diffractometer (SED), Single Crystal Diffractometer (SXD) and the
General Purpose Diffractometer (GPD). The inset shows the T-
shaped target and moderator configuration.

prototype time of flight machines at present, and the following paper deals with just one of these, the general purpose diffractometer.

DESIGN CONSIDERATIONS

The basic requirement was for a machine suitable for both liquid and high resolution powder diffraction. In practice these conditions are not so divergent as might appear at first sight, when it is realized that the highest resolutions are obtainable at the back scattering angles, whereas liquid diffraction is most usefully accomplished at a pulsed source at forward angles, where Placzek[1] corrections will be smallest. Our experience so far has shown that liquid diffraction at all angles will be important and the requirement for good resolution is needed in liquid studies as well as powders.

Early Monte Carlo simulations of the instrument together with the 120 Hz repetition rate and predicted neutron fluxes, indicated that the conflicting demands of good resolution, avoidance of frame overlap and high flux could be resolved quite well with an incident flight path of 10 m and a sample to detector distance of 1 m based on 1/2 inch diameter ^{3}He detectors. This basic design was subsequently modified to allow for geometric resolution focussing, a configuration of angled

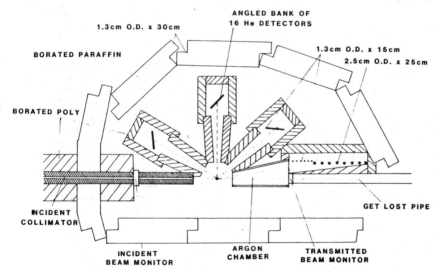

GENERAL PURPOSE DIFFRACTOMETER

Figure 2. Layout of the General Purpose Diffractometer. The incident beam has width 1.3 cm, height 5.1 cm.

banks of detectors such that the resolution for all the detectors in a given bank is the same. This condition is less sharply angled than for time focussing and so allows for more convenient detector arrangements. The layout of the resulting instrument is shown in Fig. 2.

To obtain the fully focussed condition we would need to consider all the terms in the resolution function. This can be done conveniently by Monte Carlo simulation but the results are not very transparent. Alternatively we can consider a few important terms, in particular those relating to the scattering angle. In that case the resolution focussing condition is:

$$R = \frac{\Delta t}{T} = \text{Cot}\,\theta\,\Delta\theta = \text{Constant}$$

where 2θ is the scattering angle and t is the neutron time of flight. The advantage of employing this condition as opposed to other possible configurations is that when binning a large number of detectors together there is no loss of resolution or modification of the resolution function so that the binned data are the same as those from a single detector.

Contributions to $\Delta\theta$ come from several sources and we shall consider only those due to the horizontal and vertical aspects of the detector $\Delta\theta_H$ and $\Delta\theta_V$ respectively, which are added in quadrature:

$$\Delta\theta^2 = \Delta\theta_H^2 + \Delta\theta_V^2$$

We consider two identical detectors of width w, half-height, H, separated by a distance T, and distance L from the sample, Fig. 3, (inset). One detector is allowed to rotate about the other, the angle of rotation being ϕ, and the question arises: can the second detector be placed such that both have the same resolution? We have derived expressions for $\Delta\theta_H$ and $\Delta\theta_V$ and examples of these and $\Delta\theta$ are shown in Fig. 3. We find that the condition for resolution focussing is obtained at all scattering angles. At 90 degrees and backscattering angles it is on the order of $\phi = 45$ degrees. For forward angles we find $\phi = 90 - 2\theta$ so that the bank would lie approximately parallel to the incident beam. The actual conditions are quite sensitive to the width and height of the detector, Fig. 3(b) and (c), and these have to be determined by a simulation. However, the most important result is that the condition is almost independent of the detector separation T for a large range of T, Fig 3(d), so that all the detectors in a resolution focussed bank lie in a straight line.

This principle is illustrated in Fig. 4 where a nickel powder diffraction pattern is shown. The data for a 1 micron nickel

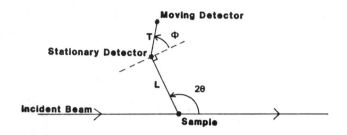

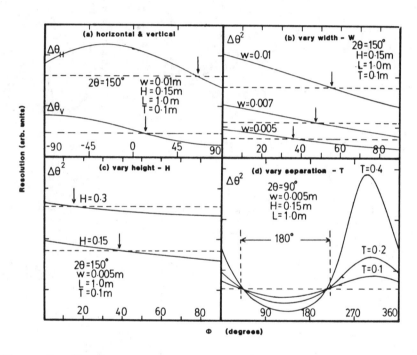

Figure 3. Resolution focussing conditions. The diagram at the
top shows the geometry of the problem. In all cases the
horizontal line shows the resolution of the stationary detector
and the curves show the resolution of the second detector which
rotates about the first. Resolution focussing occurs where the
two curves meet. In (a) the quantities $\Delta\theta_H$ and $\Delta\theta_V$ are shown:
they coincide with the stationary detector at different values
of ϕ. In (b) and (c) variation of the detectors width and
height is shown and it is seen that the focussed condition is
sensitive to these quantities. In (d) however it is found to be
quite insensitive to the detector separation.

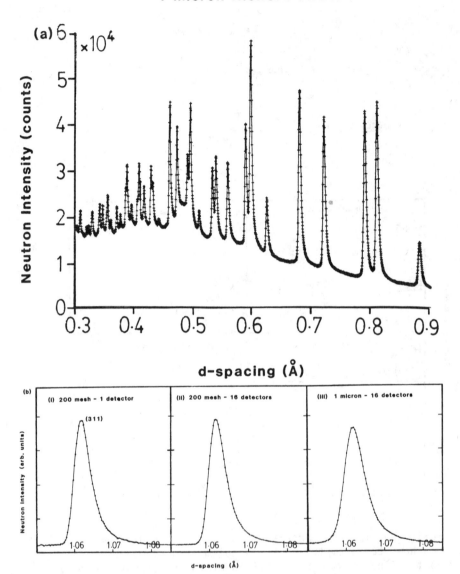

Figure 4. Demonstration of resolution focussing. In (a) is shown the diffraction pattern at a scattering angle of 150 degrees for 1 micron nickel powder. In (b) we compare a single detector ((311) peak) (i) for 200 mesh nickel powder, with 16 detectors binned for the same sample, (ii). In (iii) the same peak is shown for the 1 micron powder, and significant broadening can be seen particularly at the initial rise of the peak.

powder sample were taken in about 12 hrs of beam time, Fig. 4(a). We compare these with a 200 mesh nickel powder (particle size 100 microns) in Fig. 4(b). In particular we see that the peak shape for a single detector (i) is identical to that from 16 detectors binned together (ii)--there is no loss in resolution by binning the detectors together. Also we are able to see a significant broadening in going to the 1 micron powder (iii), where the (311) peak is broadened by about 20 percent. This is believed to be strain broadening with some particle size broadening.

The application of this instrument to powder profile refinement will be discussed in a separate paper at the Ottawa meeting.[2] We are also analyzing data taken on samples of amorphous nickel-titanium and nickel-zirconium alloys, but shall concentrate here on studies of liquid structure, in particular an experiment on liquid water.

LIQUID WATER

The purpose of this experiment is to exploit the variation of coherent neutron scattering length with hydrogen isotope to obtain the H-H partial structure factor, as well as to compare the performance of this instrument with equivalent instruments at other sources. Liquid water has of course been studied extensively by both neutron and X-ray diffraction[3,4] and an essentially equivalent experiment has been performed by Narten[5] on a reactor source, although the results are not yet published. The present experiment will extend these data to larger momentum transfers.

The isotope experiment is complicated by the large inelasticity corrections which occur with hydrogen. However, this problem was partly overcome in a recent experiment on liquid hydrogen chloride where a similar situation occurs.[9] In particular it was shown that by assuming that the dynamics of a molecule are unaffected by whether it is surrounded by H or D containing molecules then the H-H structure factor, (which includes the applicable interference inelasticity correction) could be obtained directly without using a correction procedure, and the method appeared to work well. For water the situation is complicated by the formation of HDO molecules. However, simulations based on the formulation of Rahman[6] for an assymmetric rotor, freely translating and rotating, showed that the approximation would still work very well*, particularly for time-of-flight diffraction.

* A detailed description of this experiment and the models will be published elsewhere.

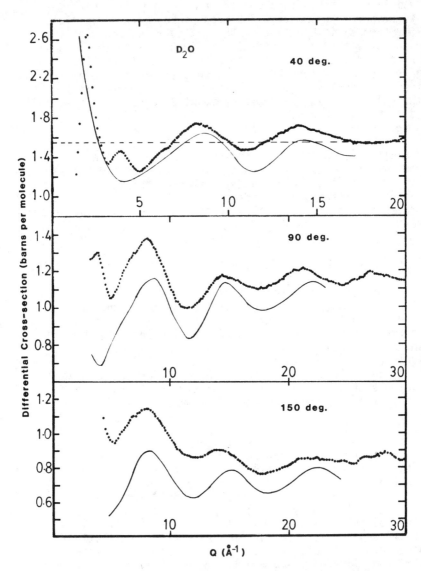

Figure 5. Neutron differential cross-sections for liquid heavy water measured at scattering angles of 40, 90, and 150 degrees. The excellent statistical quality of the data will be noted. The dashed line in (a) is the static cross-section for the self terms. The continuous lines are the result of a simulation of freely translating and rotating molecules, which however ignores vibrational motion. The OD bond length in the model was 0.958 A.

The data from the present experiment are so far incomplete but a reliable D_2O spectrum has been recorded at scattering angles of 40, 90, and 150 degrees, Fig. 5. These data have been corrected for background, attenuation and multiple scattering and normalized to a vanadium spectrum, but no inelasticity corrections have been applied. The data were taken in about 48 hours of beam time and we should note the excellent statistical quality of the results. For example at the thermal peak of the incident neutron distribution the combined data for the D_2O sample from sixteen detectors when binned in 'Q' bins of width $\Delta Q = 0.1$ A^{-1} amounted to over 2×10^6 counts.

Another point to note is that at 40 degrees the data appear to oscillate about a flat line--they do not have the characteristic 'droop' of a reactor experiment. This suggests inelasticity corrections to the self terms are rather flat at this angle. As we proceed to larger angles however the data has a pronounced fall at the smaller Q values and then flattens out, but clearly does not oscillate about a flat line. This phenomenon was apparently observed in the time-of-flight data of Clarke and Dore[7] but not commented on.

DISCUSSION

In an attempt to understand this behavior we are proceeding to develop a dynamical model which incorporates the translational, rotational, and vibrational motions. The vibrational correction has received poor treatment in the literature since all the available expressions use a simple Debye-Waller type factor which is not applicable at large Q values. Zemach and Glauber[8] give an exact expression for the vibration matrices, which has to be averaged over orientations. If we consider only the 'self' terms and one vibrational mode and assume the molecules do not translate or rotate but are randomly orientated, this average can be performed quite easily without making the usual approximation of expanding the exponential in only its first two terms, to yield a differential cross-section for transfer of n quanta of vibrational energy. The result is

$$\left(\frac{d\sigma}{d\Omega}\right)_n = b^2 \frac{k_f}{k_i} e^{\frac{-nw}{2k_BT}} F(Q,|n|)$$

where b is the scattering length, k_f, k_i are the initial and final wavevectors, Q is the momentum transfer, w is the vibrational frequency, T the temperature, and $F(Q,n)$ is given by a recurrence relation

$$F(Q, n+1) = \frac{1}{2} [(2n+1) F(Q,n) - (QA)^{2n} \exp(-Q^2 A^2)]$$

with $F(Q,0) = \frac{\sqrt{\pi}}{2} \frac{erf(QA)}{QA}$

Here $A = \hbar c / \sqrt{(2w\hbar M)}$ is the amplitude of the vibrational motions. c is the normal coordinate displacement, M is the mass of the particle and $\hbar$ is Plancek's constant divided by 2π. It will be noted that for the zero point motion (n=o) the correct expression contains an error function, rather than the exponential function which is usually assumed. Obviously as the incident energy rises, more levels can be excited and the total differential cross-section will be a sum over the available levels. The result is shown in Fig. 6, where it is assumed that the atoms have the same

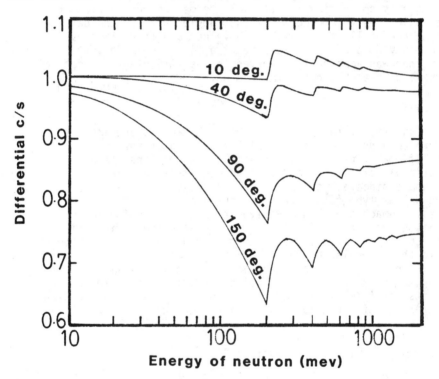

Figure 6. Total differential cross-section for a randomly orientated harmonic vibrator, at various scattering angles. The amplitude of the motions for this calculation was chosen to be the same as for the H atom in theO-H bending mode in light water.

amplitude as the hydrogen bending mode in light water (A = 0.069 A). It will be seen that at small angles the result is rather flat but as we go to larger angles the zero point motion dominates the initial part of the spectrum. The spectrum flattens out once vibrational levels start to be excited. These qualitative features appear to be born out by the D_2O data.

Obviously this model is too highly simplified since the real vibrational levels will be broadened considerably by the rotational and translational motions. We are currently developing a model which incorporates all three effects as well as the interference scattering between different atoms on the same molecule. The D_2O data were actually recorded at incident energies in excess of 1 keV. At energies of 1 eV and above, not shown in Fig. 5, the differential cross-section exhibits a few, broad features, which suggest higher energy excitations. These are not understood at present, particularly since there is currently no theory for $S(Q,w)$ in the 1 - 1000 eV energy region.

CONCLUSION

We have described a time-of-flight diffractometer capable of both good resolution (0.5%) powder diffraction and liquid and amorphous diffraction over a complete range of scattering angles. The present limited survey has stressed the ability to perform resolution focussing, and to obtain high quality diffraction patterns on liquids on this instrument. Future experiments at pulsed sources will be able to explore the structure of materials at energies far in excess of those currently in use. However, new theories will be required. We hope to publish a complete description of the instrument in the near future.

ACKNOWLEDGEMENTS

The principal collaborator in this project is R. B. Von Dreele of Arizona State University. We would like to thank J. D. Jorgensen and J. Faber for much help in early design considerations and for a copy of their Monte Carlo simulation program. This work was performed under the auspices of the U. S. Department of Energy.

REFERENCES

1. G. Placzek, Phys. Rev. 86, 377 (1952).

2. R. B. Von Dreele, A. K. Soper, 12th Congress of International Union of Crystallography, Ottawa, Canada, 1981.

3. A. H. Narten, J. Chem. Phys. 56, 5681, (1972).

4. G. Walford, J. H. Clarke, J. C. Dore, Molec. Phys. 33, 25 (1977).

5. A. H. Narten, Presented at Spring Meeting of American Chemical Society, Atlanta, Georgia, 1981.

6. A. Rahman, J. Nucl. Energy 13A, 128 (1961).

7. J. H. Clarke, J. C. Dore, referred to by J. G. Powles, Figure 9. Molec Phys. 26, 1325 (1973).

8. A. C. Zemach, R. J. Glauber, Phys. Rev. 101, 118 (1956).

9. A. K. Soper, P. A. Egelstaff, Mol. Phys. 42, 399, 1981

COLLECTION AND ANALYSIS OF SINGLE CRYSTAL TIME-OF-FLIGHT NEUTRON DIFFRACTION DATA*

A. J. Schultz,[†] R. G. Teller,[†] S. W. Peterson, and J. M. Williams
Chemistry Division, Argonne National Laboratory
Argonne, Illinois 60439, U.S.A.

ABSTRACT

This paper describes a single crystal diffractometer (SCD) which uses a position-sensitive area detector for collecting time-of-flight (TOF) neutron diffraction data at Argonne's Intense Pulsed Neutron Source (IPNS). The analysis of data obtained during operation of the prototype ZING-P' pulsed neutron source is also presented.

INTRODUCTION

With the development of pulsed neutron sources at Argonne and elsewhere, it should soon be feasible to collect single crystal Bragg data on a routine basis with higher data rates or smaller samples than are typical at steady state reactors. In order to utilize as much of the available spectrum from a pulsed source as possible, the TOF Laue method[1-4] is used for measuring Bragg intensities. In this method, neutrons of variable wavelength are used to measure diffracted beam intensities at fixed 2θ angles and maxima occur whenever the Bragg equation $\lambda = 2d\sin\theta$ is satisfied. The neutron wavelength λ is given by the de Broglie expression $\lambda = h/mv = (h/m)(t/\ell)$, where h is Planck's constant, m is the neutron mass, v is the neutron velocity, and t is the neutron time-of-flight for a path length ℓ.

At Argonne, a single crystal diffractometer with a position-sensitive area detector has been developed. This instrument samples a three-dimensional portion of reciprocal space which may contain hundreds of Bragg data in addition to possible superlattice peaks or diffuse scattering maxima. Because each measurement is obtained with both the crystal and detector stationary, experiments employing sophisticated special environment equipment should also be more easily accomplished.

INSTRUMENT DESIGN

A detailed description of the instrument design as of 1978 has been published.[5] As shown schematically in Fig. 1, angular movements of the crystal and detector are provided by means of a four-circle diffractometer utilizing a Huber goniostat with the χ circle offset from the ϕ axis. The sample-to-detector distance can also be varied.

*Work supported by the Office of Basic Energy Sciences, Division of Materials Sciences, U. S. Department of Energy, under Contract W-31-109-Eng-38.
†Authors to whom inquiries about the paper should be addressed.

36

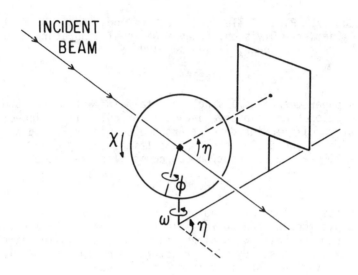

Fig. 1. A schematic representation of the instrument showing the X,
φ, ω, and η angles.

        The data presented in this paper were obtained at the ZING-P'
prototype source using a ³He-filled, multiwire proportional counter
of the Borkowski-Kopp[6] type with an active area of 20×20 cm. Data
acquisition was accomplished with a DEC LSI 11/23 microprocessor, located next to the instrument, which read the x, y, and time analog to
digital convertors, stored the events in a buffer, and transmitted
the data to a remote Xerox Sigma 5 computer for histogramming on disk
memory. The histogrammed data was then available for display and
manipulation by the user from terminals located at the Sigma 5 computer and at the instrument.
        A project at Argonne to develop a new detector with higher ef-
ficiency and resolution resulted last year in a test of the first
prototype two-dimensional position-sensitive scintillation detector
for neutrons.[7] The detector contains a close-packed array of photo-
multipliers (PM's) coupled via a light guide to a thin (1-2 mm) ⁶Li-
loaded, Ce-activated glass scintillator. Since each scintillation
event is viewed by several PM's, the position of the event is determined by calculating the centroid in the x and y directions. A second detector with an active area of 30×30 cm has been constructed and
will be used on the SCD at IPNS. The major components of a new SCD-
IPNS data acquisition system are a PDP 11/34 computer, with 20 Mbytes
of disk storage, which is linked to a Z8000 microprocessor with 2 M-
bytes of memory for building the histogram during data collection.

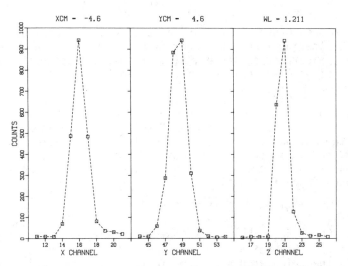

Fig. 2. Intensity profiles of a Bragg peak ($\overline{4}\,\overline{4}\,2$) in each of three directions. The peak was observed at x = -4.6 cm and y = +4.6 cm from the center of the detector (0.0,0.0) and at a wavelength of λ = 1.211 Å.

DATA COLLECTION AND ANALYSIS[8]

During the last period of ZING-P' operation 39 histograms of data from a NaCl crystal were obtained with the ^{3}He detector. The data nearly encompassed a full hemisphere of data. The crystal was 850 cm from the moderator and the detector was positioned at an angle of 90° (η in Fig. 1) to the incident beam and at a distance of 25.3 cm from the crystal. Each histogram was obtained with the diffractometer χ and ϕ at different settings such that nearly a full hemisphere of data was measured. Each event was coded with a spatial resolution of 64×64 channels and wavelength resolution of 128 channels in the range 0.701 to 3.936 Å ($\Delta\lambda$ = 0.025 Å). A peak search routine was used to locate a few Bragg peaks, which were then used in an auto-indexing routine to obtain an orientation matrix. The positions of Bragg reflections in each histogram were calculated from the orientation matrix, and raw intensities were obtained by a three-dimensional integration procedure. The intensity profiles of one Bragg peak are shown in Fig. 2.

Conversion of integrated intensities to structure factor amplitudes is based on the Laue formula:[1]

$$I_{hk\ell} = kT\phi(\lambda)\varepsilon(\lambda)A(\lambda)y(\lambda)|F_{hk\ell}|^2\lambda^4/\sin^2\theta$$

where k is a scale factor, T is the normalized monitor count, $F_{hk\ell}$ is the structure factor, θ is the Bragg angle, $\phi(\lambda)$ is the incident flux, $\varepsilon(\lambda)$ is the detector efficiency, $A(\lambda)$ is the absorption correction, and $y(\lambda)$ is the extinction correction at wavelength λ. The two factors, $\phi(\lambda)$ and $\varepsilon(\lambda)$, were obtained as a single normalization factor from the absorption-corrected spectrum obtained from the incoherent scattering from a cylinder of vanadium.[9],[10] In order to calculate the correction factor, $A(\lambda)$, the linear absorption coefficient $\mu(\lambda)$ of NaCl was obtained from the equation

$$\mu(\lambda) = 0.177\lambda + 0.333$$

The absence of neutron resonances for Na and Cl in the wavelength range 0.7 to 4.0 Å permits the use of this linear equation.

The thermal parameters for Na and Cl, a scale factor for each histogram, and an isotropic extinction factor were refined by a full-matrix least-squares procedure. The extinction correction is based on the Zachariasen formulation[11],[12]

$$|F_o|_{cor} = |F_o|(1 + 2\ g Q \bar{T})^{-\frac{1}{4}}$$

where $Q = \lambda^2 F_c^2/V^2 \sin^2\theta$, g is the refined isotropic extinction parameter, $\bar{T}$ is the absorption weighted mean path length through the crystal, F_c is the calculated structure factor amplitude, and V is the unit cell volume. The results of the final least-squares cycle are presented in Table I.

Table I. Results of Final Least-Squares Cycle[a]

Number of data:	356
Number of variables:	42
Number of scale factors:	39
Range of scale factors:	1.00(2) - 1.19(2)
Extinction factor, g:	$0.078(6) \times 10^4$
Range of extinction correction factors:	1.00 - 3.26

$$B(Na) = 1.35(2)\ \text{Å}^2$$
$$B(Cl) = 1.12(2)\ \text{Å}^2$$

$$R(F_o) = 0.033$$
$$R(F_o^2) = 0.055$$
$$R_w(F_o^2) = 1.106$$

$$^aR(F_o) = \Sigma||F_o|_{cor} - |F_c||/\Sigma|F_o|_{cor}$$
$$R(F_o^2) = \Sigma|(F_o^2)_{cor} - F_c^2|/\Sigma(F_o^2)_{cor}$$
$$R_w(F_o^2) = [\Sigma w_i|(F_o^2)_{cor} - F_c^2|^2/\Sigma w_i(F_o^4)_{cor}]^{\frac{1}{2}}$$

The function minimized in the least squares process was $\Sigma w_i|F_o^2 - S^2 F_c^2|$ where S is the scale factor.

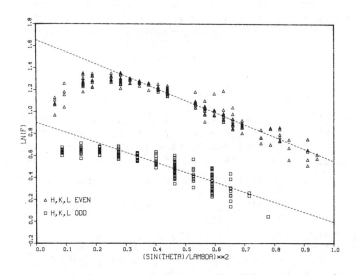

Fig. 3. A plot of $\ln|F_0|$ vs $\sin^2\theta/\lambda^2$ for data uncorrected for extinction and with $F_0^2 \geqslant 3\sigma(F_0^2)$. The data were scaled by refining a separate scale factor for each histogram based on data with $\sin^2\theta/\lambda^2 \geqslant 0.3$. The least-squares straight lines are also based only on these data.

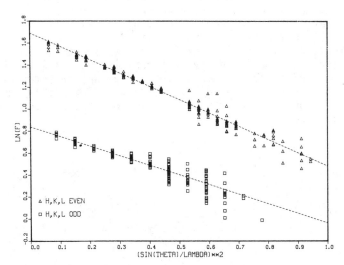

Fig. 4. A plot of $\ln|F_0|_{cor}$ vs $\sin^2\theta/\lambda^2$ after correcting for secondary extinction. Only data with $F_0^2 \geqslant 3\sigma(F_0^2)$ are plotted, but there is no $\sin^2\theta/\lambda^2$ cut-off.

Plots of $\ln|F_0|$ vs $(\sin^2\theta)/\lambda^2$ before and after correcting for extinction are shown in Figs. 3 and 4, where F_0 is the observed structure factor. The dependence of the structure factor on $(\sin\theta)/\lambda$ is

$$|F_0| = |F_0(0)|\exp(-B\sin^2\theta/\lambda^2)$$

or

$$\ln|F_0| = \ln|F_0(0)| - B\sin^2\theta/\lambda^2$$

where $F_0(0)$ is the structure factor amplitude at $(\sin\theta)/\lambda = 0$ and B is the thermal parameter for a set of $hk\ell$ planes with the same $F_0(0)$. For NaCl, with a face-centered cubic crystal structure, there are only two structure factor types:

$$|F_{hk\ell}(0)| = b_{Cl} + b_{Na}; \quad h,k,\ell = 2n$$

$$|F_{hk\ell}(0)| = b_{Cl} - b_{Na}; \quad h,k,\ell = 2n + 1$$

where b_{Na} and b_{Cl} are the scattering lengths for sodium and chlorine. Without correcting for extinction, the plots of $\ln|F_0|$ vs $(\sin^2\theta)/\lambda^2$ showed a significant variation from linearity at small $(\sin\theta)/\lambda$ values (Fig. 3). After correcting for extinction, the data clearly form two nearly straight lines (Fig. 4).

CONCLUSIONS

A separate scale factor for each histogram was included in the least-squares refinement because of relatively poor agreement of some of the very intense reflections using only a single scale factor. This can be attributed to problems encountered with the monitor counter during data collection, or to the difficulty of correcting for large extinction effects with the Zachariasen formula. Improvement in the accuracy of the intensity data can also be obtained by improving the characterization of the incident flux and detector efficiency. However, the results presented here clearly indicate that accurate Bragg data can be obtained using TOF techniques with a position-sensitive area detector.

REFERENCES

1. B. Buras, K. Mikke, B. Zebech, and J. Lecicjewicz, Phys. Stat. Sol. 11, 567 (1965).
2. D. H. Day and R. N. Sinclair, Acta Cryst. B26, 2079 (1970).
3. A. M. Balagurov, E. Borea, M. Dlouha, Z. Gheorghia, G. R. Mironova, and V. B. Zlokazov, Acta Cryst. A35, 131 (1979).
4. N. Niimura, T. Kubota, M. Sato, M. Arai, and Y. Ishidawa, Nucl. Instr. Meth. 173, 517 (1980).
5. S. W. Peterson, A. H. Reis, Jr., A. J. Schultz, and P. Day, Advances in Chemistry Series, No. 126, Solid State Chemistry: A Contemporary Overview, edited by S. L. Holt, J. B. Milstein, and M. Robbins, 1980, pp. 75-91.

6. C. M. Borkowski and M. K. Kopp, Rev. Sci. Instrum. 46, 951 (1975).
7. M. G. Strauss, R. Brenner, F. J. Lynch, and C. B. Morgan, IEEE Trans. Nucl. Sci. NS-28, No. 1, 800 (1981).
8. For a detailed discussion of the data analysis, see: A. J. Schultz, R. G. Teller, S. W. Peterson, and J. M. Williams, J. Appl. Cryst., submitted for publication.
9. D. H. Day, D. A. G. Johnson, and R. N. Sinclair, Nucl. Instr. Meth. 70, 164 (1969).
10. B. Zebech, K. Mikke, and D. Sledziewska-Blocka, Nucl. Instr. Meth. 79, 51 (1970).
11. W. H. Zachariasen, Acta Cryst. 23, 558 (1967).
12. For other discussions of the Zachariasen extinction correction to TOF data see: N. Niimura, S. Tomiyoshi, J. Takahashi, and J. Harada, J. Appl. Cryst. 8, 560 (1975); S. Tomiyoshi, M. Yamada, and H. Watanabe, Acta Cryst. A36, 600 (1980).

NEUTRON PHYSICS RESEARCH AT DUBNA PULSED NEUTRON SOURCE

L. Cser
Central Research Institute for Physics
H-1525 Budapest, P.O.B. 49, Hungary

ABSTRACT

The Time-of-Flight (TOF) method is a very powerful method for investigation of neutron diffraction phenomena. However, many efforts have to be made for solving special problems arising in a TOF diffraction experiment. As a result of recent years activity in this field at JINR pulsed reactor significant success were achieved in data collection, accumulation and evaluation techniques (e.g. new methods were proposed for orientation of monocrystals; position-sensitive detectors were used for more effective registration of scattered neutrons; by transformation of coordinates of each events into reciprocal space coordinates the number of necessary memory cells can be about hundred times decreased, etc.).

The effectiveness of TOF diffraction methods is demonstrated by determination of structure of some ferroelectric crystals. The TOF method is a useful tool also for investigation of perfect single-crystals and for texture analysis.

The TOF small-angle neutron scattering was used for measuring of the shape and dimensions of different biological macromolecules (e.g. immunoglobulins). Their conformational change was also determined. The neutron diffraction investigations in Dubna will appreciably developed with the start of the regular operation of IBR-2 pulsed reactor. Present state and future prospects of scientific program at IBR-2 reactor are briefly reviewed.

INTRODUCTION

During the several last years in Dubna, at the Neutron Physical Department of the Joint Institute for Nuclear Research (J.I.N.R.) a pulsed fast neutron reactor, IBR-30, has been operated. This pulsed source was succesfully used in many nuclear and solid state physics experiments and also serves as a good tool for some applied research e.g. in biology and medicine. At this moment a new variant of the pulsed reactor - IBR-2 - is in state of testing run.

The main characteristics of both these reactors are given in the Table I. The planned scientific activity at the IBR-2 reactor will be a direct continuation of the research program carried out at IBR-30 reactor.

In the following a brief list of selected topics will be given.

ISSN:0094-243X/82/890043-10$3.00 Copyright 1982 American Institute of Physics

Table I The main characteristics of IBR-30 and
IBR-2 pulsed reactors on fast neutrons

	IBR-30	IBR-2
Mean power	25 kw	4 Mw
Pulse repetition rate	5 p.p.s.	5 p.p.s.
Burst length	90 μsec	180 μsec
Total neutron yield	$1.3 \cdot 10^{15}$ n/sec	$1.75 \cdot 10^{17}$ n/sec
Thermal neutron flux on the moderator surface a) peak flux	10^{14} n/cm$^2 \cdot$s	$1.3 \cdot 10^{16}$ n/cm$^2 \cdot$s
b) average	$5 \cdot 10^{10}$ n/cm$^2 \cdot$s	$5.8 \cdot 10^{12}$ n/cm$^2 \cdot$s
Cooling medium	air	liquid sodium
Fuel material	metallic Pu	PuO_2

a.) Investigation of the fundamental properties of
the neutron using ultracold neutrons (such as the lifetime or electric dypole moment.
b.) Nuclear physics experiments to study processes
with very small neutron cross-sections such as (n,α);
$(n;\gamma d)$; $(n,\gamma f)$ etc.
c.) Condensed matter investigations.
- Elastic scattering measurements are aimed at the
investigation of structures of magnetic and non-magnetic
crystals, liquids, amorphous alloys and biological macromolecules.
- For investigation of dynamical properties of condensed matter a set of different spectrometers are constructed. Measurements of dispersion curves of quasi-
particle excitations (e.g. in metal hydrids, organic
crystals, quantum liquids) are in progress.
- A spectrometer of polarized thermal neutrons based
on polarizing mirror neutron guides is operating for
investigation of magnetic dynamical properties of magnetic materials.
d.) Applied research.
A neutron beam is equipped with curved mirror neutron
quide to perform microelement analysis by using capture
γ-ray at very low level of destroying sample irradiation.
Pneumotransport channels serves for studying the
kinetics of radiation damages.

Using various filters a neutron beam with average
energy of 6 MeV will be prepared. Such beam may serve for
medical neutron diagnostic and may be used for neutron
therapy at reasonable doses of 50 rad/min.

Finally, texture analysis and small-angle scattering
material testing for applied purposes are also in preparation.

REPRESENTATIVE EXPERIMENTS

1. Observation of the ferroelectric domain structure
by using a time-of-flight neutron diffractometer[1]. Various
structural phase transitions (ferroelectric, ferromagnetic, ferroelastic, etc.) are accompanied by a division
of the crystal into domains in the low-symmetry phase[2].
In general this is followed by a diffraction line splitting. The analysis of such changes in the diffraction
pattern allows to obtain information about the domain
packing, the relative volume of the domains in the crystal, and the spontaneous deformation (arising during the
phase transition) of the crystal lattice. TOF diffractometer which employs the "white" (Maxwellian) neutron
spectrum, allows one to get information about a large part
of the crystal reciprocal space. The observed volume and
consequently the information acquisition rate depend on
the intensity limits of the neutron spectrum from short
and long wavelengths.

As an illustration the simultaneously observed reg-
gion of the reciprocal space is shown in Fig. 1 for the
case when a one-dimensional position sensitive detector
covering the angle internal ($2\theta_{min}$; $2\theta_{max}$) and a wave
vector interval (K_{min}; K_{max}) are used.

The neutron TOF idffractometer has at least two more
advantages as compared with the conventional neutron
diffractometer. Firstly, all the spots of the observed
region are investigated simultaneously, and secondly, the
TOF diffractometer allows to study rather well transition
processes of short duration occurring in the crystal under the action of an external (electric or magnetic)
field[3,4].

The investigations were carried out by using the
time-of-flight diffractometer installed at the IBR-30
reactor[5]. The reactor was operated in the 5 Hz frequency
regime with an average power of about 20 kW. The sample
investigated was a parallelepiped with dimensions
8x8x10 mm³ along the crystallographic axes of the tetragonal unit cell (space group I42d), placed in a cryostat
with regulated temperature.

For the registration of scattered neutrons two
types of detectors were used: ^{3}He filled single counter
and position sensitive detector[6]. For the last one a
CHM-41 ^{3}He-counter with resistive electrode was used. The

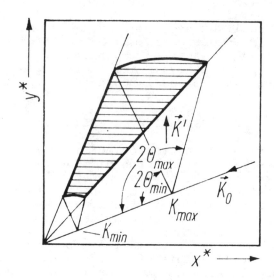

Fig. 1. Observed are of the $(x^*;y^*)$ plane in the crystal reciprocal space. The spectral interval used is $K_{min} \leq K \leq K_{max}$; the detector covers the scattering angle interval $2\theta_{min} \leq 2\theta \leq 2\theta_{max}$.

position of the neutron capture event was determined from the ratio of charges arising at the two ends of the resistive anode. The angular resolution was 10 minutes. The collected data were selected into 64 angular channels, and the whole detector covers an angular interval of $\sim 10^\circ$. For both detector the nominal value of the scattering angle was 86°.

As is known, the domain twinning in DKDP occurs in the $(x;y)$ plane[2]; therefore, we carried out the scanning in [hk0]- type directions of the reciprocal space of the crystal.

For saving of the memory capacity an essential decrease of the number of hystogram elements was achieved with the help of the preliminary transformation of the experimental coordinates (θ, φ and t) of each event (i.e. detector spatial coordinates and the time of flight) into the scattering vector space coordinates (x^*, y^*, z^*).[7] In our case we dealt with a two dimensional scan and only a connection between the variables (θ,t) and ($x^*;y^*$) should be considered.

When only the neighbourhood of a certain diffraction peak is important the intensity distribution can be represented in relative coordinates

$$\gamma = \frac{K-K_o}{K_o} \quad \text{and} \quad \xi = 2(\theta - \theta_o)$$

The ferroelectric phase transition in DKDP determined by diffraction line splitting was observed at $T_c \approx 214$ K. It corresponds to 92 % deuteration[8]. The typical shape of the diffraction lines in para- and ferroelectric phase is given in Fig. 2.

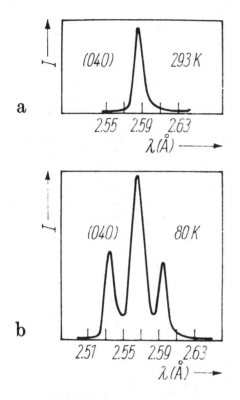

Fig. 2.
Diffraction line splitting during the phase transformation
a) T = 293 K; b) T = 80 K

The section of the reciprocal space spot (080) is given in Fig. 3 (in (ξ,γ) coordinates). The splitting of the (080) spot into four components is clearly seen in this figure. There were observed two of three possible variants of the domain packing, namely the variants corresponding to simple shear along the [110] and [110] directions. The crystal continuity remained the same, in accordance with the proposed scheme[9].

Determining the peak positions for the sake of simplicity one has to pass from the two-dimensional distribution to the one-dimensional integral sections

$$I(\xi) = \int Q(\xi,\gamma)d\gamma \qquad \text{and } I(\gamma) = \int Q(\xi,\gamma)d\xi \qquad (1)$$

where $Q(\xi,\gamma)$ is the observed two dimensional intensity distribution.

These integral sections $I(\xi)$ and $I(\gamma)$ usually consisting of superposition of several peaks were processed using a computer program[10]. This program employs a method which does not require parametric functions for the description of the spectrum components. The shape of the peak is determined experimentally and is introduced as a model into the program. The diffraction line model of DKDP in the paraelectric phase was used for the $I(\xi)$ section procession. For the $I(\gamma)$ integral section a Gaussian was used as a model function dispersion of which is composed both from the angular divergence of the primary neutron beam and from the mosaic spread of the crystal.

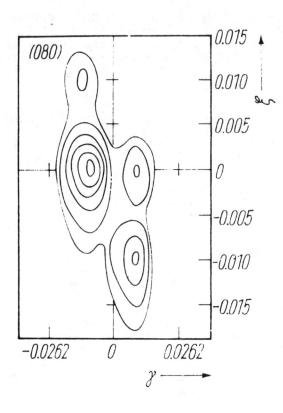

Fig. 3.
Cross section of the (080) spot for DKDP at $T = 80$ K in (ξ, γ) coordinates

Sets of (ξ_i, γ_i) derived from $I(\xi)$ and $I(\gamma)$ were transformed into reciproc lattice coordinates. The results for several reciprocal lattice points are given in Table II.

The spontaneous shear angles are for (040) and (080)
$$U_{xy} = (2x_i^* - x_o^*)/\tau_o = (31.0 \pm 0.8); \text{ for}$$
($\bar{4}40$) and ($\bar{6}60$)
$$U_{xy} = \sqrt{2}[(x_i^* - x_o^*) + (y_i^* - y_o^*)]/\tau_o = (31.6 \pm 0.8)'; \text{ on}$$
the average $U_{xy} = (31.3 \pm 0.6)'$ $(T = 80$ K). This value is in agreement with $U_{xy} = 31$ at $T = 207$ K[11].

Table II The $(x_i^* - x_o^*)/\tau_o$ values (left columns) and $(y_i^* - y_o^*)/\tau_o$ values (right columns) for several spots of the reciprocal lattice at $T = 80$ K in 10^{-3} units, x_o^* and y_o^* stands for the coordinates of unsplitted peak.

(040)		(080)		($\bar{4}40$)		($\bar{6}60$)	
-4.5(1)	-4.9(2)	-4.51(5)	-4.83(8)	-6.57(7)	-6.4(3)	-6.7(1)	-6.9(2)
-4.5(1)	4.6(2)	-4.51(5)	5.1(1)	0.0(1)	0.0(3)	0.0(1)	0.0(2)
4.5(2)	4.9(2)	4.51(6)	4.83(8)	0.38(7)	6.6(3)	6.2(1)	6.8(2)
4.5(2)	-5.2(2)	4.51(6)	-4.8(1)				

The determined value of U_{xy} is in agreement with the value of spontaneous shear, which may be found from the data on the changes in the linear dimensions of the DKDP unit cell during the phase transition[12].

2. Small-angle scattering study of the immunoglobulin conformation[13]. Immunoglobulins of IgG type constitute an important group of antibodies. About 75 % of immunoglobulins in normal human serum are of IgG type. Their functions are: to recognize foreign from the point of view of a given organism macromolecules (antigens), to trigger the events leading to elimination of these macromolecules (effector functions). The molecular weights of IgG are common (~150 000). The primary structure of a human IgG was determined by Edelman. To elucidate the structure of these molecules may help us to understand their functions.

Several physical methods (e.g. hydrodynamic measurements, electron microscopy, X-ray cyrystallography, X-ray and neutron small-angle scattering (SASX and SASN) were involved in the investigation of the real conformation of these antibodies.

According to the results of electron microscopy investigations the shape of the IgG molecules was like the letter Y.[14] X-ray crystallography for two different IgG molecules gives rise to two different shapes (Y and T). Results of small-angle scattering of X-rays used to determine the size and shape of human and rabbit IgG molecules were in accordance with T-shaped form[15,16]. The cross section plot of the measured intensity distribution ($\ln(Ix_\varkappa)$ versus $\varkappa^2$, where $\varkappa = 4\pi/\lambda \sin \theta/2$) regularly consisted of two-straight-line parts with different shapes (see Fig. 4). This characteristic feature is in a tight connection with the shape of the IgG molecule. A hapten induced conformational change was also observed[16].

Investigating pig anti-Dnp antibodies we tried to answer the following questions: Are the shape and sizes of various IgG molecule different? Does the change of the conformation due to hapten binding also arise in the case of pig anti-Dnp antibody molecules? The neutron scattering experiments were performed at the 40 meter long TOF small-angle facility with axialsymmetric geometry equiped with a ^{3}He filled detector consisting of eight ring-shaped counters[17]. This arrangement provides high data collection rate at good resolution. The last circumstances makes easier to perform correction for resolution distortion of measured data.

The obtained data show that the general pattern of cross-section curves of pig antibodies are similar to patterns yielded by IgG of other species, so one may conclude that the principal geometric features of the molecular shape are similar. Only small differences in the values of the radius of gyration and in the slopes of the straight-line parts of the cross-section curves could be observed.

The measured intensity distribution can be compared with the calculated one if a model of the molecule was constructed.

50

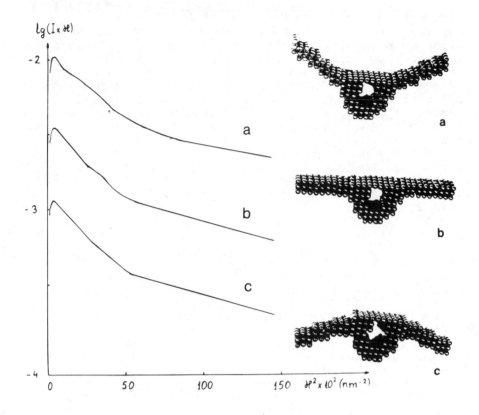

Fig. 4. Calculated cross-section curves and the corresponding models
of pig IgG molecules. The dashed line at curve "c" corresponds to the measured intensity distribution

Utilizing a generalization of the Debye formula

$$I(\varkappa) = \sum_{ik} f_i f_k \frac{\sin \varkappa r_{ik}}{\varkappa r_{ik}} \qquad (2)$$

where f_i and f_k are the scattering amplitudes of elements
of the molecule situated at points i and k, respectively,
r_{ik} is the distance between the above points, the intensity distribution may be computed[18].

The body of the molecule can be constructed in a
good approximation, from a number of homogeneous identical
spheres which are arranged in a simple way e.g. their
centres may form a primitive cubic lattice.

Three models depicted in Fig. 4 are essentially
possible. The first of them (a) corresponds to the Y-shaped
model. The second model (b) is a modification of the
T-shaped model.

Corresponding computed cross-section curves (Fig. 4) display a shoulder between the two straight-line parts of the curve. This shoulder is very pronounced in the case of the Y-shaped form and less pronounced for the T-shaped model. If we continued bending the "arms" of the model downwards to its central part ("shirt-shaped" model) the computed and measured curves almost coincide (see Fig.4c).

From these data it follows that the Y-shaped form may be excluded as a possible model of the IgG molecule in solution. But to make a choice between the T-shaped and "shirt-shaped" model is difficult.

The parameters (e.g. radius of gyration etc.) characterizing the linear sizes of the molecules decreases when the pig IgG molecules form complexes with small antigens. As the shape of the scattering curves of free and liganded antibodes are similar the suggestion that the whole antibody molecule contracts via similarity transformation seems to be correct.

The above data indicate that IgG-type antibodies of different species have quite similar conformation. Since the shape and dimensions of the IgG molecule in solution and in dry state are different, their real conformation may be effectively investigated only by small-angle scattering.

REFERENCES

1. A. M. Balagurov, I. D. Dutt, Z. Gheorghiu, B. N. Savenko and L. A. Shuvalov, Phys.Stat.Sol.(a) 51, 367 (1979)
2. F. Jona and G. Shirane, Ferroelectric Crystals, Pergamon Press, Oxford 1962.
   L. A. Shuvalov, J.Phys.Soc.Japan (Suppl.) 28, 38 (1970)
3. N. Numura and M. Muto, Nuclear Instrum. and Methods 106, 87 (1975).
4. V. V. Nitts, JINR 3-5372, Dubna 1970.
5. B. H. Ananev, A. M. Balagurov, I. P. Barabash, Z. Georgiu, V. D. Shibayev, JINR Rep. 13-11113, Dubna (1977)
6. A. M. Balagurov, V. I. Gordeliy, M. Z. Ishmuhametov, V. E. Novozhilov, Yu. M. Ostanevich, B. N. Savenko and V. D. Shibayev, JINR Rep. P13-80-440 (1980)
7. A. M. Balagurov, V. E. Novozhilov, Yu. M. Ostanevich and V. D. Shibayev, Comm. of JINR P14-12840 (1979)
8. R. J. Nelmes and V. R. Eirikson, Solid State Commun. 11, 1261 (1972)
9. I. S. Zheludev and L. A. Shuvalov, Trudy Inst. Krist. (Moskva) 12, 49 (1956)
10. A. M. Balagurov, M. Dlouha, V. B. Zlokazov and G. M. Mironova, Comm. of JINR P10-11107 (1977)

52

11. C. M. E. Zeyen and H. Meister, Ferroelectrics $\underline{14}$, 731 (1976)
12. J. Nakano, I. Shiozaki and E. Nakamura, Ferroelectrics $\underline{8}$, 483 (1974)
13. L. Cser, F. Franek, I. A. Gladkih, A. B. Kunchenko and Yu. M. Ostanevich, Eur. J. Biochem. $\underline{116}$, 109 (1981)
14. R. C. Valentine and N. M. Green, J. Mol. Biol. $\underline{27}$, 615 (1967)
15. I. Pilz, Allg. Prakt. Chem. $\underline{21}$, 21 (1970)
16. I. Pilz, O. Kralky, A. Licht and M. Sela, Biochemistry $\underline{12}$, 4998 (1973)
17. I. A. Gladkih, A. B. Kunchenko, Yu. M. Ostanevich and L. Cser, Comm. JINR, Dubna, P3-11487 (1978)
18. Yu. A. Rolbin, R. L. Kayoushina, L. A. Feigin and B. M. Schedrin, Kristallografiya $\underline{18}$, 701 (1973)

HIGH RESOLUTION POWDER DIFFRACTION
BY WHITE SOURCE TRANSMISSION MEASUREMENTS

R. G. Johnson and C. D. Bowman
National Bureau of Standards, Washington, D.C. 20234

ABSTRACT

Neutron powder diffraction has been studied by measuring the total neutron cross section using neutron time-of-flight in transmission geometry. This method is equivalent to measurements in scattering geometry of powder diffraction at $2\theta = 180°$. Measurements on iron samples were conducted using the NBS 100 MeV electron linac as a pulsed neutron source and using flight paths of 20 and 60 meters. The resolution at 60 m for 25-meV neutrons was limited to $d\lambda/\lambda=0.2\%$ primarily by moderator hold-up. Although the change in cross section at the Bragg edges may be quite small, counting rates are high permitting the recording of data with a 0.1% statistical precision in about one day. For the Fe samples, diffraction edges were distinguished as high as $n = 196$ (where n is the sum of the squares of the Miller indicies) with all edges distinguishable below $n = 90$.

INTRODUCTION

Although neutron powder diffraction is conventionally studied by observing the scattered neutrons, exactly the same information is available from the transmitted neutrons. When the wavelength of the neutron matches the spacing of crystal planes in the sample there will be a sudden increase in the cross section, the Bragg edge. The change in the transmission is equivalent to the integral of the equivalent scattering peak. Transmission geometry may have several advantages over scattering geometry. 1) In transmission geometry $2\theta = 180°$ for which the contribution of angular resolution to the wavelength resolution is minimized. 2) The useful area of the neutron beam is very large. 3) Sample thickness does not affect resolution so that the sample thickness may be optimized for counting statistics. 4) In transmission geometry information on scattering as a function of position in the sample is easily preserved.

In the following, primary emphasis will be placed on a 60-m measurement, but in the discussion some results from a 20-m measurement will be mentioned. The two experiments were quite similar with the major difference being in the appropriately tailored beam parameters and, of course, the higher count rate but poorer resolution at 20 m.

EXPERIMENTAL DETAILS

A beam of electrons from the NBS 100-MeV linac pulsed at 15 Hz and with a time spread of 4 μs were incident on a water-cooled tungsten target. The subsequent photoneutrons were moderated by a polyethylene disk 2.5-cm thick and 15 cm in diameter. With an

ISSN:0094-243X/82/890053-03$3.00

54

average electron current of 4.4 μA the flux near thermal at 60 m was
5 x 10³ n/cm²-s-eV.

An Fe sample which was 1.27-cm thick was placed in a neutron
beam collimated to 15 cm in diameter at a distance of 8 m from the
source. Neutrons transmitted by the sample were detected at 60.67 m
by a ⁶Li-glass scintillator which was 12.7 cm in diameter and 1.27-
cm thick. Data were collected by a time digitizer and computer
system into 8192 channels with a channel width of 8.192 μs covering
the neutron energy range from 4.4 meV to 950 meV.

Two measurements are required to determine the total cross
section, i.e., the incident neutron flux and the flux transmitted
by the sample. In this case the run with the sample in the beam
took approximately 20 hr while the sample-out run took only 2 hr
(since no structure was expected in the flux this data could be
heavily smoothed.) Note that a detector with a diameter of 150 cm
could be used without sacrificing resolution but providing a factor
of 100 in count-rate.

RESULTS AND DISCUSSION

A portion of the total cross section for Fe as determined in
this measurement is shown in Fig. 1 between 9 and 100 meV. Note
the suppressed zero on the cross-section axis. Each diffraction
edge is labeled by the sum of squares of the Miller indicies
$(h^2+k^2+l^2 = n)$.

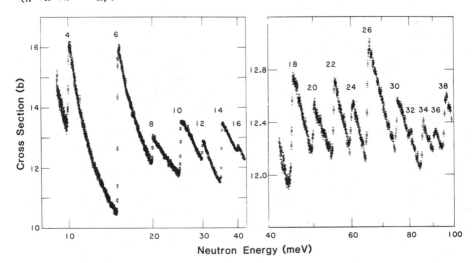

Fig. 1. Total neutron cross section for Fe. (Note the suppressed
zero on the cross section axis.)

In the data shown all diffraction edges from n = 4 to n = 38
are seen. Over the full range of the data all diffraction edges are
distinguishable below n = 50 and the stronger edges with n well over
100 are distinguished. For the higher statistical precision data

obtained in the 20-m measurement the same statement applied to n-values of 90 and 190, respectively.

It is difficult to extract the basic resolution ($\Delta\lambda/\lambda$) in this measurement since the shape of the resolution function is determined primarily by moderator hold-up which has not been carefully measured and since no special care was taken in the sample preparation (e.g., annealing to remove residual strain). However, if the resolution is taken as the interval between 25 and 75% of full magnitude for each edge, it is nearly 0.2% throughout. This estimate is in reasonable agreement with other measurements for a moderator of this type.[1]

The statistical precision on the total cross section is better than 0.3% per half resolution width over most of the energy range shown with the exception of the lowest energies where it approaches 1%. Thus the statistical precision in measuring the size of a diffraction step which is 10% of the total cross section is approximately 4%. However, if a least squares fit is performed above and below the diffraction edge, the statistical precision can be improved by at least a factor of five.

It is useful to compare the performance of the proposed high resolution diffractometer (HRD1)[2] for IPNS-II with a transmission diffractometer designed for the same resolution. Briefly, for the transmission diffractometer, the sample would be placed near the source (e.g., a 14-cm diameter sample placed 5 m from the source) and a 140-cm diameter detector would be 52 m from the source. For a $\Delta\sigma/\sigma$ of 1% the statistical precision on the integral of the scattering peak will be from 10 to 50 times better for the transmission diffractometer. This analysis assumes no background for the HRD1. For a $\Delta\sigma/\sigma$ of 10% another factor of three may be realized using the transmission geometry. Note that this result is primarily due to the use of a much larger sample and detector in transmission geometry. (For HRD1 a sample 1.0 cm in diameter and 0.5 cm thick is required.) Thus if large samples are available, the transmission geometry is apparently superior. Alternatively some of the increased count rate may be sacrificed for even better resolution.

Finally, because position information is preserved in transmission geometry several new applications can be considered. For example, the measurement of strain as a function of position for a nonuniformly stressed sample and isotope and chemical selective radiography using the strongest diffraction edges.

REFERENCES

1. K. F. Graham and J. M. Carpenter, Nucl. Instrum. Methods, 85, 163 (1970).
2. J. M. Carpenter, D. L. Price, and N. J. Swanson, IPNS--A National Facility for Condensed Matter Research, ANL-78-88 (Argonne National Laboratory, IL, 1978), p. 145.

NEUTRON SCATTERING EXPERIMENTS AT PULSED SPALLATION
NEUTRON SOURCE (KENS)

Y. Ishikawa and Y. Endoh
Physics Department, Tohoku University, Sendai 980

N. Watanabe
National Laboratory for High Energy Physics,
Tsukuba, 330-32

K. Inoue
Department of Nuclear Engyneering, Hokkaido University,
Sapporo, 060

ABSTRACT

This paper briefly summarises the results obtained with five spectrometers HIT, MAX, LAM, SAN and TOP installed at the KENS spallation neutron source during the preceding half year's operation. The results indicate that the KENS facility is quite promising, providing several powerful new tools in the fields of physics, chemistry and biology.

INTRODUCTION

The KENS neutron source is a pulsed spallation neutron source using the 500 MeV proton beams from the KEK booster synchrotron. The proton beams are stopped at a tungsten target ($78^H \times 57^W \times 120^D$ mm^3) and the fast neutrons are moderated by two kinds of moderators; a polyethylene moderator at ambient temperature and a solid methane moderator at 18.5 K. The 4π equivalent peak flux intensities of the cold (3 meV), thermal (81 meV) and epithermal (1 eV) neutrons at these moderator surfaces are estimated to be 1.6×10^{15}, 5.7×10^{14} and 2.6×10^{14} (n/cm^2, sec,ev) respectively at full power operation (6×10^{11}ppp$\times$ 38 p/2.5 sec). The time averaged beam intensities at the beam hole exit (4 m from the moderators) are 1.4×10^6n/cm^2 sec.for thermal neutrons and 6×10^5 n/cm^{2}sec.for cold neutrons respectively. The time averaged intensity of cold neutron beams at the exit of the guide tubes is about 1×10^5 n/cm^{2}sec.. Special care has been taken to reduce the fast neutrons back grounds.

The facility started to operate in June 1980 and measurements with the five spectrometers HIT, MAX, LAM, SAN and TOP have been performed since October 1980. This paper briefly summarises the results obtained during these seven months operation. The details of the KENS facility and of the spectrometers are described in a previous paper[1] as well as in a number of papers in KENS

ISSN:0094-243X/82/890057-10$3.00 Copyright 1982 American Institute of Physics

58

Report I[2] and in the Proceedings of ICANS-IV[3].

EXPERIMENTAL RESULTS

1. HIT
 HIT is the total scattering spectrometer constructed
to measure the precise structure factor, $S(Q)$, of liquids
and amorphous solids over a wide range momentum transfer
($0.1 \leq Q \leq 100$ Å^{-1}) at a high rate of data collection[4]. The
machine has fifty ^{3}He counters and is equipped with a
sample changer which can accomodate up to six different
samples. One of the main areas of study with the HIT is
the determination of atomic correlations in the amorphous
state. Some metal-metal alloy glasses such as Pd-Zr,
Cu-Ti, Ni-Zr and Ni-Ti have already been measured.
Neutron zero alloys ($<b>=0$) are particularly interesting,
because they enable the direct observation of pure con-
centration-concentration correlations. An example is

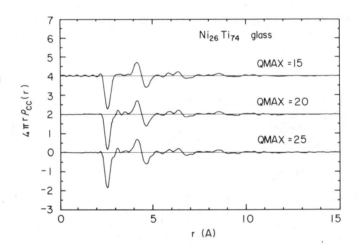

Fig. 1. $G_{cc}(r)$ of Ni-74at%Ti zero alloy glass

shown in Fig. 1 where the pair distribution functions of
chemical short range order $G_{cc}(r)$ of $Ni_{26}Ti_{74}$ glass are
plotted for different cut off wave vectors Q_{max}[5]. The
results show clearly that chemical short range order does
exist in this metal glass. Note that this kind of
measurement is quite difficult to perform with conventional spectrometers because of poor scattering intensity.
 The high data collection capability of HIT makes it
possible to measure many samples covering a wide range of
compositions, while the capability of measuring high

momentum transfers Q can provide information on highly
resolved space correlations. By taking advantage of
these merits, deuterium absorbed in metal-metal alloy
glasses were extensively studied over a wide range of
deuterium concentration[6]. In Fig. 2 are displayed the
$S(Q)$ and $G(r)$ of $Pd_{35}Zr_{65}D_{17}$ (solid lines) compared with
those of $Pd_{35}Zr_{65}$ (broken lines).

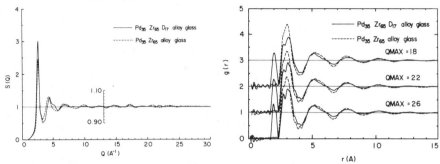

Fig. 2. $S(Q)$ (left) and $G(r)$ (right) of $Pd_{35}Zr_{65}D_{17}$ (solid
          line) and those of $Pd_{35}Zr_{65}$ (broken line).

A split peak is clearly observed at small r in the deute-
rium-containing sample which corresponds to the metal-
deuteron correlation in this glass. Systematic studies
are now in progress.
    Measurement of small and/or absorbing samples is one
of the most promising fields of application for this
spectrometer. The $S(Q)$ of amorphous alloys containing
natural Boron such as $Ni_{60}B_{40}$ was found to be measured
with a good S/N ratio[7].

2. MAX
    MAX is the multi-analyzer crystal spectrometer
equipped with fifteen separate analyzer crystals[8]. The
spectrometer was constructed to test the feasibility of
the TOF machine for studying collective excitations.
Fig. 3 shows the results of scattering from a single
crystal of Fe(Si) measured along <111> around the (110)
reciprocal lattice point. Fig. 3(a) is the TOF spectra,
while Fig. 3(b) shows the magnon and phonon dispersions
along <110> that were simultaneously obtained. The
broken lines are the dispersions previously measured with
triple axis spectrometers. We found that the magnon
scattering could be detected well up to 70 meV. The
magnon dispersions in the antiferromagnetic $\dot{\gamma}Fe_{0.7}Mn_{0.3}$
have also been measured up to 50 meV.
    Although these measurements still represent a stage
of technical development, the results indicate that MAX
is quite promising for studying collective excitations
which had long been considered a weak point of the TOF

machines. The measurements of collective excitations in non-equilibrium states will be the most promising field of application for the MAX, and such measurements will soon be realized.

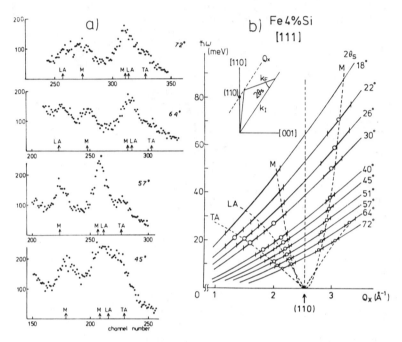

Fig. 3. Coherent scattering from a single crystal of Fe(Si) along <111> around (110) measured with MAX; a) typical TOF spectra of scattered neutrons at selected scattering angles, b) the magnon and phonon dispersions simultaneously measured.

3. LAM

The LAM is the inverted geometry quasielastic spectrometer with conventional energy resolution[9]. By virtue of the inverted geometry and the nearly flat cold source spectra, the distortion of the observed spectra from the theoretical form is quite small. Futhermore, the S/N ratio is extremely high because of the pulsed nature of the source. These features make it possible to perform accurate profile analysis of the spectra[10].

The study of the localized motion of hydrogen in water is a good application of LAM. Although extensive investigation has been performed on this subject, no definitive conclusion has as yet been obtained. Fig. 4 illustrates the energy spectra of scattering from water at ambient temperature measured over a wide range of Q.

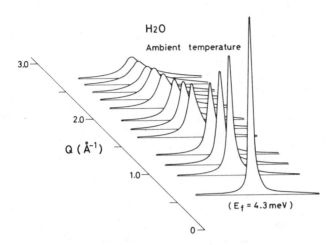

Fig. 4. Energy spectra at different Q of the quasi-
         elastic scattering from water.

The elastic incoherent structure factor, EISF was determined by separating carefully the superimposed broad
Lorentzian from the total spectra and this is plotted
against Q in Fig. 5. The EISF gives information on
localized motions such as diffusive rotation or jump of

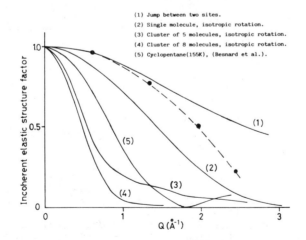

Fig. 5. The Q dependence of EICF of water and its com-
         parison with calculations based on various
         model.

hydrogen between adjacent sites. The solid lines in
Fig. 5 are those calculated based on different models
assuming (1); jump between two sites, (2)-(4); isotropic
rotation of a single molecule and that of clusters composed of five and eight molecules respectively. The line
labelled (5) is the EISF for cyclopentane calculated by
Besnard et al..

The observed curve does not agree with any of these
models, suggesting that the cluster motion models seem
not to be appropriate for describing the hydrogen motion
in water. Futhermore, the single molecule ratation model
should be rejected if the quasi-crystalline structure of
water is considered. The result strongly suggests that
the localized motion of hydrogen atoms consists of jump-
ing over a small distance or rotation with a small gyration radius.

4. SAN

The SAN is the small angle scattering spectrometer
equipped with a movable two dimensional PSD inside a
vacuum chamber[12]. A characteristic of the spectrometer
is its capacity for measuring simultaneously a wide range
of momentum transfers from 3×10^{-3} to 4Å^{-1}. This characteristic is particularly useful for studying non-
equilibrium phenomena such as spinodal decomposition or
magnetic response in spin glass systems. A typical example of the spinodal decomposition is presented in Fig. 6,
where the time evolution of scattering from $Fe_{60}Cr_{40}$
alloy annealed at 515 C for different times is displayed.

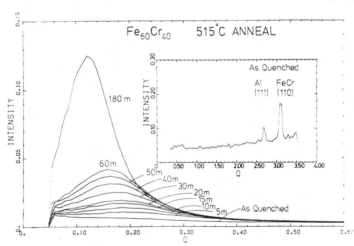

Fig. 6. Small angle scattering from $Fe_{60}Cr_{40}$ in the
process of decomposition at 515 C. The data in
the inset was obtained with fixed counters for
the as quenched alloy.

63

The sample was annealed in a molten tin metal bath at
515 C,followed by quenching in ice water. The measurement was performed at ambient temperature. The data at
low Q values were obtained by the 2D-PSD at 1m position,
while the high Q data shown in the inset was measured
with the fixed counters. Note that small angle scattering and Bragg scattering are simultaneously observed. A
careful examination of Fig. 6 suggests that the decomposition at 515 C cannot be interpreted by a simple linear
spinodal process originally proposed by Cahn[13]. The detailed analysis is now in progress.

Another characteristic of SAN is its convenience for
studying long period structures or long disturbances in
single crystals, which is discussed in detail in a separate paper[14]. The results of the small angle scattering
from a single crystal of MnSi with a long perild of 180 Å,
purple mombrane and hen collagen are also described in
that paper[14]. Two spin glass systems, $Cu_{75}Mn_{25}$ and
$88FeTiO_3$-$12Fe_2O_3$, have also been studied using single
crystals.

Fig. 7 shows the small angle scattering from 1%
polystyrene latex solution (90%D_2O). The measurement
was performed to examine the feasibility of SAN for the

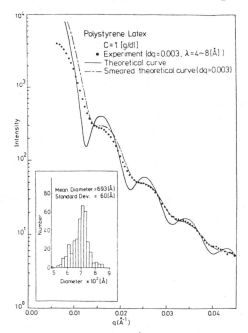

Fig. 7. Small angle scattering from 1% polystyrene-
latex solution. The distribution of diameters shown in the inset was determined by
the electron micro scope.

64

study of polymer solutions. The average diameter of latex determined by the electro-microscope is 693 Å with a standard deviation of 60 Å as shown in an inset to the figure. The solid line is a theoretical curve calculated using the determined distribution, while the chain line is a smeared theoretical curve which takes into account the instrumental resolution (dq=0.003). The agreement between observation and calculation is fairly satisfactory except for the low Q data for which more accurate instrumental corrections such as that due to the converging Soller slit would be necessary. The result suggests that SAN can also be used for this type of experiment.

5. TOP

TOP is the cold polarized neutron spectrometer which polarises cold neutrons with wave lengths between 3.5 Å and 12 Å by means of a curved Soller slit type opical mirror[15]. The mirror is made of films of $Fe_{45}Co_{65}$ 2000A thick, evaporated on commercially available polypropylene films 50μ thick, which are stretched and bent so that all neutrons in the incident beam (50×20mm^2) suffer from total reflection within the mirror. The polarizability determined by Cu_2MnAl analyzer was found to be satisfactory in the shorter wavelength region as shown in Fig. 8 and is uniform across the beam cross

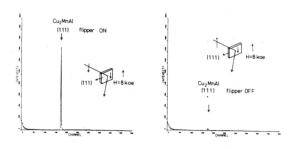

Fig. 8. Wavelength dependence of the polarizability of TOP measured by Cu_2MnAl analyzer.

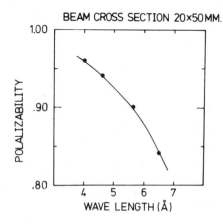

section. The polarizability as well as the uniformity, however, drop significantly at longer wavelengths, a problem which will be overcome by improving the polarizer.

TOP is now used to study the magnetic diffuse scattering from ferromagnetic alloys, magnetic fine particles, magnetic amorphous alloys and so forth. Here we present the recent results of an investigation on the interface magnetism in multi-layer films. In Fig. 9 is shown the flipping ratio of reflections from FeSb and FePd bilayer structures of about 100 Å thickness, plotted against the applied magnetic field. The difference in the flipping ratio for different interfaces and for different reflections indicates that the interface state of ferromagnetic iron evidently depends on the junction element. The effect of the magnetic field is also be interesting and a detailed discussion on this problem will be published elsewhere.

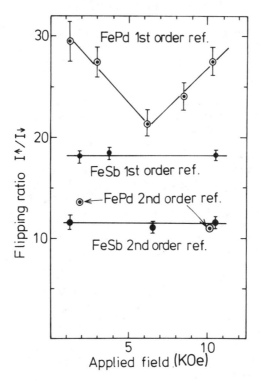

Fig. 9. Flipping ratios of the first and second reflections from different bilayer materials plotted against external magnetic field.

CONCLUSION

Although the KENS neutron source provides, for a moment, a medium neutron flux which does not exceed the flux of the medium flux reactor, we have been able to obtain many interesting data with a good S/N ratio. In the case of the established spectrometers HIT and LAM, the results exceed those of a high flux reactor. MAX, SAN and TOP are still in a stage of developemnt, but we have already got results which are different in quality compared with those obtained using a conventional reactor source. Since technical developments are now progressing rapidly and the flux intensity will increase nearly one order of magnitude in the near future, the KENS facility may contribute significantly to the research in many different feilds. Three new spectrometers PEN (polarized epithermal neutron spectrometer), FOX (four circle single crystal spectrometer)and RAC (resonance detector spectrometer) are now under constuction.

The experimental results summarised in this paper were obtained by many collaborators, the contributions of all of whom the authors gratefully acknowledge. Their thanks are particularly due to H. Sasaki, Director of the Booster Synchrotron Utilization Facility for his continuous help and encounragement and to J.M. Newsam for a critical reading of manuscript.

REFERENCES

1. Y. Ishikawa and N. Watanabe, KEK report-19 (1978).
2. Y. Ishikawa, KENS Report I., KEK Internal-1 (1980).
3. Y. Ishikawa, N. Watanabe, Y. Endoh, N. Niimura and J.M. Newsam, Proceedings of ICANS-IV (1981).
4. N. Watanabe et al., Proc. ICANS-IV, 539 (1981).
5. T. Furusaka et al., to be presented at the 4th International Conf. Rapidly Quenched Metals (August 1981).
6. K. Kai et al., ibid.
7. N. Watanabe et al., Proc. ICANS-V (1981) to be published.
8. K. Tajima et al., Proc. ICANS-IV, 600 (1981).
9. K. Inoue et al., Proc. ICANS-IV, 592 (1981).
10. K. Inoue et al., Nuclear Inst. Meth. 178 (1980) 459.
11. M.E. Besnard et al., Neutron Inelast.Scattering 1977 IAEA 1, 361 (1978).
12. Y. Ishikawa et al., Proc. ICANS-IV 563 (1981).
13. J.W. Cahn, Act. Met. 9 795 (1961).
14. N. Niimura et al., present Proceedings.
15. Y. Endoh et al., Proc. ICANS-IV (1981).

HOW GOOD A CHOICE ARE THE 5oo µs-PULSES IN THE SNQ-PROJEKT?

G.S. Bauer and B. Alefeld
Institut für Festkörperforschung, Kernforschungsanlage Jülich GmbH
D-517o Jülich, FRG

ABSTRACT

The new high power spallation neutron source proposed in the Federal Republic of Germany is designed to be competitive with a high flux reactor in its time average flux of $\bar{\phi}$ = 7·1o^{14} cm^{-2}s^{-1} and will operate in an intensity-modulated mode with 1oo Hz repetition rate and 5oo µs pulse width to yield a peak flux of $\hat{\phi}$ = 1.3·1o^{16} cm^{-2}s^{-1}. This source will serve conventional cw-instruments, allowing background reduction and elimination or simultaneous use of higher order monochromator reflections by a time gate on the detector. If equipped with synchronized choppers, it provides excellent conditions for conventional and inverted time-of-flight techniques as well as correlation techniques. Special instruments to exploit this time structure can be designed.

1. INTRODUCTION

The choice of a linear accelerator to feed the German high power spallation source implies that only a limited peak proton current (1oo mA) will be available. Together with the time average proton beam power of 5.5 MW (5 mA at 1.1 GeV), required to achieve the design goal of $\bar{\phi} \geq$ 6·1o^{14} cm^{-2}s^{-1} with a lead target, this sets the duty cycle to 1/2o. The combination of 1oo Hz repetition rate and 5oo µs pulse width has been selected to provide good experimental conditions for a large variety of instrument types.

2. TIME STRUCTURE OF THE NEUTRON FIELD

Two different moderators will be provided, a large D_2O-tank above and a Pb-reflected H_2O moderator, with a grooved surface viewed by the beam holes, below the target[1]. The size and position of the H_2O moderator is optimized for high integrated flux. The time structure of its thermal neutron field, as inferred from measurements[2], is shown in Fig. 1. Possible improvements in peak flux and pulse width in the H_2O-moderator by shortening the proton pulse at constant integrated intensity are indicated in Fig. 2. As can be seen, the maximum possible improvement is a factor of 3.4 in peak flux. The intrinsic pulse width would still not be sufficient for TOF-experiments without a chopper. Shortening of the neutron pulse by decoupling and poisoning the moderator results in serious penalties in the integrated thermal flux (see Fig. 3). Moreover, the maximum number of protons per pulse so far proposed for a rapid cycling synchrotron (6o Hz) is a factor of o.16 lower than in the SNQ-linac pulse and therefore the peak flux, even in an undecoupled moderator, would be a factor of o.55 lower, with a reduction factor of o.1 in $\bar{\phi}$.

ISSN:0094-243X/82/890067-05$3.00 Copyright 1982 American Institute of Physics

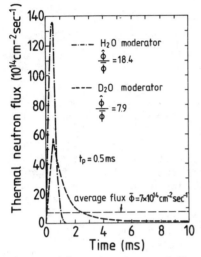

Fig. 1 Time structure of the thermal neutrons in the SNQ-moderators.

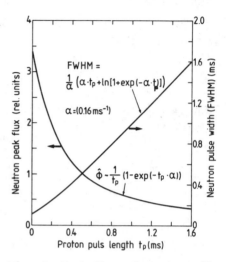

Fig. 2 Peak thermal neutron flux and pulse width (FWHM) in an unde-coupled H_2O-moderator as a function of proton pulse length. $1/\alpha$ =.16ms [4]

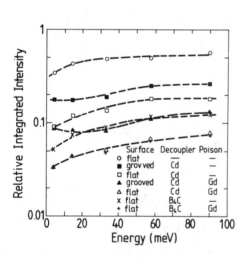

Fig. 3 Energy dependent integrated neutron intensity for an H_2O moderator with various decouplers and poisons relative to the proposed SNQ-moderator [3,4].

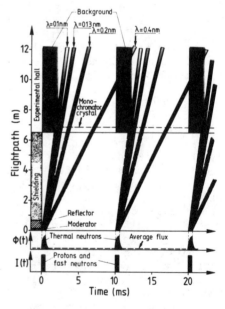

Fig. 4 Space-time diagram for neutrons of different velocities for the SNQ-pulse sequence.

3. USES OF THE SNQ-TIME STRUCTURE

a.) <u>Traditional cw-instruments</u> (TAS and 4-circle diffractometers), although essentially depending on ϕ will profit from the time
structure by the use of a time gate on the detector to reduce background (through reduced counting time, different flight times and its
intrinsic time structure) and to eliminate higher order reflections
from the monochromator crystal. With a detector placed at an effective distance of 1o m from the source, this is easily accomplished
with the proposed time structure (see Fig. 4).

b.) <u>Traditional TOF-instruments</u> with a monochromator crystal and
a chopper close to the sample will fully benefit from the peak-to-
average flux ratio. Synchronisation of the choppers is relatively
easy because of the "long" source pulse. By the same token as before,
higher order monochromator reflections will not pass through the
chopper. The repetition rate of 100 Hz may be low in some cases, but
constitutes a good compromise for the bulk of the instruments.

c.) <u>Inverted time-of-flight instruments</u> are likely to require
pulse repetition rates lower than 1oo Hz to avoid frame overlap, if
the full white spectrum is allowed to impinge on the sample. Apart
from the risk of causing unneccesarily high activation of the sample
and its environment, this also reduces the possible data collection
rate. Using a chopper in the beam will alleviate both of these problems. In Fig. 5, the energy band transmitted through a chopper
(whose opening time is neglected relative to the 5oo µs pulse width)
is shown as a function of the delay time Δt for a chopper at 2 m from
the moderator in the target shield. For an overall flight path of
1o m, the fast background from the following pulse will interfere
only with incident energies between 4 and 5 meV which are of low intensity in a thermal spectrum anyway. Usually the analyser energy
will even be higher than that, which makes also very small energy
transfers accessible. By varying the delay time Δt, the energy band
incident on the sample can be tuned to the phenomenon under investigation. At longer flight paths, which might be required for good resolution, the "forbidden" window is shifted to higher energies, but
a variable moderator-to-sample distance will allow to cover the
whole range. Although the integrated intensity cannot be used in full
in this technique, the usable peak flux will be higher than in any
other facility. Since the time interval of neutron arrival at the
sample is D2/D1 · T (Fig. 5), the full interval between two pulses
will only be covered with a detector at 4o m from the moderator, if
the chopper is at 2 m. For shorter distances it will be possible to
determine the background between the pulses of "useful" neutrons.
These considerations show that, although not optimized for inverted
TOF experiments, the SNQ will provide good conditions for this technique. Development of internally cooled choppers with friction-free
magnetic bearings for use in relatively high radiation fields will
therefore be an important issue. The fact that a repetition rate as
high as 1oo Hz can be used in inverted TOF, is an additional advantage of this approach.

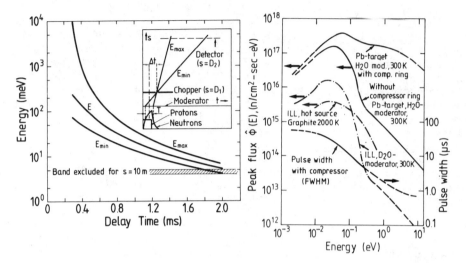

Fig. 5 Energy band as a function of chopper delay time for a
chopper at 2 m from the moderator. The inset shows the space-
time diagram.

Fig. 6 Peak flux as a function of
neutron energy for the SNQ without
and with proton compressor as compared to the time average flux of
the HFR Grenoble.

d.) Specially adapted instruments have been proposed[5] to exploit the time-structure of the SNQ. These include multiple-order
back-scattering spectrometers designed to use up to three orders of
the monochromator crystal simultaneously, as well as a technique,
where neutrons of slightly different energies are reflected from a
stack of monochromator crystals with a temperature gradient in such
a way as to arrive in adjacent time intervals at the sample. This
latter technique avoids the need for a Doppler shift on the monochromator and makes it possible to use the peak flux during the full
time interval between two pulses, while still retaining the high
energy resolution of the back-scattering technique.
     Finally, it should be noted that a superposition of the pseudo-
statistical pulse sequence of a correlation chopper on the time
structure of the SNQ will alleviate the well-known problem of correlation techniques, that a strong line in the spectrum gives rise to
a high background everywhere. The crude separation by conventional
TOF leads to a situation, where the background under each line is
determined by its own intensity only and hence full benefit can be
drawn from the good duty cycle of the correlation chopper, using the
integrated intensity.

4. CONCLUSIONS

The 1oo Hz - 5oo μs time structure of the SNQ basic concept has a
potential to satisfy an extraordinarily broad spectrum of experimental techniques, some of which are traditional cw-instruments but
have reached a high degree of reliability and perfection and will

continue to play an important role also in the future. At the same time, it offers most of the advantages of a pulsed source as far as thermal neutron utilisation is concerned and therefore makes the source an unusually versatile one. In order to exploit the properties of the slowing-down regime at energies above the thermal range, it will be necessary to supplement the facility by a proton pulse compressor[6], which would lead to tremendous gains in peak flux in the epithermal energy range, as shown in Fig. 6.

REFERENCES

1. G.S. Bauer, Proc. ICANS IV, report KENS-2 pp 154-180 (1981) and G.S. Bauer, H. Sebening, J.E. Vetter and H. Willax, eds. report JÜL-SPEZ-113 and KfK 3175
2. G.S. Bauer, W.E. Fischer, F. Gompf, M. Küchle, W. Reichardt and H. Spitzer, Proc. ICANS V, to be published as JÜL-Conf report (1981)
3. G.S. Bauer, J.P. Delahaye, H. Spitzer, A.D. Taylor and K. Werner Proc. ICANS V, to be published as JÜL-Conf report (1981)
4. G.S. Bauer, H.M. Conrad, H. Spitzer, K. Friedrich and G. Milleret, Proc. ICANS V, to be published as JÜL-Conf report (1981)
5. B. Alefeld, Proc. ICANS-IV, report KENS-2 pp 678-689 (1981)
6. G. Schaffer, ed.: IKOR, JÜL-SPEZ-114 (1981)

NEUTRON DIFFRACTION STUDIES OF $CeVO_4$, $PrVO_4$, and $ScVO_4$

J. Faber, Jr., and A. T. Aldred
Materials Science Division, Argonne National Laboratory
Argonne, Illinois 60439, U.S.A.

ABSTRACT

High resolution neutron powder diffraction studies (T = 300 K) have been carried out on $CeVO_4$, $PrVO_4$, and $ScVO_4$. Bond distance determinations show that although the rare earth ion size varies over a wide range, the characteristic V-O distance remains the same. The rare-earth-oxygen distances, however, approach an isotropic configuration with increasing rare earth ion size.

INTRODUCTION

Rare earth vanadates, AVO_4, crystallize with the tetragonal zircon-type structure[1,2], $I4_1/amd$. Figure 1 illustrates the unit cell arrangement in which each V atom is surrounded by an oxygen tetrahedron and each rare earth atom is eight-fold coordinated by nearest neighbor oxygen atoms[2] (see the dashed line in Fig. 1). An interest in Jahn-Teller interactions, centered mainly on the mid-rare earth series elements (e.g., $GdVO_4$ and $TbVO_4$), has led to several detailed structural investigations. However, vanadates compounds with extremes in A atom size have not been extensively studied. The purpose of our neutron experiments is to begin a systematic examination of bonding interactions in AVO_4 compounds, using high resolution neutron powder diffraction techniques. The stability of these oxide materials (and related structure-types) and the wide range of A atom size that can be accommodated in the structure may suggest potential usefulness as hosts for radioactive wastes.

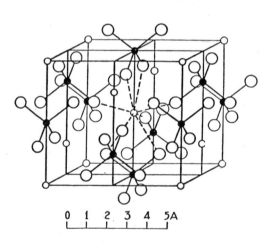

0 1 2 3 4 5A

Fig. 1. Zircon-type crystal structure for the rare earth vanadate compounds, AVO_4. The small open circles represent rare earth atoms, the small closed circles are V atoms, and the large open circles are oxygen atoms.

EXPERIMENT

Polycrystalline samples of $CeVO_4$, $PrVO_4$, and $ScVO_4$ have been examined with neutron time-of-flight (TOF) techniques at ZING-P', the Argonne prototype pulsed source. The experiments were carried out at T = 300 K using the high resolution TOF powder diffractometer[3]. The mechanically time focused back-scattering detector banks gave $2.5 < Q = 4\pi \sin\theta/\lambda < 25$ Å^{-1} with $\Delta Q/Q$(fwhm)=0.0033. Rietveld least-squares structure refinement techniques, modified for TOF data, were used to obtain values for the structural parameters. The model assumed in the analysis was the zircon structure, $I4_1/amd$. Approximately 300 Bragg reflections were collected for each sample. In Table 1 we show the results obtained for the lattice and atom positional parameters. An indication of goodness of fit, wR, based on the weighted sum of residuals is also given in the table. Small estimated standard deviations associated with the structural results demonstrate that neutron scattering experiments are ideal for determining the A-O and V-O atomic distances in AVO$_4$ compounds because the oxygen scattering length, b_0, is comparable to b_A, and the V atoms are at high symmetry positions in the unit cell.

DISCUSSION

To obtain insight into the general systematics of bonding in rare earth vanadate compounds, the parameters in Table 1 were used to calculate nearest neighbor interatomic distances. These results are illustrated in Fig. 2 where we show A-O and V-O nearest neighbor bond distances plotted as a function of eight-fold crystal radii, r_A. The radii values illustrated in Fig. 2 assume that the A atoms are tripositive ions in the lattice. To span the entire range of A ion vanadate compounds that crystallize in the zircon-type structure, we treat Sc as a small pseudo-rare earth atom. Several important observations can be drawn from the results illustrated in Fig. 2. Whereas the A atom radii span a wide range of values, the V-O bond distance is essentially independent of rare earth ion size. Moreover, the experimental values lie

Table 1. Lattice and atom positional parameters obtained from profile refinements of CeVO$_4$, PrVO$_4$, and ScVO$_4$. In the centro-symmetric representation for $I4_1/amd$, atom fractional coordinates are: A: 0,3/4,1/8; V: 0,1/4,3/8; Ox: 0,y,z. The values in parenthesis are estimated standard deviations applied to least-significant digits.

Compound	a_0	c_0	y_{0x}	z_{0x}	wR(%)
CeVO$_4$	7.4017(2)	6.4984(2)	0.4280(2)	0.2065(2)	3.9
PrVO$_4$	7.3636(2)	6.4661(1)	0.4290(2)	0.2059(2)	3.0
ScVO$_4$	6.7785(1)	6.1354(1)	0.4430(1)	0.1965(1)	2.2

74

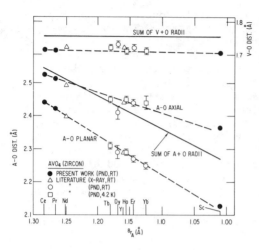

Fig. 2. Nearest neighbor interatomic distances as a function of
crystal radii, r_A. Note that V-O
distances at the top of the
figure require the right hand
scale. Our results are given by
the closed circles; the error
bars are the size of the
points. Open symbols are values
extracted from the literature.[4]

systematically below the V+O
radii sum. Since these radii
sums imply hard sphere interactions, it is reasonable to
attribute the differences to
covalent bonding effects in the
VO_4 tetrahedra. As shown in
Fig. 2, two characteristic A-O
nearest neighbor distances
exist in this structure; as
the rare earth ion size increases, the differences in
these characteristic dis-
tances decrease. However,
only the A-O axial component
in the bonding approaches
the hard sphere A+O radii
sum. For the largest stable
rare earth vanadate, $CeVO_4$,
we see that considerable
anisotropy in A atom bonding
environment persists.

Although the residuals
given in Table 1 are quite satisfactory, we have detected
systematic differences between
observed and calculated Bragg
line shapes (not shown here)

that suggest a small distortion of the tetragonal structure to
orthorhombic symmetry. Experiments are planned to characterize the
magnitude of the distortion as a function of A atom size, and a
more detailed account of these observations will be published
elsewhere.[5]

REFERENCES

1.   R. W. G. Wyckoff, Crystal Structures, Interscience, N.Y.
      1965, Vol. 3, p. 15.
2.   L. Bragg and G. F. Claringbull, Crystal Structures of
      Minerals, G. Bell and Sons, London, 1965, p. 185.
3.   J. D. Jorgensen and F. J. Rotella, J. Appl. Cryst., 1981 (in
      press).
4.   H. Fuess and A. Kallel, J. Sol. St. Chem. 5, 11 (1972); J. A.
      Baglio and O. J. Sovers, J. Sol. St. Chem. 3, 458 (1971), and
      references therein.
5.   J. Faber and A. T. Aldred, to be published.

*This work was supported by the U.S. Department of Energy.

CRYSTAL STRUCTURE AND THERMAL EXPANSION OF
α-SiO₂ AT LOW TEMPERATURES

G. A. Lager*
University of Louisville, Louisville, KY   40292

J. D. Jorgensen and F. J. Rotella
Argonne National Laboratory, Argonne, IL   60439

ABSTRACT

The crystal structure of α-SiO$_2$ (low-quartz) has been refined at 296°, 78° and 13°K from time-of-flight neutron powder diffraction data. The major effect of temperature from 296° to 78°K is a nearly rigid body rotation of the SiO$_4$ tetrahedra about the two-fold axis of symmetry. Below 78°K the mechanism of thermal expansion is tetrahedral distortion rather than rotation. When compared to published thermal expansion data, the unit cell expansion at low temperatures exhibits a pronounced nonlinearity which can be ascribed to the invariance of the intertetrahedral (Si-O-Si) angle.

INTRODUCTION

This study reports the first refinement of the α-SiO$_2$ structure at temperatures below 94°K. The analysis of interatomic distances and angles at 13° and 78°K illustrates the relation between structure and thermal expansion at low temperatures.

EXPERIMENTAL

A sample of crystalline α-SiO$_2$ powder (Fisher Scientific Co., S-152: 99.7% SiO$_2$, 0.1% Al$_2$O$_3$, 0.2% [Fe$_2$O$_3$ + CaO + MgO + H$_2$O]) was sealed in a thin-walled vanadium can and cooled by a closed-cycle DISPLEX refrigerator (Air Products, Inc.). Diffraction data were collected at 296°, 78° and 13°K in the back-scattering detector banks ($2\theta = \pm$ 160°) of the high-resolution time-of-flight powder diffractometer (HRPD) at Argonne National Laboratory's ZING-P' pulsed spallation neutron source. The design and operation of the HRPD have been described elsewhere.[1] The data at all three temperatures were fit over the range of d-spacings from 0.55 Å to 2.28 Å, which included 1840 observed profile points and 442 allowed Bragg reflections. Rietveld refinements[2] in space group P3$_1$21[r(+) setting] with 21 variable parameters converged to discrepancy indices (R_{wp})[1] of 3.11% (296°K), 4.14% (78°K) and 3.87% (13°K). The structural

*Work performed at Argonne National Laboratory while a Faculty Research Participant. Program administered by the Argonne Division of Educational Programs, and supported by the U. S. Department of Energy, Office of Energy Research, through its University/National Laboratory Cooperative Program.

parameters and interatomic distances and angles are reported as a function of temperature in Tables I and II, respectively.

Table I. Structural Parameters (x 10^4) for α-SiO_2*

Parameter		T (°K) 13	78	296
Si	x	4680(2)	4684(2)	4700(2)
	β_{11}	33(3)	30(3)	68(3)
	β_{22}	41(5)	34(6)	54(6)
	β_{33}	15(3)	16(3)	44(3)
	β_{13}	2(2)	2(2)	6(2)
O	x	4124(2)	4125(2)	4131(2)
	y	2712(1)	2707(1)	2677(2)
	z	-1163(1)	-1163(1)	-1189(1)
	β_{11}	79(3)	84(3)	195(4)
	β_{22}	57(3)	59(3)	120(3)
	β_{33}	30(1)	30(1)	72(1)
	β_{12}	42(3)	48(3)	104(3)
	β_{13}	4(2)	6(2)	23(2)
	β_{23}	11(1)	13(2)	45(2)
a (Å)		4.9021(1)	4.9030(1)	4.9141(1)
c (Å)		5.3997(1)	5.3999(1)	5.4060(1)
V ($Å^3$)		112.37(1)	112.42(1)	113.06(1)

*Thermal parameters are of the form: $\exp[-(h^2\beta_{11} + k^2\beta_{22} + \ell^2\beta_{33} + 2hk\beta_{12} + 2h\ell\beta_{13} + 2k\ell\beta_{23})]$. Si is at position (x, 0, 0) with $\beta_{12} = \beta_{22}/2$ and $\beta_{23} = 2\beta_{13}$. Numbers in parentheses in Tables I and II are standard deviations of the last significant figures.

Table II. Selected Distances and Angles for α-SiO_2

Quantity	T (°K) 13	78	296
Si-Si(Å)	3.0530(1)	3.0532(1)	3.0577(1)
Si-O(Å)	1.612(1)	1.611(1)	1.609(1)
Si-O(Å)	1.613(1)	1.614(1)	1.611(1)
O-Si-O(°)	110.71(5)	110.67(5)	110.63(5)
O-Si-O(°)	108.57(1)	108.61(1)	108.75(1)
O-Si-O(°)	109.38(8)	109.48(8)	109.32(8)
O-Si-O(°)	108.89(8)	108.80(8)	108.75(8)
Si-O-Si(°)	142.41(6)	142.48(6)	143.45(6)
δ(°)*	17.25(8)	17.25(8)	16.73(8)

*Tetrahedral tilt angle defined as $\tan \delta = \frac{2}{9}\sqrt{3}\,(c/a)[(6z-1)/x]$ where x and z refer to the oxygen parameters.[7]

DISCUSSION

Unit-cell parameters determined at 13°, 78° and 296°K (Table I) are plotted in Fig. 1 together with thermal expansion data from the Thermophysical Properties Research Center[3] (TPRC) and the low-temperature X-ray study of Le Page et al.[4] The TPRC data, reported as percent thermal expansion, have been normalized to the 296°K unit-cell parameters. As a check on their thermocouple measurements, Le Page et al. extrapolated the cell parameter expression of Lindman[5] (experimentally determined between 283° and 573°K) to 94°K. Judging from Fig. 1, the temperature of their experiments may be lower than reported.

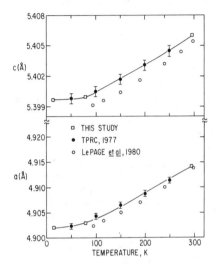

Fig. 1. Temperature dependence of unit-cell parameters. Curves were visually fit.

The effect of temperature on the quartz structure is well known.[6] The major effect of cooling is a nearly rigid body rotation of the SiO_4 tetrahedra with a corresponding decrease in the intertetrahedral angle (Si-O-Si). The small change in the cell parameters from 78° to 13°K (Fig. 1) reflects the almost insignificant change ($\sim 1\sigma$) in the Si-O-Si angle over this temperature range (Table II). Below 78°K the tetrahedra no longer rotate (Table II) so that any change in the Si-O-Si angle must result from distortion of the tetrahedra, i.e. the mechanism of thermal expansion is tetrahedral distortion rather than rotation. In terms of the O-Si-O angular variation per degree, tetrahedral distortion increases in the temperature range 78° to 13°K.

REFERENCES

1. J. D. Jorgensen and F. J. Rotella, J. Appl. Cryst., in press.
2. H. M. Rietveld, J. Appl. Cryst. 2, 65 (1969).
3. Thermophysical Properties of Matter, Vol. 13, edited by Y. S. Touloukian (Plenum Publishing Corp., 1977), p. 350.
4. Y. Le Page, L. D. Calvert and E. J. Gabe, J. Phys. Chem. Solids 41, 721 (1980).
5. F. K. Lindman, Acta Soc. Fenn. 46, 1 (1916).
6. H. D. Megaw, Crystal Structures: A Working Approach (W. B. Saunders Co., 1973), p. 453.
7. H. Grimm and B. Dorner, J. Phys. Chem. Solids 36, 407 (1975).

NEUTRON DIFFRACTION INVESTIGATION OF
ORDERED OXYGEN VACANCIES IN $Pb_2Ru_2O_{6.5}$

R. A. Beyerlein, H. S. Horowitz, and J. M. Longo
Exxon Research and Engineering Co., Linden, NJ 07036

J. D. Jorgensen and F. J. Rotella
Argonne National Laboratory, Argonne, IL 60439

INTRODUCTION

A pyrochlore identified as $Pb_2Ru_2O_6$ was originally reported by Longo, et al.[1] A recent investigation[2] has shown[2] the correct stoichiometry to be $Pb_2Ru_2O_{6.5}$. In this paper we report the results of a neutron diffraction investigation of this oxygen deficient pyrochlore.

The general pyrochlore formula may be written as $A_2O' \cdot B_2O_6$, emphasizing the fact that this structure may be viewed as two interpenetrating networks.[3] The B cations are located at the center of oxygen (O) octahedra which share only corners so as to form a system of interconnected cages. The special oxygen O', located at the center of these cages, may be partially or totally absent and is the basis for the anion nonstoichiometry observed in pyrochlores. The A_2O' sublattice may be viewed as a system of intersecting $-A-O'-A-O'-$ zig zag chains (cuprite structure) or as a corner shared network of A_4O' tetrahedra. In the random vacancy model of $Pb_2Ru_2O_{6.5}$ with A = Pb and B = Ru, each O' site is 1/2 occupied. This work confirms the oxygen stoichiometry of this model while at the same time demonstrating that the vacancies are ordered.

EXPERIMENTAL

The $Pb_2Ru_2O_{6.5}$ sample was prepared by solid state reaction similar to a procedure previously described.[1,2] Powder x-ray diffraction showed a single phase pyrochlore with a unit cell edge of a_0 = $10.252 \pm 0.001Å$. Thermogravimetric reduction in hydrogen gave a formula $Pb_2Ru_2O_{6.5 \pm .04}$. Time-of-flight neutron diffraction data were collected at the High Resolution Powder Diffractometer[4] (HPRD) at Argonne's ZING P' pulsed neutron source using the $2\Theta = 160$ detector bank. Data collection required about 52 hours of machine time for a $4 cm^3$ powder sample. A full description of the diffractometer and the modified Rietveld profile analysis which was used in analyzing the data is given elsewhere.[4,5]

RESULTS AND DISCUSSION

The data were first refined in the pyrochlore space group Fd3m, origin at $\bar{3}$m, with Pb at 16c, Ru at 16d, O at 48f, and special oxygen O' at 8a. The refinement included 3790 data points, 437 allowed reflections and 19 refined parameters, including the cubic cell constant, occupation factors and full anisotropic thermal parameters for all atoms. The best fit profile is shown in Figure 1. The weighted R

ISSN:0094-243X/82/890078-03$3.00 Copyright 1982 American Institute of Phys

value calculated point by point with background included was 3.99% while the Rietveld R, calculated after removing the background, was 6.06%. The value for the oxygen position parameter was x_O = 0.4268 (1) in agreement with the value determined previously by x-ray diffraction[1], x_O = 0.429(3). The fractional occupancy of the anion defect site, O' at 8a, was determined to be 0.50(1), in agreement with the stoichiometry determined from our thermogravimetric data.

While our initial refinement appeared successful, close examina-tion of the neutron data and our own powder x-ray data showed several weak lines of the type hk0 where h+k=2n which are forbidden in Fd3m. The presence of a weak but definite (420) in the neutron data and (420) and (640) in our x-ray data is consistent with the loss of the center of inversion symmetry at the cation sites. This loss of symmetry is expected if the oxygen vacancies are ordered. A second refinement was performed using the space group F$\bar{4}$3m in which the oxygen vacancies are ordered and there is the possibility of associat-ed cation displacement along [111]. Refined values for the position, thermal, and occupation parameters are given in Table I. For the oxygen atoms O_1 and O_2 which form the octahedral cage around each Ru, the Ru-O bond lengths are 1.954(4) and 1.969(4)Å, respectively. These bond lengths agree with the Ru-O bond length in Fd3m, 1.961(1) Å, within two standard deviations. The occupation parameter for the special oxygen at 4d refined to a value 0.96(2) with the alternative oxygen defect site at 4a vacant. This yields a stoichiometry parameter for the O' site of 0.48(1). The refined values for the Pb and Ru atom positions at 16e (x,x,x), x_{Pb} = 0.8772(2) and x_{Ru} = 0.3754(2) show that Pb has moved off its position of inversion symmetry, x_{Pb} = 0.875, in the Fd3m structure while Ru has remained fixed at x_{Ru} = 0.375 within two standard deviations. Each Pb atom is

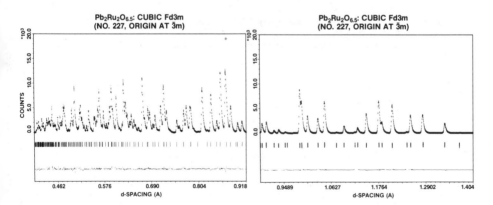

Fig. 1. Refinement profile for $Pb_2Ru_2O_{6.5}$. Plus marks (+) are the raw data points. The solid line is the best-fit profile. A differ-ence (observed-calculated) curve appears at the bottom. Tick marks below the profile indicate the positions of all allowed reflections included in the calculation. Background has been removed prior to plotting. The portion of the profile for 1.4Å<d<2.38Å is omitted.

displaced by 0.04(1)Å toward its associated vacancy as shown in Fig. 2, so that, along the [111] direction, the Pb-vacancy distance is 2.180(1)Å while Pb-0 distance is 2.260(1)Å. These results show that the anion vacancies in $Pb_2Ru_2O_{6.5}$ are ordered with an associated ordering of the Pb cations. To our knowledge this is the first example of anion/vacancy ordering in a defect pyrochlore.

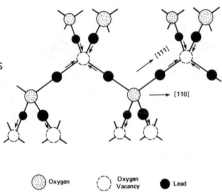

◎ Oxygen   ○ Oxygen Vacancy   ● Lead

Fig. 2 $Pb_2O_{.5}$ sublattice in the ordered vacancy model for $Pb_2Ru_2O_{6.5}$.

TABLE I.   Structural Parameters for $Pb_2Ru_2O_{6.5}$
(Space Group F$\bar{4}$3m, Ordered Vacancy Model)

a_o = 10.2519(2) $R_{weighted}$ = 3.63% $R_{Rietveld}$ = 5.43%

Atom	Site	Fractional Occupation	x^a		$B(\overset{\circ}{A}{}^2)^b$
Pb	16e (X X X)	0.99(1)	0.8772(2)	[7/8]	0.51(4)
Ru	16e (X X X)	0.97(1)	0.3754(2)	[3/8]	0.18(3)
O_1	24f (X 0 0)	1.0	0.3028(2)	[0.3018(1)]	0.68(10)
O_2	24g (X 1/4 1/4)	1.0	0.4492(2)	[0.4482(1)]	0.50(10)
O	4d (3/4 3/4 3/4)	0.96(2)	3/4	—	0.47(9)

a. Values in brackets [] are refined positions in Fd3m with a change of origin by (1/8 1/8 1/8) from $\bar{3}$m. X_{O1} and X_{O2} are related by symmetry in Fd3m.

b. Full anisotropic temperature parameters were refined, but only B_{eff} (isotropic) = $4a_o^2\beta_{11}$ is given here.

REFERENCES

1. J. M. Longo, P. M. Raccah, and J. B. Goodenough, Mat. Res. Bull. 4, 191 (1969).
2. H. S. Horowitz, J. M. Longo, and J. T. Lewandowski, Mat. Res. Bull. 16, 489 (1981).
3. A. W. Sleight, Inorganic Chem. 7, 1704 (1968).
4. J. D. Jorgensen, and F. J. Rotella, J. Appl. Cryst., in press, April, 1981.
5. J. D. Jorgensen, D. G. Hinks, and F. J. Rotella in Ternary Superconductors, Shenoy, Dunlap, Fradlin, eds., Elsevier North Holland, 69-73 (1981).

HYDROGEN CONTENT AND LEAD VACANCIES IN β-PbO₂ FROM ACTIVE BATTERY PLATES*

J. D. Jorgensen, R. Varma, F. J. Rotella, G. Cook and N. P. Yao

Solid State Science Division and Chemical Engineering Division
Argonne National Laboratory, Argonne, IL 60439

Numerous chemical analyses of lead dioxide, PbO_2, have shown it to be a nonstoichiometric compound with a deficiency of active oxygen and sufficient adsorbed or lattice water or hydroxl ions to give an overall Pb:O ratio near 1:2.[1-4] The oxygen deficiency and the amount of hydrogen (assumed to be in the form of H_2O or OH) was found to depend on the method of preparation. Of course, from the chemical analyses alone it was impossible to determine whether oxygen vacancies actually existed in the PbO_2 structure or H_2O or OH groups occupied some of the oxygen sites.

Attempts to understand mechanisms of failure of lead-acid battery electrodes have recently focussed attention on the incorporation of hydrogen in electrochemically formed PbO_2. In a lead-acid cell the positive electrode (in the charged state) is PbO_2. During discharge the reaction can be represented as

$$PbO_2 + 3H^+ + HSO_4^- + 2e^- \rightarrow PbSO_4 + 2H_2O.$$

The reverse reaction occurs during charging. One of the proposed mechanisms of failure after repeated cycling is a reduction in the electrochemical activity of the PbO_2; i.e., a reduction in the ability of PbO_2 to participate in the discharge reaction. A recent NMR study indicates that freshly-formed, electrochemically active PbO_2 contains a hydrogen species not present in PbO_2 cycled to failure or PbO_2 prepared by chemical methods.[5,6] This result suggests that hydrogen incorporated in electrochemically formed PbO_2 may play an important role in its electrochemical activity.

For the neutron diffraction studies reported in this paper two 10 gm samples of β-PbO_2 were obtained from cycled commercial lead-acid battery plates. One cell contained H_2SO_4 electrolyte while the other contained D_2SO_4 electrolyte. The diffraction data were obtained on the back scattering detectors (2θ = ± 160°) of the high resolution powder diffractometer at Argonne's ZING-P' pulsed neutron source.[7]

The presence of hydrogen (deuterium) in the samples was immediately obvious from the difference in incoherent background for the

---

*Work supported by the U.S. Department of Energy.

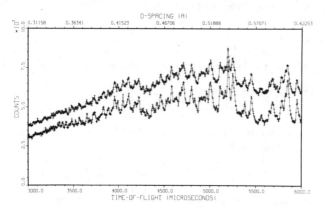

Fig. 1. Normalized raw time-of-flight data
in the high neutron energy region
for PbO_2:H (top curve) and PbO_2:D
(bottom curve).

sample cycled in H_2SO_4 versus that cycled in D_2SO_4. Figure 1 shows normalized raw time-of-flight data for the two samples, PbO_2:H and PbO_2:D, in the low d-spacing (high neutron energy) region. In the high energy regime where coherence is lost and Bragg peaks are no longer observed the scattering cross section is simply the sum of the coherent and incoherent cross sections for the individual atoms. Thus the fractional concentration of hydrogen can be calculated directly from the ratio of the observed cross sections. At ~ 220 meV (corresponding to the low d-spacing limit in Fig. 1) the total cross sections for hydrogen and deuterium were taken as 27 and 4.6 barns respectively.[8] This resulted in a calculated hydrogen concentration in the PbO_2:H sample of 0.21 H atoms per PbO_2 unit.

The data for both samples were refined by the Rietveld method over the region 0.78 A $\leq$ d $\leq$ 2.4 A which included 62 Bragg reflec-

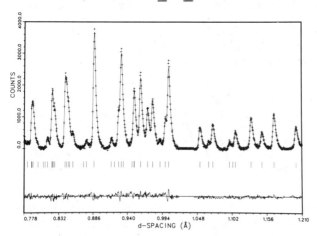

Fig. 2. Portion of the observed (+) and
calculated (continuous line)
diffraction profile minus background for PbO_2:D.

tions. Figure 2 shows the raw data and calculated profile for PbO_2:D over a portion of the refined region. For both samples weighted profile R values near 5% were obtained using space group $P4_2/mnm$. Structural parameters including anisotropic temperature factors agreed well with a recent powder refinement of chemically prepared β-PbO_2.[9] Complete structural results are published elsewhere.[10] As a test for oxygen vacancies, the Pb:O stoichiometry was refined. Both samples showed a small but significant <u>lead</u> defficiency--a 3±2% deficiency for PbO_2:H and a 5±1% deficiency for PbO_2:D.

From these results it is clear that there are not large numbers of oxygen vacancies in electrochemically formed $\beta-PbO_2$ as had been previously suggested. If hydrogen is incorporated as lattice water or hydroxl ions the oxygens of these species occupy regular oxygen sites. The observed existence of 3-5% lead vacancies suggests that charge balance may be achieved by incorporation of H in the $\beta-PbO_2$ lattice. Within experimental uncertainties, the 0.21 H atoms per PbO_2 unit observed in the bulk sample is in general quantitative agreement with the refined lead (Pb^{4+}) vacancy concentration. Unfortunately, the present results do not rule out the possible existence of an amorphous component or poorly crystallized surface layers. More work will be required to conclusively demonstrate that the hydrogen is actually in the $\beta-PbO_2$ lattice and to locate its site.

REFERENCES

1. G. Butler and J. L. Copp, J. Chem. Soc., 725 (1956).
2. N. E. Bagshaw, R. L. Clarke and B. Halliwell, J. Appl. Chem. <u>16</u>, 180 (1966).
3. J. A. Duisman and W. F. Giauque, J. Phys. Chem. <u>72</u>, 562 (1968).
4. A. B. Gancy, J. Electrochem. Soc. <u>116</u>, 1496 (1969).
5. S. M. Caulder, J. S. Murday and A. C. Simon, J. Electrochem. Soc. <u>120</u>, 1515 (1973).
6. A. C. Simon, S. M. Caulder and J. T. Stemmle, J. Electrochem. Soc. <u>122</u>, 461 (1975).
7. J. D. Jorgensen and F. J. Rotella, J. Appl. Cryst., in press.
8. D. J. Hughes and J. A. Harvey, <u>Neutron Cross Sections</u>, BNL 325, (Brookhaven National Laboratory, New York, 1958).
9. P. D'Antonio and A. Santoro, Acta Cryst. B36, 2394 (1980).
10. J. D. Jorgensen, R. Varma, F. J. Rotella, G. Cook and N. P. Yao, J. Electrochem. Soc., submitted.

MEASUREMENTS OF INELASTIC SCATTERING OF eV NEUTRONS

C. D. Bowman and R. G. Johnson
National Bureau of Standards, Washington, D.C. 20234

ABSTRACT

A technique has been demonstrated for studying the inelastic scattering of eV neutrons using a pulsed white source. Measurements have been completed on benzene for incident energies in the range 1.5 to 15 eV and for q values from 13 to 120 $A^{\circ -1}$. Details of the method and possibilities for improvement and extension are presented.

INTRODUCTION

One of the often stated most promising possibilities for inelastic neutron scattering research are studies using eV incident neutrons.[1] Such studies offer the possibility of studying high lying molecular rotational-vibrational states, molecular electronic excitations, electronic levels in solids, etc. which presently can only be studied using electromagnetic probes and even then often not in bulk material. Two types of neutron interactions appear to be promising. The kind studied here is excitation arising from hard sphere nuclear scattering. The other is direct interactions between the neutron and electron through their magnetic moments. The experimental problem in these measurements is the simultaneous measurement of both the incident and scattered neutron energies with the time-of-flight method. Of the several possibilities for such measurements, the approach chosen here is the detection of the scattered neutrons by measuring capture γ-rays from an absorber with a strong isolated eV resonance.[2]

EXPERIMENT

The geometry for the experiment is shown in Fig. 1. The source of neutrons is the NBS 100-MeV electron linac using a tungsten target and operating with a pulse width of 2 μsec, an average power of 3.5 kW and a pulse rate of 360 pps. The total average 4π neutron production rate is about 7×10^{12} n/sec. The neutrons are moderated in CH_2 and separated in energy by time-of-flight measurements along a 21-m drift tube. The beam is collimated at the sample to a 5×10 cm^2 rectangle. Scattered neutrons travel about 15 cm to an In or Au foil placed against the surface of a C_6D_6 liquid scintillator surrounded on five sides with lead shielding. By measuring the flight time from source to scatterer to detector the incident energy can be determined since the final energy is fixed by the resonance. The resolution ultimately will be no better than the FWHM of the gold and indium resonances which are 72 and 140 meV, respectively. Backgrounds are significantly reduced by using the organic scintillator loaded with deuterium to eliminate bothersome

ISSN:0094-243X/82/890084-03$3.00

neutron capture in hydrogen. Most of the background can be measured by absorber-in, absorber-out comparisons.

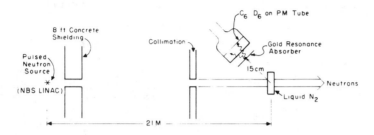

Fig. 1. Arrangement for eV inelastic scattering experiments.

RESULTS

The results of our experiments on benzene using an indium resonance absorber are given in Fig. 2. Spectra are shown for scattering angles of 15°, 30°, and 45°. The incident neutron energy is
shown on the abscissa and the neutrons detected in the indium resonance per timing channel is given on the ordinate. The increased
probability of higher molecular excitation energy with increasing
scattering angle is clearly evident. For the 30° spectrum the
momentum transfer varies from q = 10 to q = 50 $A^{°-1}$. In the 30°
spectrum which was run the longest (24 hours) the statistical
accuracy is sufficient to clearly show structure in the scattered
spectrum. The structure might arise from carbon-hydrogen stretch-
ing vibrations at about 0.37 eV. Successful measurements also have
been completed on liquid nitrogen using a gold absorber. For neutrons in the 5-15 eV range the q value for a 142° scattering angle
ranges from 90 to 120 $A^{°-1}$.

FUTURE STUDIES

The results presented here, apparently the first successful
inelastic neutron scattering measurements in the eV range, demonstrate the practicality of these measurements with reasonable run-
ning times at established white sources. The data collection can be
greatly improved by more efficient detector design and by more complete coverage of scattered neutron solid angle. However, there are
other methods for these experiments which should be explored before
commitment to a particular measurement system. Two are shown in
Fig. 3. The upper figure uses a resonant absorber but in a transmission geometry. This method has been explored to some degree but
has not been compared carefully with the capture measurement used
here. The lower geometry shows a threshold method probably offering
improved resolution. It uses a thin ^6Li glass for a 1/v detection
efficiency in the 10-50 meV range and takes advantage of the

v-dependent rise in s-wave interaction thresholds to give a distinct step at each new interaction threshold.

There are several other attractive geometries which we wish to
explore and which promise an array of versatile experimental
methods for studies in the previously inaccessible eV range of
neutron interaction studies.

REFERENCES

1. For example, J. M. Carpenter,
   D. L. Price, and N. J. Swanson,
   "IPNS-A National Facility for
   Condensed Matter Research,"
   ANL-78-88, p. 41, (1978).
2. J. M. Carpenter, T. H. Blewitt,
   D. L. Price, and S. A. Werner,
   Phys. Today, p. 42 (Dec. 1979).

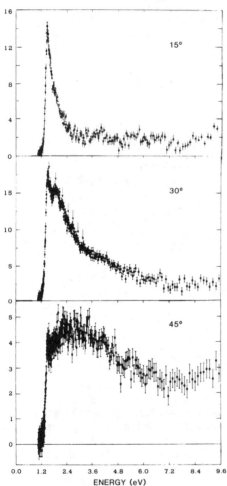

Fig. 2. Inelastic neutron scattering from benzene for 1-10 eV
incident neutrons. Scattering
angles top to bottom are 15°,
30°, and 45°.

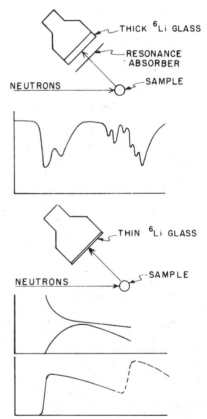

Fig. 3. Alternative geometries
for eV neutron inelastic scattering spectrometry.

PULSED NEUTRON SCATTERING STUDY ON AMORPHOUS
ZIRCONIUM-NICKEL HYDRIDES AND DEUTERIDES

M. Hirabayashi, H. Kaneko, T. Kajitani,
K. Suzuki and M. Ueno

The Research Institute for Iron, Steel and Other Metals,
Tohoku University, Sendai 980, JAPAN

ABSTRACT

Localized hydrogen vibration spectra have been measured for
crystalline $ZrNiH_{2.8}$ and $Zr_2NiH_{4.6}$ and amorphous $ZrNiH_{1.8}$ and
$Zr_2NiH_{4.4}$ by means of the time-of-flight neutron inelastic
scattering experiment. Elastic scattering intensity has been also
measured for amorphous ZrNi and $ZrNiD_{1.8}$. It is concluded that the
amorphous matrix mainly consists of tetrahedrons and hexahedrons,
and the solute hydrogen and deuterium atoms are located in these
polyhedrons.

INTRODUCTION

Hydrogen or deuterium atoms are typical interstitial solutes in
metals and alloys either in the crystalline state or presumably in
the amorphous state. In the amorphous state, larger number of interstitial holes in a unit volume may be involved than in the
crystalline state because of the topological disorder. On the other
hand, the possibility to find a specific configuration of metal
atoms, in which a hydrogen atom can sit, decreases because of the
quantitative disorder or the fluctuation of interatomic distances.
Recently, Rush, Rowe and Maeland [1] have disclosed that the neutron
inealstic scattering can provide an unusual probe for studying the
local environment around the hydrogen atom in the amorphous hydrides.
The present work is aimed to elucidate the dynamical behavior and
the location of the solute hydrogen and deuterium atoms in amorphous
Zr-Ni alloys by means of pulsed neutron scattering experiments.

EXPERIMENTAL PROCEDURE

Crystalline $ZrNiH_{2.8}$ and $Zr_2NiH_{4.6}$ and amorphous $ZrNiH_{1.8}$,
$ZrNiD_{1.8}$ and $Zr_2NiH_{4.4}$ were prepared by the process described in our
previous paper [2]. Neutron inelastic scattering experiments were
carried out utilizing a time-focused crystal analyzer [3] with the
pulsed neutron source at the Tohoku University 300 MeV Electron
Linac. The energy resolution of this spectrometer falls $\Delta E/E \simeq 0.04$
in the energy range from 50 to 200 meV. The data acquisition time
for each specimen was about 36 hours. A high-resolution total-
scattering spectrometer was employed to obtain the structure factor,
$S(Q)$, in the wave vector region up to $|Q| = 40$ Å^{-1} [4]. All the
experiments were carried out at room temperature.

RESULTS

Fig. 1 and 2 show the frequency distribution function, $g(\omega)$, of the zirconium-nickel hydrides in crystalline and amorphous states, respectively. For the thermal vibration correction, the mean square amplitude of hydrogen was assumed as 0.01 Å^2. For the crystalline $ZrNiH_{2.8}$ and amorphous $ZrNiH_{1.8}$, pronounced peaks are seen at almost identical energy of 120 meV which are due to the primary vibration modes of the solute hydrogen. Note a striking difference in the peak width of the two spectra; the peak width for the amorphous hydride is broader than the crystal.

It is known that there are two types of hydrogen sites in the crystalline $ZrNiH_{2.8}$; tetrahedral site (so called c-site) and hexahedral site (b-site). On account of the local symmetry, it was concluded [2] that three modes may be excited at the b-site and one mode at the c-site, and hence four modes are overlapped in the peak of the $g(\omega)$ curve. Using the RANDOM SEARCH LEAST SQUARES program [5], the $g(\omega)$ curve was separated into four Lorentzian peaks by fitting three parameters, i.e., peak center, peak height and full width at half maximum (FWHM). The $g(\omega)$ curve of the amorphous $ZrNiH_{1.8}$ was also separated in a similar way as shown in Fig. 2. Indices Ty, Rz, A" and Tx indicate the irreversible representations for each vibration mode; Tx, Ty and Rz modes are due to the hydrogen motion at the b-site, and A" mode is due to the motion at the c-site.

In the amorphous hydride the peak centers shift only a little from those of the crystal. The peak broadening is ascribed to the fluctuation of the interatomic distances between hydrogen and metal atoms. The results suggest that the environment of hydrogen atoms in the amorphous alloys is very similar to that in the crystalline hydrides. The enhancement of the central peak, A", in the amorphous hydride indicates high population of hydrogen in the c-like sites.

Unfortunately the $g(\omega)$ curves of crystalline $Zr_2NiH_{4.6}$ as well as amorphous $Zr_2NiH_{4.4}$ cannot be separated in a similar way to the cases of $ZrNiH_{2.8}$ and $ZrNiH_{1.8}$, because the hydrogen sites in Zr_2Ni are not known yet. However, the similar profiles of the two $g(\omega)$ curves in Fig. 2 indicate that the environment of hydrogen atoms is notably similar in the amorphous $Zr_2NiH_{4.4}$ and $ZrNiH_{1.8}$.

Quite consistent results are obtained by the elastic scattering measurements. Fig. 3 shows the radial distribution function (RDF) for the amorphous ZrNi and $ZrNiD_{1.8}$. Remarkable change is noted after the deuterium charging of the amorphous ZrNi; two peaks centered at 1.7 Å and 2.2 Å appear for the amorphous deuteride. The first peak at 1.7 Å is due to the Ni-D correlation, and the second peak at 2.2 Å is due to the Zr-D correlation. The coordination numbers are approximately one for the Ni-D and three for the Zr-D. Accordingly it is concluded that the structures of amorphous $ZrNiH_{1.8}$ and $ZrNiD_{1.8}$ consist mostly of many tetrahedrons and hexahedrons of the host metal atoms, which are similar to the c- and b-sites in the crystalline $ZrNiH_3$, and major part of these sites are occupied by the hydrogen or deuterium atoms.

AKNOWLEDGEMENT

Authors are indebted to Dr. K. Aoki and Professor T. Masumoto for valuable discussions.

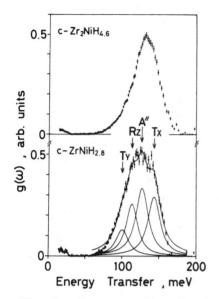

Fig. 1. Frequency distribution functions for crystalline hydrides.

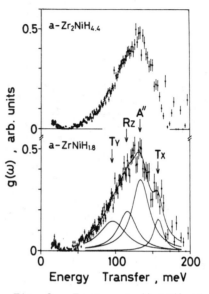

Fig. 2. Frequency distribution functions for amorphous hydrides.

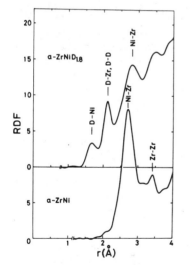

Fig. 3. Radial distribtuion functions for amorphous ZrNi and ZrNiD$_{1.8}$.

REFERENCES

1. J. J. Rush, J. M. Rowe and A. J. Maeland, J. Phys. F: Metal Phys., 10, L283 (1980).
2. T. Kajitani, H. Kaneko and M. Hirabayashi, Sci. Rep. Res. Inst. Tohoku Univ., A-29, 210 (1981).
3. N. Watanabe, M. Furusaka and M. Misawa, Res. Rep. Lab. Nucl. Sci., Tohoku Univ., 12, 72 (1979).
4. K. Suzuki, M. Misawa, K. Kai and N. Watanabe, Nucl. Instr. and Meth., 147, 519 (1977).
5. M. Kitamura, Y. Takeda, K. Kawase and K. Suguyama, Nucl. Instr. and Meth., 136, 363 (1976).

90

NEUTRON DIFFRACTION STUDY OF LIQUID BENZENE

T. Matsumoto

University of Hokkaido, Sapporo, Japan 060

ABSTRACT

This paper describes a neutron diffraction study of liquid benzene. The experiment described herein was made by the Time-of-Flight method using an electron linear accelerator. The molecular structure factor $S_m(Q)$ with the wide range of Q at about 1 - 30 $\overset{\circ}{A}^{-1}$ was obtained from the measurements at the scattering angles of 45° and 150°. The dynamical effect, which causes problems in neutron diffraction experiments of light nuclei liquids, was successfully corrected by our previous method.[1] In this study the parameters for the intramolecular structure were derived from $S_m(Q)$ in the high Q region. Furthermore, a simple model calculation method, which can effectively include the coupling effect between the spatial and the orientational correlations, was proposed for the intermolecular structure of nonspherical molecular liquids. The calculation done on the model with three types of arrangements for the two benzene molecules can explain well the characteristical structure in the low Q region of liquid benzene.

INTRODUCTION

The structure of liquid benzene has been studied by the X-ray diffraction method[2,3] and Narten has recently remeasured it by the same technique.[4,5] Liquid benzene is known to show a characteristical structure factor in X-ray diffraction which has two, well-resolved peaks of about 1 or 2 $\overset{\circ}{A}^{-1}$ in the low Q region. Narten demonstrated two calculations for the structure factor using a six-site and a 12-site model. Only the latter model, which had 12 interaction sites on atoms C and H, could explain the existence of the two peaks, which indicated that the hydrogen atoms located on the outside of the molecule took on an important role in the molecular configuration, and that the neutron diffraction method could be useful for obtaining information.

The neutron diffraction method needs some correction for the dynamical effect, which is due to the inelastic scattering of the neutrons. As this correction is difficult to make in light nuclei liquids, few studies of the neutron diffraction method have been made on hydrogeneous liquids. The author has recently proposed a method for this dynamical correction which is available to light nuclei liquids.[1] The method was applied to the liquid benzene.

EXPERIMENT

The measurements were made using the neutron diffraction equipment of a 45 MeV electron linear accelerator maintained at Hokkaido

University. The details of this equipment are presented in Ref. 6. The path lengths were 693 cm and 23 cm for the incident and the scattered neutrons, respectively. The deuterated liquid benzene (D: 99.6 %) was kept in a cylindrical container (10 mmϕ x 100 mm) made of vanadium (0.025 mm thickness). A series of measurements was made at room temperature using the following: (1) a copper poly-crystalline plate for calibrating the scale of Q; (2) a vanadium rod (10 mmϕ) for cancelling the distortions due to the energy depend-ence of the neutron source and detector; (3) liquid benzene; and (4) the vanadium container for the background counts.

After the diffraction pattern in the time space was transformed to that in the Q space in this study, several corrections were made. The effects of the absorption in the sample and in the vanadium were corrected by following the method described in Ref. 6. The dynamical correction was made
separately in the high and low Q regions. In the high Q region the ratio of the count-ing rate of the sample to the vanadium at 150° was fitted by the least square method to Eq. (14) in Ref. 1, where, for simplicity, the third term at the right hand side was neglected. The dynami-cal correction in the low Q region was made by Eq. (18) in Ref. 1

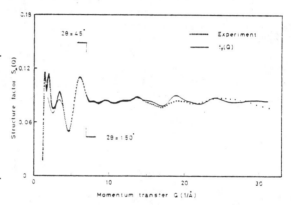

Fig. 1 Structure factor of liquid benzene

for 45°. The molecular structure factors obtained from the measure-ments at 45° and 150° were in good agreement at around Q=5 - 10 Å^{-1}. Figure 1 shows the corrected experimental result of the molecular structure factor for liquid benzene. The measurements at 45° and 150° were connected at Q= 6.9 Å^{-1}. The neutron diffraction method using the electron linear accelerator gives the value of Q ranging widely from about 1 - 30 Å^{-1}.

INTRAMOLECULAR STRUCTURE

The molecular structure factor $S_m(Q)$ is dominated by the intra-molecular function $f_0(Q)$ in the high Q region, which enabled us to determine the parameters for the intramolecular structure of benzene in the liquid state. Assuming that the planar molecule has the angle ∠$C_1C_2C_3$=120°, the experimental $S_m(Q)$ in the Q region 7 - 18 Å^{-1} was fitted to $f_0(Q)$. Table 1 shows the determined parameters related to the single benzene molecule in the liquid state. The neutron diff-raction method allowed for an accurate determination of the distance D - C. The distance C - C may be compared with that obtained by other methods. The value derived by the neutron method was closer

92

to the vapour value than to that of
the X-ray. The function $f_0(Q)$, which
was calculated by using the parameters
given in Table 1, was compared with
the neutron diffraction experimental
results, shown in Fig. 1.

distance	Å
D - C	1.081 ± 0.005
C - C	1.395 ± 0.005
	1.41 liquid (XD^4_7)
	1.393 vapour (ED^7)

Table 1 Atom - atom distance

INTERMOLECULAR STRUCTURE

The structure factor of the liquid benzene obtained from the
neutron diffraction method showed two well-resolved peaks in the low
Q region, whose positions corresponded well with the ones in the X-
ray diffraction.[5] The second peak at Q= 1.96 Å^{-1} was more significant in the neutron diffraction than the corresponding peak in the

X-ray study. Since the intramolecular function $f_0(Q)$ has no peaks
in the low Q region, the two peaks
reflected the intermolecular
structure of liquid benzene.

We also similar preliminary
neutron diffraction experiments
using liquid toluene, nitrobenzene
and pyridine. The two well-
resolved peaks occurred with the
pyridine, but only one peak appeared in the corresponding range
of Q when the toluene and nitrobenzene were used. This finding
suggested that the two peaks resulted from the disk-like shape of

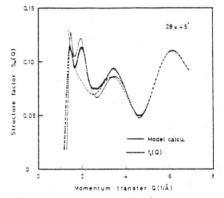

Fig. 2 Model calculation

the molecule, and that the benzene molecule could be treated effectively by a circular disk.

The structure factor of the liquid benzene for the low Q region
was analyzed by a simple model,[8] which can effectively include the
coupling effect between the spatial and the orientational correlation of nonspherical molecular liquids. Herein the benzene molecule
was treated as a circular disk, and $S_m(Q)$ was represented by three
arrangements of two molecules. A calculational result is shown in
Fig. 2, where the diameter and the thickness of the disk was 6.00
and 3.40 Å, respectively.

REFERENCE

1. T. Matsumoto, J. Nucl. Sci. Technol. 16, 401 (1979).
2. S. Katzoff, J. Chem. Phys. 2, 841 (1934).
3. W. C. Pierce, J. Chem. Phys. 5, 717 (1937).
4. A. H. Narten, J. Chem. Phys. 48, 1630 (1968).
5. A. H. Narten, J. Chem. Phys. 67, 2102 (1977).
6. T. Matsumoto, J. Nucl. Sci. Technol. 15, 863 (1978).
7. I. L. Karle, J. Chem. Phys. 20, 65 (1952).
8. T. Matsumoto, to be published.

EXPERIMENTAL STATUS AND RECENT RESULTS
OF NEUTRON INTERFERENCE OPTICS

A. Zeilinger
Department of Physics
Massachusetts Institute of Technology, Cambridge, MA 02139, USA
and
Atominstitut der Oesterreichischen Universitaeten
A-1020 Wien, Austria

R. Gaehler
Institut Laue-Langevin, F-38042 Grenoble, France

C. G. Shull
Department of Physics
Massachusetts Institute of Technology, Cambridge, MA 02139, USA
and
Hahn-Meitner-Institut, D-1000 Berlin 39, Germany

W. Treimer
Fritz-Haber-Institut, D-1000 Berlin 33, Germany

ABSTRACT

Up to the present time, perfect crystal interferometers of both
two-crystal and three-crystal Laue-case variety have been used to
explore the optical characteristics of neutrons of diffracting wavelength. In addition, very recently the classical optical experiments
on the diffraction by an absorbing edge, by an absorbing wire and by
single and double slit assemblies have been performed for cold
neutrons. These experiments show with high precision the validity
for matter waves of the standard Fresnel-Kirchhoff approach to diffraction.

PERFECT CRYSTAL NEUTRON INTERFERENCE OPTICS

After its invention in 1974, the three-crystal or LLL interferometer[1] has found numerous interesting applications. The experiments performed up to 1978 have been reviewed at the International
Workshop in Neutron Interferometry.[2] Since the time of this conference, measurements with this type of interferometer of the coherent scattering lengths of various gases including tritium have been
performed by the Vienna-Dortmund[3] group at the ILL, and an improved
version of the experiment on the effect of Earth's rotation on neutron
interferences was performed by the Missouri[4] group. Thus, we can
state that LLL neutron interferometry has become in the few years
since its invention a very well established technique.
The LLL interferometer corresponds topologically to the standard Mach-Zehnder interferometer for light if one modifies the particular type of beam splitter used, i.e. if the mirrors of the Mach-
Zehnder interferometer are considered to be replaced by the perfect

ISSN:0094-243X/82/890093-07$3.00 Copyright 1982 American Institute of Physics

crystals in Laue geometry (Fig. 1). An important advantage of the LLL interferometer is that one can, as in the Mach-Zehnder interferometer, use large apertures in the incident beam and still obtain macroscopically well separated beams.

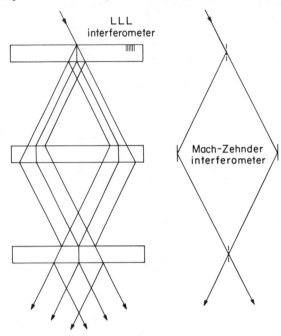

Of other types of possible perfect crystal interferometers discussed in the literature[5], only the two-crystal or LL interferometer has found experimental application[6,7] up to now. For the operation of this interferometer type the mutual coherence properties of radiation following different optical paths within the Borrmann fan are crucial. Even without possible coherence between different components in the incident radiation, there will still be coherence left between rays which follow symmetric paths within the crystal with respect to the lattice planes. If the thickness of the second crystal plate is identical to that of the first one, corresponding rays meet at the exit face of the second crystal plate forming a focus where they coherently superpose (Fig. 2). Therefore this interferometer type can be viewed as being a thermal neutron analog of the standard Rayleigh interferometer. This interferometer type has been successfully employed in experiments[8] aimed at limiting the size of hypothetical nonlinear terms in the Schroedinger equation by three orders of magnitude as compared to Lamb-Shift experiments[9] and for a search for an Aharonov-Bohm effect for neutrons.[10]

Fig. 1. LLL-interferometer showing central ray propagation and the Borrmann fan limitation within the system is topologically equivalent to the Mach-Zehnder interferometer.

In such experiments, spatially well separated beams are used which necessitates a limitation of the entrance aperture. But it is interesting to note that the operation of the LL interferometer does not depend on selecting just two symmetric rays at the exit face of the second crystal. In contrast, because any pair of such symmetric rays exhibits essentially the same coherence properties the whole beam can be used between the crystals. An advantage of the LL interferometer stems from the property that only two crystal plates are encountered by the beam. This implies that the operation of that

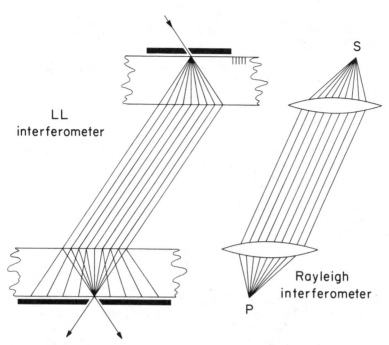

Fig. 2. Ray propagation in the two-crystal or LL-interferometer exhibits focusing action at the backface of the second crystal plate in analogy to the Rayleigh interferometer for light.

interferometer type is insensitive to parallel lateral displacements of the crystal plates relative to each other. On the other hand, it is very sensitive with respect to the relative angular position of these crystal plates the sensitivity being in the range of 10^{-3} arcsec.[11] For experiments sensing a deflection of the whole beam on its way from the first to the second crystal plate a very wide entrance aperture can be used. An experiment exploiting that angle sensitivity feature is at present in progress at MIT. It concerns the neutron analog of the classical Sagnac experiment, an experiment sensing the effect on the interference fringes of active rotations of the interferometer.

Another interesting feature of the LL-crystal set-up is its very high sensitivity (10^{-8}eV) to energy changes of the neutron on its way from the first to the second crystal plate. This has been used in an experiment demonstrating the wavelength change of the neutron when entering a modest magnetic field.[12] The very existence of the two related focal points (Fig. 2) allows one to conclude that the flight time of the neutron from one to the other focal point has to be the same since generally the optical path is of equal length along any path connecting two focal points in an optical system. This property together with symmetry considerations leads to the conclusion that the flight time for a neutron from the entrance face is the same independent of the actual path followed by the neutron

within the crystal plate. The experimental result[13] demonstrates
this property.

NEUTRON DIFFRACTION AT ABSORBING EDGE,
SINGLE SLIT AND DOUBLE SLIT ASSEMBLIES

One of the motivations of introducing a nonlinear term into the
Schroedinger equation[9] was to prevent wave packets from spreading
without limit. Therefore, it is reasonable to expect that effects
of such a nonlinear term may be seen when looking closely on the
free space propagation of neutrons. It can be shown[14] that a non-
linear term of the type

$$F = F(\rho), \quad \rho = |\psi|^2 , \tag{1}$$

would lead to a bending of the wavefront of

$$\frac{d^2y}{dz^2} = \frac{1}{2E} \frac{dF}{d\rho} \frac{d\rho}{dy} , \tag{2}$$

where z is measured along the propagation direction of the neutrons
and y orthogonal to it. As the effect increases with increasing
lateral gradient of the probability density, a precision measurement
of the Fresnel diffraction of slow neutrons at a straight absorbing
edge was performed.[14] The apparatus used in that experiment and in
those mentioned later made use of an optical bench arrangement[15]
located at one of the neutron beam lines emerging from the cold
source of the High Flux Reactor of the ILL (Fig. 3). Monochromatic

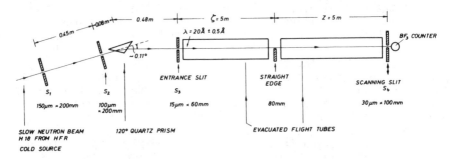

Fig. 3. Cold neutron optical bench set-up (not to scale).

neutrons of wavelength 20.0 Å with a spread of ±0.5 Å were used in
the experiment and these were prepared by prism refraction. Fig. 4
shows the experimental Fresnel pattern together with a computer
generated pattern as calculated with the standard linear Schroedinger
equation. As can be seen from these data, any effect due to non-
linear action was found to be below experimental resolution and this
permitted a lowering of the limit on the size of nonlinear terms by
another two orders of magnitude as compared with the neutron interferometer result. Specifically, for a logarithmic nonlinearity type

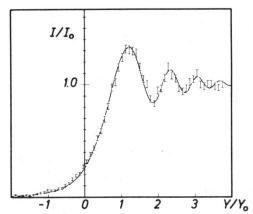

Fig. 4. Experimental Fresnel diffraction pattern of an absorbing straight edge, together with pattern calculated from standard linear Schroedinger equation[14] ($Y_0 = 100\mu m$).

of term $F = -b \ln(\alpha|\psi|^2)$, the quantity b which measures the strength of the nonlinear term cannot be larger than 3.3×10^{-15}eV in order to be in agreement with the experiment.

Further experiments which have been performed using the same experimental arrangment concerned the diffraction of neutrons at single and double slit assemblies. In slit diffraction, experiments have been performed using slits of width 96µm and 22µm in the single and double slit configurations respectively. Fig. 5 shows the result of the 96µm single diffraction experiment obtained in a measuring time of 11¾ days for the whole pattern which amounts to 192 minutes per point. Also displayed is the computer generated solution of the

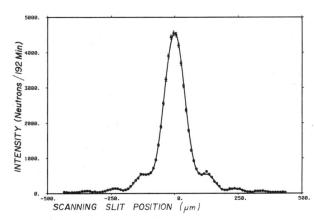

Fig. 5. Experimental 96µm single slit diffraction pattern with pattern calculated from the Schroedinger equation

standard linear Schroedinger equation. This solution was obtained using a Fresnel-Kirchhoff type approach[16] where the entrance slit of the optical bench was assumed to be a plane wave coherent over the full width of that slit. Different wavelength contributions and different incident direction contributions were assumed to be incoherent with each other. It should be noted that in the single slit diffraction pattern the interference maxima up to third order are clearly visible with some indication of the fourth-order maximum.

By mounting a Boron wire of 104µm thickness into the gap of a 148µm wide single slit a double slit assembly was obtained. Fig. 6 shows a drawing to scale of that arrangement. The absorbing edges were made of a borate glass with 10% Gd_2O_3 added to ensure a high neutron absorption action. The flat parts of the edges opposite to

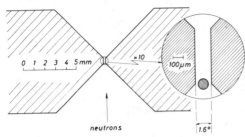

Fig. 6. Double slit arrangement: Boron wire mounted in the gap between absorbing glass edges.

each other were arranged to be slightly divergent as seen from the side of the incident neutrons in order to avoid contamination of the diffraction pattern by surface reflected neutrons. This same approach was also used for obtaining the single slit diffraction patterns. In order to bring once again to attention the scale of the experiment we note that the wavelength of the neutrons (λ = 18.45 Å) was nearly five orders of magnitude smaller than the center-to-center distance (d =126 μm) of the two slit openings. Hence the angular distance between two adjacent maxima in the double slit diffraction pattern is only 3 arcsec. In order to obtain acceptable statistical accuracy within a reasonable time, the wavelength band used in that experiment had been increased to ±1.4 Å. This still implied an overall measuring time of 7500 sec per point or of 7½ days for the whole pattern. The slight asymmetry of the diffraction pattern (Fig. 7) can be fully accounted for by the existence of slight asymmetry in the slit widths.

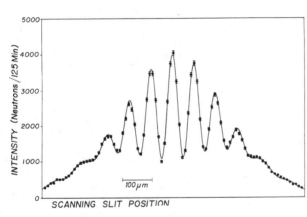

Fig. 7. Neutron diffraction pattern of a double slit with pattern calculated from the Schroedinger equation

This double slit interferometer set-up has also been recently used for a measurement of the Fizeau-effect for neutrons,[17] i.e. the effect of the motion of the phase shifter on the phase of a neutron wave. The result, in agreement with quantum mechanics, may be viewed as a demonstration of the transformation laws of both the deBroglie wavelength and the frequency of a massive particle.

Another interesting group of experiments performed using the optical bench arrangement was performed by Klein et al.[18] In these experiments the operation of both circular and cylindrical Fresnel lenses for thermal neutrons was successfully demonstrated. Furthermore, it could be shown that a Billet split-lens type interferometer[19] using Fresnel lenses can be expected to be operated at a high flux reactor in a sensible intensity range. Furthermore, the introduction of Fresnel lenses into perfect crystal neutron optics[20] seems to open an interesting new field for experimentation.

This work was supported by the National Science Foundation (USA), the U.S. Department of Energy, the Fonds zur Förderung der Wissenschaftlichen Forschung (Austria) project no. 4230 and the Bundesministerium für Forschung und Technologie (Germany) project no. 41E06P.

REFERENCES

1. H. Rauch, W. Treimer and U. Bonse, Phys. Lett. A 47, 369 (1974); W. Bauspiess, U. Bonse, H. Rauch and W. Treimer, Z. Phys. 271, 177 (1974).
2. U. Bonse and H. Rauch (Eds.), Neutron Interferometry (OUP, Oxford, 1979).
3. H. Kaiser, H. Rauch, G. Badurek, W. Bauspiess and U. Bonse, Z. Phys. A 291, 231 (1979); S. Hammerschmied, H. Rauch, H. Clerc and U. Kischko, to be published.
4. S. A. Werner, J.-L. Staudenmann and R. Colella, Phys. Rev. Lett. 42, 1103 (1979); J.-L. Staudenmann, S. A. Werner, R. Colella and A. W. Overhauser, Phys. Rev. A 21, 1419 (1980).
5. W. Graeff, p. 34 in Ref. 2.
6. A. Zeilinger, C. G. Shull, M. A. Horne and G. L. Squires, p. 48 in Ref. 2.
7. S. Kikuta, p. 60 in Ref. 2.
8. C. G. Shull, D. K. Atwood, J. Arthur and M. A. Horne, Phys. Rev. Lett. 44, 765 (1980).
9. I. Bialynicki-Birula and J. Mycielski, Ann. Phys. (NY) 100, 62 (1976).
10. D. M. Greenberger, D. K. Atwood, J. Arthur, C. G. Shull and M. Schlenker, Phys. Rev. Lett. (in press).
11. U. Bonse, W. Graeff and H. Rauch, Phys. Lett. 69A, 420 (1979).
12. A. Zeilinger and C. G. Shull, Phys. Rev. B 19, 3957 (1979).
13. C. G. Shull, A. Zeilinger, G. L. Squires, M. A. Horne, D. K. Atwood and J. Arthur, Phys. Rev. Lett. 44, 1715 (1980).
14. R. Gaehler, A. G. Klein and A. Zeilinger, Phys. Rev. A 23, 1611 (1981).
15. R. Gaehler, J. Kalus and W. Mampe, J. Phys. E 13, 546 (1980).
16. M. Born and E. Wolf, Principles of Optics (Pergamon Press, New York, 5th ed., 1980) p. 382.
17. A. G. Klein, G. I. Opat, A. Cimmino, A. Zeilinger, W. Treimer and R. Gaehler, Phys. Rev. Lett. 46, 1551 (1981).
18. P. D. Kearney, A. G. Klein, G. I. Opat and R. Gaehler, Nature 287, 313 (1980); A. G. Klein, P. D. Kearney, G. I. Opat and R. Gaehler, Phys. Lett. 83A, 71 (1981).
19. A. G. Klein, P. D. Kearney, G. I. Opat, A. Cimmino and R. Gaehler, Phys. Rev. Lett. 46, 959 (1981).
20. V. L. Indenbom, JETP Lett. 29, 5 (1979).

A NEUTRON SPIN ECHO DEVICE TO IMPROVE
THE ENERGY RESOLUTION OF TRIPLE AXIS SPECTROMETERS

Claude M.E. Zeyen

Institut Laue-Langevin, 156X, 38042 Grenoble Cédex, France

ABSTRACT

The design of a spin-echo option to the D10 thermal neutron Triple axis spectrometer at I.L.L. is described. The option consists of vertically focussing Heusler crystal polarizer/monochromator and analyser systems, transverse field electromagnets for the Larmor processions and the necessary spin turn devices. The polarized neutron intensity at the sample position is reduced by a factor of 5 with respect to the standard using P.G. crystals. The resolution obtained is 1 μeV at 14 meV incident energy.

INTRODUCTION

The resolution limitation of classical triple-axis spectrometers is well known to be due to the necessary strict collimation and monochromatisation required to measure the incoming and outgoing neutron energies and momenta precisely. Thus the practical limit in relative energy resolution is usually about one percent. In order to get the best from a given spectrometer one usually adapts the incident neutron energy to the resolution required and the spectrometers become very much specialised to certain types of studies. For example in order to reach the μeV spectroscopy region triple axis spectrometers with perfect crystal monochromator and analysers set to backscattering have been developed. They take advantage of the fact that in the backscattering geometry the reflected wavelength band does not in first order depend on the beam divergency. By using large solid angle focussing analysers this method is very powerful for incoherent or momentum-independent weakly inelastic scattering (the scanning range is limited because the Bragg angles on monochromator and analyser have to be kept near 90°) but is difficult for coherent quasi-elastic scattering because of intensity.

An alternative to this resolution/intensity dilemma is the spin echo method introduced by Mezei[1]. By measuring experimentally a quantity directly related to the energy transfer of the neutron, this method is capable within certain limits to decouple the resolution performance of a neutron spectrometer from its intensity. As a drawback this method requires the use of polarized neutrons and a rather delicate use of special magnetic field configurations which renders its use somewhat more cumbersome experimentally. The polarization analysis used in NSE, results in a signal loss of a factor of three for spin incoherent scattering but may on the other hand prove to be very useful for the separation of coherent and incoherent effects as well as for the study of magnetic properties.

It should be said too that for given precession magnets certain limitations may have to be imposed on beam cross sections and sample dimensions for the sake of magnetic field homogeneity. The resolution limits in NSE are usually conditioned by the field homogeneities along the neutron trajectories.

The I.L.L. cold neutron NSE spectrometer IN11 [1] more specially designed for the study of quasi-elastic scattering has demonstrated the power of this new method.

The use of NSE together with a classical triple axis spectrometer has been proposed long ago [2] and its potential possibilities discussed both for quasi-elastic work, the separation of elastic and inelastic scattering [3] and the study of phonon lifetimes [2,4].
In this paper we will describe the design of the first triple-axis spin echo spectrometer. We will show how the handicap of the need for polarised neutrons can be well overcome by the use of especially designed vertically focussing Heusler crystal polariser/analyser systems.

Some particular techniques to keep the spectrometer practically as compact as a normal TAS will also be described. We will give the measured performances of the machine. Practical limitations will be discussed.

Several possible applications of the NSE-TAS will shortly be outlined and some will be illustrated by very recent experimental data which illustrate its good performed. It will be seen that the new NSE device is very well suited for the separation of elastic and inelastic scattering with a resolution more than two orders of magnitude better than the classical TAS with the extra possibility of separating the incoherent and coherent components as well, if relevant. The detector count rates are reduced by a factor of four to five only as compared to the classical use of the machine. For quasi-elastic scattering even at high momentum transfers (up to 7 Å^{-1}) lines have been measured down to a width of less than 1 µeV at 14.6 meV incident energy.

For phonons only preliminary data could be taken until now. For the non dispersive case (locally flat phonon branches) no particular problems arise and recent experiments seem to show that linewidths down to a few µeV can be measured.

For the case of excitations with dispersion where NSE focussing is required the proposed "tilted-magnet" technique [2] turns out to be very cumbersome experimentally and its practical usefulness is still under question.

THE SPECTROMETER

This basis of our NSE TAS is the existing thermal neutron triple-axis spectrometer D10 [5] of the I.L.L. situated at the end of a 100m long guide tube of large vertical cross section (3 x 12 cm). In its classical mode of operation the spectrometer is equipped with sets of both PG and copper vertically focussing monochromator/ analysers. The useful wavelength band is inbetween 1 and 3.5 angstrom and below 1.5 Å the cut-off of the curved neutron guide is indeed very useful because λ/2 filters are not needed. This is a particularly relevant feature where Heusler crystals are used as the second order reflexion is very strong.

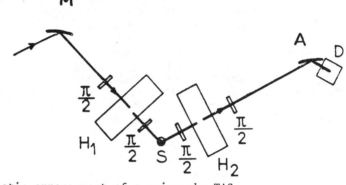

Fig. 1. Schematic arrangement of a spin-echo TAS
M : focussing crystal monochromator/polarizer
A : focussing crystal analyser/spin analyser
D : detector
S : sample
$\frac{\pi}{2}$: spin-turn coils (the $\frac{\pi}{2}$ coils around the sample can be replaced by a single π coil)
H_1, H_2 : precession fields.

a) Polarized Neutrons

In order to transform the spectrometer into a polarization analysis spectrometer vertically focussing composite Heusler systems were especially developed for this purpose. Good Heusler crystals were cut into slabs of 75 x 15 x 7 mm with the front 15 x 7 mm surface parallel to the (111) crystallographic planes so that the magnetic (111) reflexion could be used in Bragg geometry. The front surfaces were well polished to avoid surface depolarisation. These plates were tested and aligned using thermal neutrons. Only those with single-peaked mosaic distributions and a FWHM of about 25 minutes of arc were used to compose the polarizer systems. 8 slabs were oriented one by one to within ± 1 minute of arc and then glued onto a flexible thin Aluminium support with their slab axis horizontal to form a flexible plate of 75 x 120 mm. By bending

the support an adjustable vertical curvature with radii between
1.5 meter and ∞ can be obtained. Before designing a magnet frame to
magnetize the Heuslers, we measured the magnetization curves on small
model crystals with the applied field along and transverse to the
slab axis in order to assess the minimum field required to saturate
the Heusler crystals. At 97% of saturation the applied magnetic
field was found to be 1.2 k Oersted along the slab axis and 3.3
k Oersted perpendicular to it. As 3 k Oersted are difficult to
obtain with a compact magnet over a gap of 120 mm (not to speak of
the air gaps between the crystal slabs) the horizontal field along
the slabs was adopted[x]. Due to the large lattice constant of the
Heusler alloy (3.43 Å for the 111 planes) useful Bragg angles will be
small so that the magnet has to be kept very flat such as to allow
the neutrons to reach and leave the crystals under very small incident angles ($\Theta > 15°$). For wavelengths smaller than 1.7 Å flat
Heusler plates in transmission are used. Such a magnet with a
Θ-frame was built using permanent Samarium-Cobalt blocks. To avoid
depolarizing return fields where the outgoing polarized neutrons
pass close to the magnet blocks soft-iron shims were used. These
systems give a focussed beam of 30 x 30 mm cross section starting
from a white beam of 120 x 30 mm. The polarization efficiency
obtained is 96%. The final polarization of the spectrometer
including the flipper efficiency reaches 92% provided the sample
presents no spin incoherent scattering.

b) Precession Electromagnets

In order to keep the spectrometer rather compact an Iron core
magnet-type with the field transverse to the neutron trajectories
was chosen. The design of the magnets is a compromise to satisfy
various possible uses of the present NSE option and is therefore
not optimum for any of them. The magnets may be mounted with the
field horizontal or vertical. In the latter case a rotation around
a vertical axis (tilting) is possible. We will come back to the
use of this facility. The requirements and design calculations
have been discussed before [1] (pages 151 and 171) [3]. We will simply
recall the main design parameters and the results from their use in
practice.
    The magnets have a wide window frame soft iron core dimensioned
so that the iron does not saturate anywhere at maximum field. This
means that the field should be always proportional to the applied
current. During the experiments it was found that close to maximum
field only a small quadratic correction is needed to describe the
exact current to field relationship. The windings were chosen of
the racetrack type because they give the maximum of homogeneity.

---

[x]The vertical transverse field is possible too provided the slabs are
joined tighly at small fixed angles resulting in a fixed curvature
Furthermore the height has to be restricted to a gap size for which
the required field may be obtained.

The maximum field in the magnet gap was chosen to be 1.5 k Oersted. For particular small sample geometries extra pole pieces may be added to give a maximum field of 3 k Oersted. For NSE the relevant property of the precession magnets is the line integral of the magnetic field H along neutron path and over the distance ℓ where the neutron spins are allowed to precess around the field direction. This integral was calculated to be about

$$\int H\, d\ell \sim 5.10^4 \text{ kOe.cm} \quad \text{(without pole pieces)}$$

the value depending of course on the precise integration length ℓ. In practice and at maximum field the value of $5.63.10^4$ kOe.cm was found. This corresponds to 1000 Larmor precessions for 2.41 Å $(N = 7.37.10^{-3}\ \lambda\ H d\ell)$. This number of precessions N has to be as homogeneous as possible for all neutrons of identical wavelength. The parameters responsible for field integral homogeneity are the magnet dimensions and mainly the gap size (Our magnets being three gaps wide). The empirical formula

$$\frac{\Delta \int H d\ell}{\int H d\ell} = 0.4 \ \frac{y^2}{g(\ell + 1.4g)}$$

where 2g is the gap size, 2ℓ the length of the magnet (in the direction of the neutron path) and 2y the beam size in the gap can be used to estimate the results of the principal inhomogeneity onto the dephasing of the Larmor precessions.

For our magnet $\ell = 25$ cm, g = 15 cm. Thus for y = 0.5 cm the relative error in number of precessions will be 0.4×10^{-3} which is not a serious drawback. Experiments have shown that it corresponds to a loss of NSE signal of about 30% at maximum field which is acceptable. This example corresponds to the situation for which the magnets are mounted to have the field in the horizontal plane. The large width is then useful to permit rather large vertical beam dimensions without signal loss. This geometry was found to be very useful and gave excellent results for all work not involving NSE focussing to dispersive excitations. The vertical field geometry allows only significantly smaller beam cross sections for comparable homogeneity.

c) Flippers and Correction Coils

We use the popular flat Mezei coils in a somewhat different manner than on the IN11 NSE spectrometer. In order to keep the spectrometer as compact as possible the flippers will have to operate in the relatively high fringe fields (200 to 300 Oersted) of the precession electromagnets. Beyond the flippers this fringe field is useful as a spin guide field. The solution which consists of introducing around the flippers strong correction coils to reduce the field locally to values of typically 20 Oersted normally used introduces very large field inhomogeneities into the system and gives rather unsatisfactory results. We solved this difficulty by developing special air-cooled flippers which can develop coil fields of upto 300 Oersted. Such coils, properly oriented within the fringe fields of the magnets make very good π and $\frac{\pi}{2}$ flippers without destroying the homogeneity. With our rather narrow incident wavelength distributions ($\frac{\Delta\lambda}{\lambda} < 3\%$) they have flipping efficiencies always better than 99%. Small correction coils are then only needed to take into account variations of precession fields for the purpose of NSE scanning.

PERFORMANCE AND RESULTS

From the point of view of polarized neutron flux the present solutions are very satisfactorily. The flux loss with respect to the standard graphite crystal monochromator/analyser is of a factor 4 to 5 only. A comparison with IN11 using the identical sample gave a 30 times stronger signal on D10 (at a higher incident wavelength and therefore lower resolution).

The quantity measured in an NSE experiment is the beam average of the final NSE polarisation. As extensively discussed in [1] this quantity is related to the scattering function $S(Q,\omega)$ of the sample by

$$P_{NSE} = P_o \int S(Q,\omega) \cos \omega t \, d\omega$$

where t is a Fourier time available depending on the machine constants and the incident wavelength. P_o is a normalising polarization which contains eventual changes of polarisation by the sample. This relation can be specialised to particular scattering examples.

For example for a Lorentzian TDS distribution of width Γ in the presence of a purely elastic Bragg peak this polarisation simply gives the elastic fraction x of the scattering

$$P = 1 - x + x\exp(-\Gamma t)$$

the exponential term can be kept vanishingly small provided sufficient Fourier times t can be obtained.

For quasi-elastic Lorentzian scattering the polarisation is just given by

$$P_{NSE} = P_o \exp(-\Gamma t)$$

where Γ is the linewidth parameter of the Lorentzian.

For phonon linewidth studies the measured polarisation is given by

$$P_{NSE} = P_o \cos 2\pi (N_1 - N_2)$$

where N_1 and N_2 stand for the mean numbers of spin precessions in the initial and final beams. The quantity P_o describes the decay of the echo signal as a function of N_1 and contains both the energy width of the classical resolution function of the spectrometer as the standard deviation of the phonon lineshape (both assumed to be Gaussian).

The resolution function of an NSE spectrometer is given by the decay of the NSE polarisation with increasing precession field strength. This decay originates from field inhomogeneities which

introduce Larmor dephasing and lower the measured NSE polarisation.
Fig. 2 shows a typical resolution curve of D10-NSE for an incident
wavelength of 2.41 Å measured using a Bragg peak at low temperatures
to avoid any inelastic scattering. One sometimes defines the spectral resolution which gives rise to a 5% change in P_{NSE} at the
maximum Fourier time t_{max} obtainable. In our case this quantity
amounts to 0.3 μeV. We have verified that for quasi-elastic scattering this is indeed measurable (see example of paraterphenyle).
Deconvolution of measured data is particularly simple due to the
fact that the measured quantity P_{NSE} is the Fourier transform of
the scattering law. It is sufficient to divide the measured P
value by the polarisation of the resolution curve of Fig.2 to
obtain the effective polarisation to be used in the above formulas.

To verify these resolution characteristics we used a molecular
crystal which exhibits a quasi-elastic peak, the energy width is a
function of temperature and drops to zero near a phase transition
of the system. The sample was a 0.1 cm^3 single crystal of deuterated paraterphenyle. At T_c = 179 K this material undergoes an order-
disorder phase transition resulting in a superstructure. Above T_c
at the positions of the superstructure Bragg peaks (such as
(2.5 .5 0)) quasi-elastic scattering is observed arising from the
critical slowing-down of a rotational libration of the molecule [6].
Figure 3 shows our results obtained on D10 as compared to the IN11
(NSE) and IN10 (backscattering) data. It should be added that the
signal was a factor of 30 stronger than on IN11. On IN10 a 3 times
larger sample was used but the intensity was still considerably
lower. This stems from the fact that the present spectrometer
is well adapted to this type of studies from the point of view of
momentum resolution. We conclude that the calculated performance
can be well obtained. The spectrometer is therefore well adapted
to completely separate out TDS scattering under a Bragg peak. An
example of such an experiment is shown in figure 4. The fact that
the measured effective NSE polarization rises at both ends of the
scan is due to the fact that at these reciprocal distances from the
Bragg peak the acoustic phonons are high enough in energy that inelastic scattering is removed by the classical resolution of the
spectrometer. Thus the fraction of elastic "background" scattering
occurring within the classical resolution ellipsoid increases and
the polarization rises. Concerning the measurement of phonon lifetimes only very preliminary measurements have been made. It can
be stated that there is no experimental problem for non dispersive
branches although the data treatment and corrections are not yet
completely under control.

108

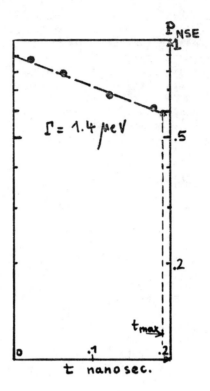

Fig. 2. Example of D10-NSE Resolution as a function of
the Fourier time ($\lambda_0 = 2.4$ Å). The maximum Fourier time
of 0.185 nanoseconds corresponds to a field of 1.5 kOe.

FUTURE TRENDS

Focussing by the "tilted magnet" scheme although possible in
principle and nice looking on paper is not compatible with a compact
spectrometer to be used in a reasonably standard fashion. The secondary effects of tilting those big magnets with their complicated
stray field pattern are very difficult to control. Any action on
one of the magnet tilt angles requires resetting of several other
machine parameters. More patient experimentation will be needed to
decide whether the method of tilting necessarily bulky precession
magnets is the good way to perform NSE focussing.

In fact we are working on the development of superconducting
precession magnets which though very strong could be made extremely
compact. By using superconducting sheets around solenoid magnet
windings the field could be completely confined to the desired beam
region avoiding any stray field problem.

Furthermore, this creates a sample area under low field

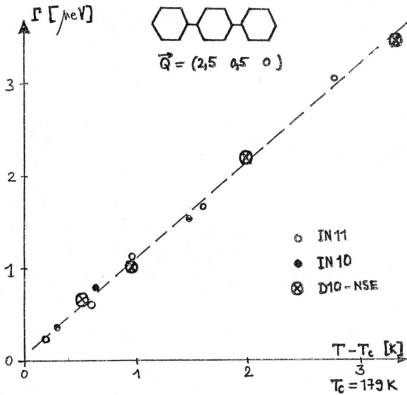

Fig. 3. Quasielastic Critical Scattering in Para Terphenyle: Γ is the measured HWHM of the quasielastic scattering arising from the critical slowing down of an intra molecular libration.

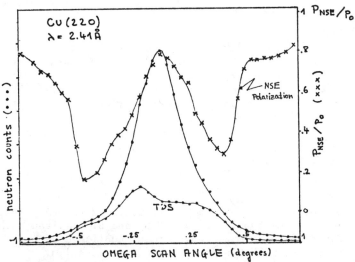

Fig. 4. Example of TDS Separation.

conditions despite a compact spectrometer. The advantages of this are that for large samples and high scattering angles one avoids the otherwise significant Larmor precession dephasing due to different possible path lengths within the sample. It also eases the use of sample orienting devices (Eulerian cradle) or more sophisticated sample environment equipment which usually contain some magnetic material.

Similar shielded superconducting magnets (field of 50 kOe) have been built [7] for other purposes. The shielding action of the super-conducting Nb-Sb sheets was found to be complete and the stability is such that no field variations could be detected over a period of a few months.

CONCLUSION

We have developed a Spin Echo option to a classical Triple-Axis Spectrometer which is capable of improving the energy resolution by two orders of magnitude for quasielastic scattering and the separation of elastic and inelastic scattering. The possibilities of the new NSE-TAS could be well demonstrated by the study of the critical slowing down near the order-disorder phase transition of parterphenyl. Previous results measured on IN11 could be reproduced with a factor of 30 more intensity. New results concerning the q-dependence of this critical scattering have already been obtained and will be reported on later.

The method is ideally suited for TDS studies. Phonon lifetime measurements have been tried very recently. For non-dispersive phonon branches no experimental problems arose although the data evaluation still leaves us with a few unanswered questions. On the other hand the practicability of the "tilted magnet" method for NSE focussing is still an open question.

REFERENCES

(1) Workshop on Neutron Spin Echo, Proceedings
    Grenoble 1979, F. Mezei Ed. Lecture Notes in Physics
    n° 128, Springer-Verlag 1980. Reference to papers by
    Mezei (p.3) Pynn (pp. 159 and 234), Zeyen (p. 151).

(2) F. Mezei, Proc. Int. Symp. on Neutron Scattering (IAEA,
    Vienna 1977)

(3) J.B. Hayter, M.S. Lehmann, F. Mezei, C.M.E. Zeyen, Acta Cryst.
    735 333 (1979)

(4) R. Pynn, J.Phys. E 11 1133 (1978)

(5) The I.L.L. "Yellow Book", I.L.L. neutron beam
    Facilities at the H.F.R. available for users, B. Maier
    I.L.L. (1980)

(6) H. Cailleau, A. Heidemann and C.M.E. Zeyen, J. Phys.C,
    Sol. State Phys.12, L411 (1979) and H. Cailleau, A.Heidemann,
    F. Mezei and C.M.E. Zeyen, to be published

(7) M. Firth, CERN Geneva, private communication.

LOCATION AND INTEGRATION OF SINGLE CRYSTAL REFLECTIONS USING AREA MULTIDETECTORS

C. Wilkinson and H.W. Khamis
Queen Elizabeth College, LONDON W8 7AH

ABSTRACT

Many techniques are currently under development for the identification of peak and background points and the subsequent positioning of a software window in the data array for the integration of Bragg reflected intensities. A direct method for the automatic determination of crystal orientation and a three-dimensional analytic minimum $\sigma(I)/I$ method for the integration of reflections which have been developed at Institut Laue Langevin, Grenoble are described in this paper.

INTRODUCTION

Area detectors for X-ray crystallography[1] and for neutron diffraction[2] are now in use. Larger detectors are planned and will be particularly important for the pulsed neutron sources and synchrotron X-ray sources which are under development. Whatever the physical mechanism of particle detection, the data from area detectors can be presented in "detector space" as a three-dimensional array of integers representing the intensity diffracted by the crystal. This array is built from frames of data on which are represented the number of particles arriving in each spatial element of the detector. Successive frames represent crystal rotation steps in the case of a monochromatic beam or units of energy discrimination for a static crystal and white beam. Bragg reflections occur in this distorted reciprocal space as high regions and the process of data analysis is therefore to identify these regions, index the reflections and evaluate their integrated intensities.

If the lattice parameters and orientation of the crystal are known a trial and error method similar to that described by Milch and Minor[3] for indexing oscillation films can be used. When the unit cell is large, however, it may be that the orientation of the crystal is not sufficiently well known for this method to succeed. We have therefore developed a direct method for the determination of crystal orientation and unit cell dimensions from the observed positions of unindexed reflections.

Several procedures [1,4-7] have been proposed for the integration of reflections from film (two-dimensional data) and area detectors.

The method described by Xuong et al[1] for integration of peaks with an X-ray area detector uses a box of fixed volume. A software window of 3x3 elements is placed at the predicted reflection position and can be adjusted by ±1 element so that its centre corresponds more closely to the centre of gravity of the distribution.

ISSN:0094-243X/82/890111-10$3.00 Copyright 1982 American Institute of Physics

112

Integration is made within this window over nine successive frames. This technique does not attempt any optimisation of peak to background measurement and can also give strong positive bias in the case of very weak reflections.

The one-dimensional profile analysis technique first described by Diamond[8] has been developed by Ford[4] for the analysis of precession photographic data. The advantage of profile fitting is that the estimated uncertainty of the intensity of the reflection is lower when a fitted function is integrated than when the raw intensity measured in each position is used for integration. The drawback of using such a technique in three dimensions is that the precise form of the profile changes more rapidly than in one or two dimensions and that the processing time increases rapidly with the number of dimensions.

Recently, Spencer and Kossiakoff[9] and Wlodawer and Sjölin[10] have described simple pattern recognition techniques which rely on the fact that the Bragg peaks are approximately ellipsoidal in the detector space. They define an elliptical boundary for a peak by consideration of the ratio of the signal in contiguous elements within the ellipse to the variance of the background outside the ellipse. These methods reduce the estimated standard deviations of the integrated peak intensities as they exclude the measurement of unnecessary background points.

The work which will be described here is also designed to minimize the standard deviations in measured intensities of weak reflections. It is based on the $\sigma(I)/I$ minimum method described by Lehman and Larsen[11] for one dimensional data and uses on a knowledge of the peak profile of strong reflections to optimise the integration of weak reflections.

AUTOMATIC LOCATION OF PEAKS IN THREE DIMENSIONAL DATA ARRAYS

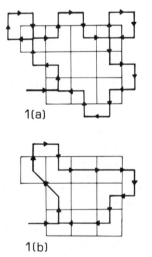

1(a)

1(b)

Fig.1. Contour tracing around peak elements.

On each frame of data, which will in general contain peaks in various stages of development, elements which contribute to a particular peak are defined to tbe those which are contiguous and are greater than a nominally constant background value B by a specified amount. This is normally a few standard deviations of the background level. The problem is then effectively reduced to a binary one which can be tackled by a contour following routine. The one which we have adopted is similar to that described by Duda and Hart[12]. Starting from the bottom of the frame, each row of data is scanned until a peak element is encountered. The algorithm for following the contour around that peak is to take one step to the left if the current element is a peak element, or one step to the right if it is a background element. This traces a contour around the peak in a clockwise direction as illustrated in Fig.1(a). All n elements within the contour are used to calculate the approximate intensity I and the centroid $\bar{x}, \bar{y}$ of the peak section on that frame according to the formulae:

$$I = \sum_{}^{n} (I_j - B) \qquad (1)$$

$$\bar{x} = \sum_{}^{n} x_j (I_j - B) / I \qquad (2)$$

$$\bar{y} = \sum_{}^{n} y_j (I_j - B) / I \qquad (3)$$

where I_j is the count falling in the detector element with coordinates x_j, y_j. The value of B is obtained iteratively for each frame of data by calculating the mean intensity per element, eliminating those elements which are more than a given number of standard deviations from the mean. Except in the case of very weak peaks the position calculated for the centroid is not very sensitive to the precise value of B.

In order to speed up the contouring algorithm and also to prevent the possibility of it becoming trapped in a region of "background" which is included within the boundary of a peak, a more economic decision scheme for tracing the contour is adopted in the computer program (SETUP). This involves the successive testing of the immediate neighbours to the left and in front of the current element, making left, right, diagonal or straight on moves according to the result found. A test is also made for the "single element" peak. The contour traced in this way is shown in Fig.1(b).

The centroids of all peaks identified on a frame are stored and compared with those measured on the previous frame. Real diffraction peaks, as opposed to background fluctuations, will occur in similar positions on many successive frames of data. Before classification as a diffraction peak, therefore, the further requirement is made that there should be at least five successive frames on which the centroids of the peak sections be within a given distance (normally 1 element) of each other. When this condition is satisfied the centroid $\bar{X}, \bar{Y}, \bar{\Omega}$ of the peak which spans m successive frames in Ω is calculated as

$$\bar{X} = \sum_{}^{m} I_k \bar{x}_k / I \qquad (4)$$

$$\bar{Y} = \sum_{}^{m} I_k \bar{y}_k / I \qquad (5)$$

$$\bar{\Omega} = \sum_{}^{m} I_k \Omega_k / I \qquad (6)$$

where I is the sum of the intensities I_k attributed to the peak on successive frames. The detector coordinates of the centroid are then converted to reciprocal lattice coordinates on the axial system defined by Busing and Levy[13]

114

DETERMINATION OF CRYSTAL ORIENTATION AND UNIT CELL SIZE

The lattice of the crystal is found by noting that real lattice points are positions in space where the sets of Bragg planes which are normal to each reciprocal lattice vector have a common intersection. (An alternative statement of this is that the Patterson function calculated from observed reflections has the periodicity of the real lattice). Since points $\vec{r}$ on the set of Bragg planes which correspond to the jth scattering vector $\vec{k}_j$ have

$$\vec{k}_j \cdot \vec{r}_j = n_j \qquad (7)$$

where n_j is an integer, the problem can therefore be reduced to one of searching real space for points $\vec{r}$ where $\vec{k}_j \cdot \vec{r}$ is integral for <u>all</u> measured reflections.

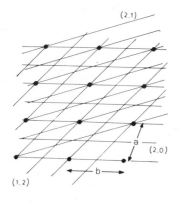

This is illustrated for the case of a two dimensional lattice in Fig.2. The intersections of the Bragg planes corresponding to the (1,2), (2,1) and (2,0) reciprocal lattice vectors are shown. The real lattice point a,2b for example has n_j equal to 5,4 and 2 respectively for these reciprocal lattice vectors.

Since measured reciprocal lattice vectors $\vec{k}_j$ are subject to experimental error, the $\vec{k}_j \cdot \vec{r}$=integer condition must be relaxed in accordance with the uncertainty in each $\vec{k}_j$. The precision with which a particular real lattice vector can be established can therefore be seen to depend on the accuracy in that direction of the measured reciprocal lattice vectors.

Fig.2. Bragg plane intersections in real lattice points

A computer program (BRAGG) has been written to effect this search of real space. The triclinic cell obtained by using BRAGG on data collected by the D17 multidetector at Institut Laue Langevin (ILL), Grenoble from a crystal of nucleosome[2] using 9.25Å neutrons was

$$a = 109.3 \pm 1.9\text{Å} \qquad b = 194.5 \pm 5.8\text{Å} \qquad c = 110.3 \pm 2.1\text{Å}$$

$$\alpha = 89.5 \pm 1.5° \qquad \beta = 90.06 \pm 1.2° \qquad \gamma = 89.5 \pm 2.0°$$

which agrees with the known or orthorhombic cell (space group $P2_12_12_1$) of a = 111Å, b = 198Å, c = 111Å[14]. The components of ten scattering vectors which were used to obtain this cell are given in Table I and are referred to the X, Y and Z axes defined by Busing and Levy[13].

TABLE I.   Observed and calculated scattering vector components in units of 10^{-5} Å^{-1}

$k_x$obs	$k_x$calc	$k_y$obs	$k_y$calc	$k_z$obs	$k_z$calc	h	k	ℓ
994(14)	990	234(7)	238	1809(28)	1826	1	1	2
1731(11)	1723	-478(4)	-477	1(11)	1	1	3	0
1733(5)	1734	-463(2)	-463	-907(5)	-906	1	3	1
3444(22)	3444	-953(11)	-954	11(13)	-2	2	6	0
3453(38)	3457	-940(18)	-940	-903(25)	-905	2	6	1
3409(20)	3425	-977(10)	-982	1810(16)	1813	2	6	2
3749(13)	3746	57(3)	56	29(6)	26	3	5	0
2398(28)	2358	129(7)	133	1852(33)	1833	2	3	2
1011(3)	1012	267(1)	266	12(4)	13	1	1	0
2004(16)	2013	515(4)	518	930(14)	932	2	2	1

Also shown are the indices produced by the program and the scattering vector components calculated from the crystal orientation matrix, which the program also gives.

THE $\sigma(I)/I$ MINIMUM METHOD FOR PEAK INTEGRATION

As originally described by Lehmann and Larsen[11] for one dimensional data scans, the ratio $\sigma(I)/I$ of the standard derivation of the integrated intensity (taking into account background subtraction) to the intensity I is calculated for points which are successively more distant from the peak of a reflection.   The boundary of the peak is defined to occur when $\sigma(I)/I$ reaches a minimum.   In this form the technique presents several difficulties for multidetector data analysis.   These are:

(i)   Although the minimisation of $\sigma(I)/I$ gives the correct statistical decision for the point at which peak integration should stop it is known that it underestimates the intensity of a reflection.   This can be particularly serious for three-dimensional data where the error can be as high as 20%.

(ii)   It fails in the case of weak peaks where a clear minimum in $\sigma(I)/I$ does not always occur.

(iii)   It is time consuming to have to numerically evaluate $\sigma(I)/I$ for every peak.

These difficulties can be overcome if the peak shape is known. Suppose that this is measured for a strong peak in term of p, the number of peak points within a certain contour level of intensity $I(p)$ and that $I(p)$ is a fraction $x(p)$ of the total intensity of the peak $I_o(p_o)$.   Consider now the application of this to a weak peak $I'(p)$ which has the same shape as $I(p)$ (ie $I'(p)/I(p) = f$, a fractional constant independent of p) and sits on the same background B.   We wish to minimise the quantity $\sigma(^1/xI)$, which is

116

also achieved by minimizing its square $\sigma^2(^1/_x\ I')$. This can be expressed as $\sigma^2\ (^1/_x fI)$ which when x is well known and an equal number of peak and background points are measured is given by

$$\sigma^2\left(^1/_x fI\right) \simeq ^1/_{x^2}\left(fI + 2_p\beta\right) \qquad (8)$$

$$= ^{I_o^2}/_{I^2}\left(fI + 2p\beta\right) \qquad (9)$$

Differentiating with respect to p and setting $\dfrac{d\sigma^2}{dp} = 0$

to find the minimum gives the relationship

$$I_c\frac{dp}{dI} - \frac{2p}{x} = \frac{fI_c}{2\beta}\ \left(= \frac{I_o'}{2\beta}\right) \qquad (10)$$

This equation is useful in that it separates the parameters of the strong peak and the weak peak. The quantities $I_o\ \dfrac{dp}{dI}$, x and $\dfrac{2p}{x}$

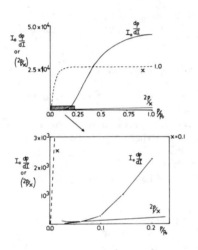

are shown in Fig.3 for a typical strong peak observed on the D17 multidetector. The curves are plotted as a function of the normalised coordinate

$p/_{po}$. In order to decide which is the statistically optimum position to stop the integration it is only necessary to observe where the difference between $\dfrac{2p}{x}$ and $I_o\dfrac{dp}{dI}$

is equal to the ratio of the intensity of the peak to twice the background intensity. (The intensity of the peak needs only to be roughly known for this purpose).

Fig.3. Peak profile parameters.

The advantages of using this analytic method are as follows:

(i)   The statistically optimum point to stop measuring the weak peak can be identified and a correction factor $1/_x$ applied to the intensity.

(ii)   Weak peaks present no difficulty as the use of standard curves from the strong peak renders unnecessary the detection of a numerical minimum in $\sigma(I)/_I$.

(iii)   It is quicker to apply as $\sigma(I)/_I$ does not need to be numerically evaluated for each peak.

WEAK AND STRONG PEAKS

A peak is defined to be strong and added to a library built for the purpose of integrating weak peaks if there is no substantial increase in the variance of the integrated intensity $\sigma^2(I_o)$ as the volume p of the peak is increased.

From Fig.4 it can be seen that this is true provided the ratio $p_o B/I < 0.1$, while for weak peaks, when there is an appreciable increase in the value of $\sigma^2 I$ with p, the noise to signal ratio $poB/I > 1$. An investigation of the full expression (involving $\sigma^2(1/x)$) for the uncertainty in I' has shown that the assumption that $1/x$ is well known is good when $f < 0.3$. Peaks which have $0.1 < \dfrac{poB}{I} < 1.0$ are therefore integrated as strong peaks, but their profiles are not added to the library.

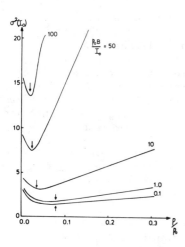

Fig.4. Variance of integrated intensity as a function of noise to signal ratio.

INTEGRATION OF STRONG PEAKS AND CONSTRUCTION OF A MODEL LIBRARY

A Fortran computer program (PEKINT) has been written to carry out the integration procedures and to construct a library of peak profile models. Since the observed intensity contours of the peaks are approximately ellipsoidal in shape the peak profile $I(p)$ is found for strong peaks by dividing the peak into concentric ellipsoidal shells. The smooth ellipsoid which most closely resembles a contour is obtained by calculating the components of the moment of inertia tensor for equally weighted elements which lie inside that contour. Typically, the one twentieth peak height contour is chosen and the smooth solid ellipsoid which has identical tensor components is used as the modelling volume. This shape is expanded and contracted in suitable steps to give the intensity profile $I(p)$, the gradient $dI/_{dp}$ and $\sigma(I)/_I$ as a function of the numbers of points p contained within an ellipsoidal volume. For a strong peak the position of the minimum in $\sigma(I)/_I$ is not critically dependent upon the precise background level chosen. For a more accurate estimate of the background the average count within an ellipsoidal shell far from the peak centre is measured. This shell is normally chosen to be twice as far from the centre of the peak as the surface at which $\sigma(I)/I$ is a minimum. The integrated intensity of the peak is then defined to be the sum of the counts in elements within the background surface after the subtraction of a constant background count per element.

INTEGRATION OF WEAK PEAKS

When a peak is classified as weak, the surface of minimum $\sigma(I)/I$ which envelopes the optimum integration volume is found from equation (10), using parameters obtained from nearby strong peaks. The background is calculated in a shell which was used to calculate the background for the model reflection. The intensity within the optimum integration volume is summed and the background subtraction made. A minimum volume for integration is set in the program and this is used when it is larger than the predicted optimum volume. This is because the effects of mis-centring the integration volume and the discrete nature of the detector elements become important at very small integration volumes.

The gain in accuracy in limiting the integration to the $\sigma(I)/I$ minimum point can be gauged from Fig.4 and is also illustrated by Table II, where the intensity of the same peak for which the profiles are shown in Fig.3. has been measured on the D17 detector with different signal to noise ratios.

The one with the largest signal to noise ratio is the "model" peak for the sequence, with a volume $p_o = 595$ elements. The

Table II

$P_o B / I_o'$	f	x	$I'(p) \pm \sigma I'(p)$	$I/fx \pm \sigma/fx$	$I'(p_o)/f \pm \sigma/f$
0.012	1.0000	1.000	$28,872 \pm 172$	$28,872 \pm 172$	$28,872 \pm 172$
0.065	0.1847	0.984	$5,355 \pm 74$	$29,464 \pm 407$	$29,241 \pm 422$
0.65	0.0185	0.936	481 ± 24	$27,777 \pm 1,386$	$27,776 \pm 1867$
6.5	0.0018	0.857	54 ± 10	$35,000 \pm 6,500$	$35,700 \pm 15,000$

fraction x of the peak included in the $\sigma I/I$ minimum volume is shown, the intensity $I'(p)$ measured within this volume and $I'(p)/fx$, the intensity scaled to compare with the "model" intensity is also shown. The last column gives the intensity measured over the full peak volume po, scaled by I/f to compare with the "model" intensity. It can be seen that when the noise becomes appreciably larger than the signal, there is a gain of more than two in the estimated accuracy of the measured intensity, when integration is terminated at the minimum $\sigma I/I$ point.

MEASUREMENT OF DIFFRACTION INTENSITIES FROM PHTHALOCYANINE CRYSTAL

Integrated intensities of reflections from a phalocyanine crystal have been measured individually with the small area detector on the present D19 diffractometer at ILL[15]. These have been analysed with the $\sigma(I)/I$ technique and also with a method devised by A. Filhol and M. Thomas of ILL, which is a statistical method similar in effect to that described by Spencer and Kossiakoff[9]. The results of this analysis are described in more detail elsewhere[15] and therefore will be mentioned only briefly here. The agreement between intensisities obtained by the two methods is good, with an overall scale factor of 0.999 taken over 1270 reflections and a merging R factor on F^2 of 0.016. Detailed examination shows that in general the $\frac{\sigma I}{I}$ technique gives slightly lower intensities for stronger reflections with $\frac{poB}{I} < 1.0$, but slightly higher intensities in the intermediate and lower intensity ranges.

DISCUSSION

It has been demonstrated that the measurement of the integrated intensities of Bragg reflections using a multidetector can be made on automatic procedure. With a large area detector which is capable of measuring many reflections simultaneously the crystal orientation matrix can be found and the reflections automatically indexed using information gained from the reflection centroids. (There is of course no reason why this approach should not also be used on positional information gained with a single detector, except that this implies some previous knowledge of the crystal orientation in order to find the reflections). Models of peak profiles can be established from strong reflections and can be stored in a library for the integration of weak peaks.

The phthalocyanine data illustrate how systematic differences in intensity measurement can arise with different measurement techniques. The differences in the intensities of strong reflections arise from different techniques of estimating the background, which did not appear to be flat within the range of measurement. The intermediate range intensities were slightly lower when measured by the statistical method as the extent of these reflections is underestimated by this technique.

ACKNOWLEDGMENTS

The authors wish to thank the Institut Laue Langevin for the provision of facilities and a research contract under the auspices of which the work was carried out.
* The nucleosome data was kindly provided by G.A. and A. Bentley.

REFERENCES

1. N.H.Xuong, S.T. Freer, R. Hamlin, C. Nielsen and W. Vernon,
    Acta Cryst. A34, 289 (1978).
2. G.A. Bentley, T.J. Finch and A. Lewit-Bentley, J.Mol.Biol.
    145, 771 (1981).
3. J.R. Milch and T.C. Minor, J.Appl. Cryst. 7, 502 (1974).
4. G.C. Ford, J.Appl. Cryst. 7, 555 (1974).
5. C.E. Nockolds and R.H. Kretsinger, J.Phys.E: Sci Instrum 3,
    842 (1970).
6. W. Kabsch, J.Appl. Cryst. 10, 426 (1977).
7. M.G. Rossman, J.Appl. Cryst 12, 225 (1979).
8. R. Diamond, Acta Cryst A25, 43 (1969).
9. S.A. Spencer and A.A. Kossiakoff, J.Appl. Cryst. 13, 563 (1980).
10. A. Wlodawer and L. Sjölin. Private communication.
11. M.S. Lehmann and F.K. Larsen, Acta. Cryst. A30, 580 (1974).
12. R.O. Duda and P.E. Hart, Pattern Classification and
    Scene Analysis (Wiley N.Y., 1973).p.290.
13. W.R. Busing and H.A. Levy, Acta Cryst. 22,457 (1967).
14. T.J. Finch, R.S. Brown, D. Rhodes, T. Richmond, B. Rushton,
    L.C. Lutter and A. Klug, J.Mol. Biol. 145, 757 (1981).
15. M. Thomas and A. Filhol, ILL Report 81TH08T (1981).
16. R. Stansfield, Ibid.p

NEUTRON SCATTERING STUDIES OF MATERIALS UNDER PRESSURE

C. Vettier
Institut Laue Langevin, Grenoble, France

ABSTRACT

A review of experimental techniques used in the study of materials under applied pressure by neutron scattering is given along with a discussion of the amount of information that can be obtained from various pressure conditions. Some experimental results are presented to illustrate both the feasibility and the interest of elastic and inelastic neutron scattering measurements of materials under pressure.

INTRODUCTION

For a long time pressure has been considered as an important thermodynamic intensive variable in physics and chemistry. Technological applications have developed as a consequence of the steadily increasing high pressure limit in research work. The most famous example is given by diamond synthesis which requires subjecting graphite simultaneously to high pressure ($P \simeq 55$ Kbar) and high temperature ($T \simeq 2000$ K). The pressure range which is commonly encountered in our everyday life is fairly narrow when compared to the spectrum of pressures that can be found or estimated in nature at large; in particular geophysicists are accustomed to thinking in terms of megabar pressures (table I). The megabar range can be achieved experimentally [1], but under such extreme conditions that only a limited number of operations can be performed (resistivity, materials synthesis ...). Neutron scattering studies of materials under pressure have been carried out up to pressures of 40 Kbar, this upper limit is determined by constraints of the neutron scattering technique (windows, background, sample size, etc.)

Table I. Typical values for high pressure

Examples	Pressure (bar)
Bottom of oceans	1. 10^3
Crust-mantle interface	1.1 10^4
Mantle-core interface	1.4 10^6
Neutron scattering exp.	$\sim$ 4. 10^4
X-ray exp.	$\sim$ 3. 10^5

The outline of this review is as follows. Firstly the type of information which may, in principle, be gained from neutron scattering experiments under conditions of high pressure will be outlined. Secondly various experimental pressure techniques and their limitations in neutron scattering will be discussed. Finally some specific examples which characterize the range of present neutron scattering studies of materials under pressure will be presented.

PROPERTIES OF MATERIALS UNDER PRESSURE

The purpose of this section is to present a condensed overiew of the known effects of pressure which can be probed by neutron scattering [2,3,4]. The essential effect of pressure is to reduce interatomic distances. Therefore, it is of fundamental importance to study the effects of pressure on structural properties.

1. Structural properties. All materials are compressible. Certainly solids are less compressible (compressibility $K \sim 10^{-6}$ bar^{-1}) than liquids ($K \sim 50 \ 10^{-6}$ bar^{-1}), but there exist exceptions (solid He at low pressure, $K \sim 1000 \ 10^{-6}$ bar^{-1}). Structural investigations under pressure have been made on liquids [5] and solids [6], using neutron diffraction. In the case of liquids, a study of the microstructure of heavy water has indicated that rigid water molecules rearrange themselves in such a way that the scale of volume varies as the inverse of the density [5].

In the case of solid materials, applied pressure leads to more complex changes in properties due to the stronger interactions between atoms. At this point it is worth pointing out that, by virtue of the fairly low compressibility of condensed matter, quite large applied pressure must be applied to observe noticeable effects. Firstly, experimental (Pressure, Volume, Temperature) relationships can be compared with various equations of state [7]. Furthermore crystallographic phases can be induced by a pressure driven softening of phonons [8]. The dependence of the interatomic forces on distance, especially the anharmonicity of interatomic potential has been studied by measuring phonon spectra as a function of temperature and volume [9]; by varying pressure and temperature, it is possible to keep the sample volume constant and therefore study the effects of the anharmonicity as a function of temperature. Another aspect of the variation with lattice constants of interactions is illustrated by incoherent inelastic neutron scattering studies of tunneling frequencies [10] under pressure which provide a measure of the volume dependence of barrier heights.

2. Phase transitions. In solid materials, when a transition occurs between a high temperature (disordered) phase and a low temperature (ordered) phase, the symmetry is lowered. In some cases, this change in symmetry gives rise to domain structure. Application of anisotropic (unaxial) stress would favour one domain orientation; indeed, neutron diffraction experiments under unaxial stress have allowed structure determinations of ordered phases on single domain samples [11]. Pressure as an external field, introduces a further

dimension to phase diagrams. Both magnetic and structural phase
diagrams under pressure have been determined with neutrons [12].
Since the order of the transition itself is the result of a delicate
balance between forces, it can be modified when applied pressure
reaches a threshold value; such multicritical points, observed
experimentally, are of great interest in the theory of critical
phenomena. In addition, the propagation wavevector $\vec{\delta}$, which
characterizes commensurate or incommensurate ordered phases is also
pressure dependent [13]; in the case where δ "locks in" to rational
values, incommensurate to commensurate transitions can be investigated.

3. Electronic properties. Reductions of lattice spacings produce
changes in band structure, which in turn can induce electronic
transitions (e.g. metal-insulator transitions, superconductivity,
magnetic order, valence changes ...). These changes in electronic
properties can be probed with neutrons [14] either directly or
indirectly by observing secondary effects such as compressibility or
phonons anomalies.

HIGH PRESSURE TECHNIQUES FOR NEUTRON SCATTERING

Pressure is defined as the ratio of force by unit area. From
this definition, it follows that very high pressure could be achieved by using very small sample dimensions. However, neutron scattering studies require fairly large sample volumes and, as a consequence, high pressure can only be obtained by applying large forces
which implies massive mechanical devices.

Applied stress can be isotropic (hydrostatic pressure) or anisotropic (e.g. unaxial stress). When subjected to unaxial stress,
a solid is simply squeezed between two pistons; therefore the
upper limit of stress depends essentially on the mechanical properties of the sample itself (covalent compounds can hold up to 20 Kbar
unixial stress) : the sample fractures ! Since the material under
investigation is in contact with external surfaces (pistons) on two
opposite faces only, it is clear that any neutron scattering experiment on solids subjected to unaxial stress can be performed
easily [15].

On the other hand, hydrostatic pressure does not destroy the
materials (a part from violently first-order pressure induced
transition) and so pressure limitations are simply the limits of
the present technology. In essence, pressure is applied to a sub-
stance by immersing it into a pressure transmitting medium which is
pressurized either by a volume reduction (fluid and solid media)
or by compression at constant volume (gaseous medium). The pressure
medium is contained in a pressure vessel whose wall thickness is a
function of the mechanical strength of the material that constitutes the vessel and is critically dependent on the sample volume.
The difficulties encountered in neutron scattering experiments
appear immediately : the neutron beams must travel through rather
thick sections of the pressure cell which are in close contact with
the sample under investigation; this gives rise to absorption and
coherent or incoherent "parasitic" scattering, this extra scat-

tering can be eliminated by using fixed scattering angle geometry.

It is necessary to balance absorption against the pressure vessel strength [2,3,4,16]. The choice of materials is confined to aluminium alloys (Al-Zn-Mg) copper alloys (Cu-Be), maraging steels (Fe-Ni-Co-Mo), Titanium-Zirconium (null matrix alloys) or sintered alumina (Table II). According to the current technology, the maximum pressure attainable is around $P \simeq 40$ Kbar with a reasonable sample volume ($V \approx .3$ cm^3). Two types of apparatus are available according to the maximum working pressure to be achieved.

Table II. Materials for high pressure cells. Approximate values for bursting pressures and absorption for a cylindrical vessel (i.d = 5mm, o.d = 20mm) with a continuous window (h = 5mm).

Materials	Al. alloys	Ti-Zr	Cu-alloy	Maraging steel	Al$_2$O$_3$
Bursting pressure (10^3bar)	25	40	65	85	45
Absorption (%)	20	40	67	82	50

At pressures below 6 Kbar, the pressure vessel materials can hold the pressure without being reinforced. Sample volume can be as large as 5 cm^3. He gas has been widely used as pressure medium; it offers good hydrostaticity even at low temperatures. However, clamped piston-cylinder devices, with a liquid pressure medium, have also been used; deuterated methanol- ethanol mixtures, or "Fluorinert" C_6F_{12} are the most common fluids. Such pressure vessels are suitable for any conventional neutron scattering experiments, except small angle scattering. In this particular case, which is of interest for polymer studies and biology, sapphire windows must be mounted in the pressure vessel.

At higher pressures, the cells must be reinforced to hold the pressure [18] (Fig. 2). Windows are cut through the opaque and strong supporting parts to let neutrons in and out of the sample volume. Sintered alumina is chosen as a material for the pressure cylinder because of its transparency to neutrons and its very high compressive strength. Fixed angle window cells, with a better support, provide a higher pressure limit or a larger sample volume; these devices are well suited to elastic and inelastic studies of polycrystalline samples at fixed scattering angles. On the other hand, continuous windows (Fig. 2) provide access to an entire scattering plane, which allows for single

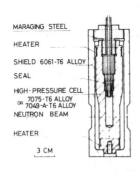

MARAGING STEEL

HEATER

SHIELD 6061-T6 ALLOY

SEAL

HIGH-PRESSURE CELL
7075-T6 ALLOY
OR 7049-A-T6 ALLOY

NEUTRON BEAM

HEATER

3 CM

Fig. 1. Aluminium alloy pressure cell.

crystal studies. The consequence is a rather small sample volume.

The actual pressure on the sample can be measured by means of standard manometers; in the case of a clamped device, the pressure is monitored by the observation of changes in lattice constants of calibrant (NaCl, CsCl) whose equation of state is known and which is mixed with the sample.

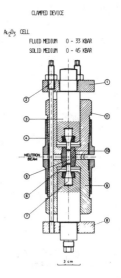

CLAMPED DEVICE

Al_2O_3 CELL
FLUID MEDIUM   0 - 33 KBAR
SOLID MEDIUM   0 - 45 KBAR

SOME NEUTRON STUDIES
AT HIGH PRESSURE

A few experimental studies are presented in order to emphasize the advantages and drawbacks of different scattering experiments under high pressure.

1. Phase transition in NiF_2 [19]. At ambient pressure, NiF_2 has the rutile structure ($P4_2/mnm$). A considerable amount of interest has been focussed on such compounds because a pressure induced softening of transverse acoustic phonons branches can produce structural transitions. Indeed ultrasonic measurements have indicated of softening of the elastic constant ($C_{11}-C_{12}$) with increa-

Fig. 2. Supported Al_2O_3 pressure cell with continuous window.

sing hydrostatic pressure; subsequent X-ray studies confirmed a tetragonal to orthorhombic transition, but were unable to determine the high pressure space group because of the lack of resolution. Time-of-flight neutron measurements were done on the powder diffractometer H-8 at Argonne's CP-5 research reactor. Pressure was applied at room temperature using a supported Al_2O_3 pressure cell with fixed angle windows. The scattering angle ($2\theta = 90$) was such that, with collimated neutron beams, no reflections from the cell contamined the diffraction pattern. Structural and anisotropic thermal parameters were determined by profile refinement techniques. The high pressure phase was found to have the P_{nnm} space group, a subgroup of $P4_2/mnm$, which is consistent with a continuous phase transition (Fig. 3). This study demonstrates that a small amount of strain can be measured under pressure with TOF techniques. In the particular case of high pressure studies, the major advantage of TOF is that reflections from the pressure cell can be completely eliminated.

2. Modulated structures of thiourea[20]. Thiourea ($SC(NH_2)_2$) is one of the earliest insulators found to undergo a commensurate-incommensurate phase transition. Above T_o, the space group of the paraelectric phase is Pnma; at low temperature ($T < T_c$), it becomes ferroelectric with space-group $P2_1ma$. Between T_o and T_c, $SC(NH_2)_2$ exhibits different modulated phases, as evidenced by (h, k ± nδ, l) satellite reflections in the diffraction patterns (δ is the modulated vector). At ambient pressure, δ was found to be temperature dependent, and

a commensurate phase ($\delta = 1/9$) was found just above the ferro-electric transition ($\delta = 0$). Pressure has been applied in order to study the variation of δ, especially close to commensurate-incommensurate transitions. In the course of this study, a new lock-in phase ($\delta = 1/7$) was discovered (Fig. 4); the range of stability of this phase is such that it suggests the existence of two distinct phases with the same translational symmetry (7b) but different structures. This would deserve further experimental verification.

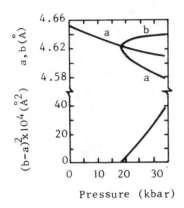

Fig. 3. Pressure dependence of lattice parameters a and b of NiF_2.

This experiment was performed on the IN2 three-axis spectrometer at the Institut Laue-Langevin, with a high strength Al alloy pressure cell. Similar experiments on other compounds have been performed at higher pressures using a supported Al_2O_3 cell.

In such studies, the main difficulty comes from background scattering due to the pressure cell : Bragg or incoherent scattering from the cell can overwhelm weak satellite reflections from the crystal.

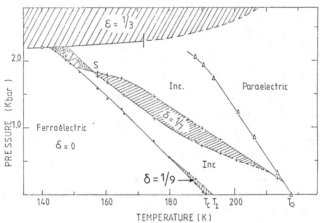

Fig. 4. Pressure-temperature phase diagram of $SC(ND_2)_2$ as determined by neutron diffraction.

3. Pressure induced magnetic order [21,22].

Praseodymium antimonide has a simple NaCl structure; the cubic crystalline electrical field lifts the degenerate free ion state of Pr^{3+} in such a way that a low temperature PrSb can be described as a singlet-triplet model. Excitations between crystal field levels can be studied by inelastic neutron scattering, and their variations with pressure [21] can be tested against current theories of crystal field splittings Δ and exchange J. Experiments were performed at Brookhaven National Lab. and I.L.L. using clamped Al_2O_3 cells at pressure up to 35 Kbar and temperatures down to 5K. The experimental data show that Δ decreases with applied pressure. At high pressure (P > 30 Kbar), Δ is sufficiently small that the ratio J/Δ exceeds some critical value leading to magnetic order. Since the softening occurs at the Brillouin zone boundary, the ordered phase is antiferromagnetic. Values of the magnetic moment and the phase diagram have been obtained.

Similar studies of the more complex properties of pure praseodymium metal have been reported [22]. In this case, the variations of the crystal field was produced by a change in symmetry due to applied uniaxial stress.

CONCLUDING REMARKS

These experimental results demonstrate that elastic and inelastic neutron scattering experiments under high pressure are both feasible and fruitful. Since neutron scattering has already had a considerable impact in the study of condensed matter, it is hoped that further technological developments will allow more sophisticated neutron studies of materials under pressure in more diversified areas of physics.

REFERENCES

1. S. Block, G. Piermarini, Physics Today, 44 (Sept. 1976).
2. D. Bloch, J. Paureau, Nato Advanced Institute on "High Pressure Chemistry", C41 (D. Reidel, Dordrecht, 1978), p.111.
3. C.J. Carlile, D.C. Salter, High Temp.-High Press, 7, 529 (1975).
4. D.B. McWhan, C. Vettier, J. Physique C5, 107 (1979).
5. Microstructure of liquids under pressure : P.A. Egelstaff, D.I. Page, C.R.T. Heart, J. Phys. C4, 1453 (1971) G.W. Neilson, D.I. Page, W.S. Howells, J. Phys. D12, 901 (1979).
6. Structure determinations of solids under pressure : R.M. Brugger, R.B. Bennion, T.G. Worlton, Phys. Lett. A24, 714 (1967); D.R. McCann, L. Cartz, R.E. Schmunk, Y.D. Harker, J. Appl. Phys. 43, 1432 (1972); D.B. McWhan, G. Parisot, D. Bloch, J. Phys. F4, L69 (1974); C. Vettier, W.B. Yelon, J. Phys. Chem. Solids 36, 401 (1975); T.G. Worlton, R.A. Beyerlein, High Temp.-High Press, 8, 27 (1976); G.A. Mackenzie, B. Buras, G.S. Pawley, Acta Cryst. B34, 1918 (1978); S.R. Srinivasa, L. Cartz, J.D. Jorgensen, J.C. Labbe, J. Apply. Cryst. 12, 511 (1979); G.M. Meyer, R.J. Nelmes, C. Vettier, J. Phys. C13, 4035 (1980)
7. D.L. Decker, T.G. Worlton, J. Apply. Physics 43, 4799 (1972).
8. D.B. McWhan, R.J. Birgeneau, W.A. Bonner, H. Taub, J.D. Axe, J. Phys. C8, L81 (1975).
9. Gruneisen parameters : W.B. Daniels, G. Shirane, B.C. Frazer, H. Umebayashi, J.A. Leak, Phys. Rev. Lett. 18, 548 (1967); G. Shirane, W.B. Daniels, Phys. Rev. B6, 4766 (1972); J. Eckert, W. Thomlinson, G. Shirane, Phys. Rev. B18, 3074 (1978); J.R.D. Coppley, C.A. Rotter, H.G. Smith, W.A. Kamitakahara, Phys. Rev. Lett. 33, 365 (1974); O.Blasko, G. Ernst, G. Quittner, J.R.D. Coppley, Sol. State. Commun. 21, 1043 (1977); D.B. McWhan, R.C. Dynes, S.M. Shapiro, Int. Conf. on lattice Dynamics (Flammarion, Paris, 1978), p. 648.
10. Tunneling frequencies under pressure : S. Clough, A. Heidemann, M. Paley, C. Vettier, J. Phys. C12, L781 (1979); J. Eckert, C.R. Fincher, J.A. Goldstone, W. Press, to be published.
11. Structure determinations under unaxial stress. D. Bloch, C. Vettier, J. Magn. Magn. Mat. 15-18, 589 (1980); B. Barbara, M.F. Rossignol, J.X. Boucherle, C. Vettier, Physica 102B, 177 (1980).
12. Phase diagrams and multicritical points : W.B. Yelon, D.E. Cox, P J. Kortmann, W.B. Daniels, Phys. Rev. B9, 4843 (1974); C. Vettier, W.B. Yelon, Phys. Rev. B11, 4700 (1975); R. Youngblood, B.C. Frazer, J. Eckert, G. Shirane, Phys. Rev. B22, 228 (1980); P. Bastie, M. Vallade, C. Vettier, C.M.E. Zeyen, H. Meister, J. Physique 42, 445 (1981).
13. Incommensurate phases : F.A. Smith, C.C. Bradley, G.E. Bacon, J. Phys. Chem. Sol. 27, 925 (1966); H. Umebayashi, G. Shirane, B.C. Frazer, W.B. Daniels, J. Phys. Soc. Japan 24, 368 (1968).

W. Press, C.F. Majkrzak, J.D. Axe, J.R. Hardy, N.E. Massa,
F.G. Ullmann, Phys. Rev. B22, 332 (1980); S. Megtert,
R. Comès, C. Vettier, R. Pynn, A.F. Garito, Solid State
Commun. 37, 875 (1981).

14. Electronic and Magnetic Properties : B.D. Rainford, B. Buras,
B. Lebech, Physica 86-88B, 41 (1977); C. Vettier, D.B. McWhan,
E.I. Blount, G. Shirane, Phys. Rev. Lett. 39, 1028 (1977);
D.B. McWhan, S.M. Shapiro, J. Eckert, H.A. Mook, R.J.
Birgeneau, Phys. Rev. B18, 3623 (1978); B. Barbara,
M. Rossignol, J.X. Boucherle, C. Vettier, Phys. Rev. Lett. 45,
938 (1980).

15. Uniaxial stress devices : A. Draperi, D. Herrmann-Ronzaud,
J. Paureau, J. Phys. E9, 174 (1976); J. Pretchel, E. Lüscher,
J. Kalus, J. Phys. E10, 432 (1977).

16. D.B. McWhan, High Pressure Science and Technology (Plenum,
N.Y., 1979), p.292.

17. Non supported pressure vessel for neutron scattering.
R. Lechner, Rev. Sci. Inst. 37, 1534 (1966); H. Umebayashi,
G. Shirane, B.G. Frazer, W.B. Daniels, Phys. Rev. 165, 688
(1968); J. Paureau, C. Vettier, High Temp.-High Press 7,
529 (1975); R.M. Moon, W.C. Koehler, D.B. McWhan, F. Holtzberg,
J. Apply. Phys. 49, 2107 (1978); J. Mizuki, Y. Endoh, J. Phys.
Soc. Japan 50, 914 (1981).

18. Supported pressure vessels for neutron scattering :
D.B. McWhan, D. Bloch, G. Parisot, Rev. Sci. Instr. 45, 643
(1974); D. Bloch, J. Paureau, J. Voiron, G. Parisot,
Rev. Sci. Instr. 47, 296 (1976); B. Buras, W. Kofoed,
B. Lebech, G. Bäckström, Danish Atomic Energy Commission,
Report N° 357 (Risø, 1977), p.32; J.D. Jorgensen, T.G.
Worlton, J.C. Jamieson, High Pressure Science and Technology
(Plenum, N.Y., 1979) p.152.

19. J.D. Jorgensen, T.G. Worlton, J.C. Jamieson, Phys. Rev. B17,
2212 (1978).

20. F. Denoyer, A.H. Moudden, R. Currat, C. Vettier, A. Bellamy,
H. Lambert, to be published.

21. D.B. McWhan, C. Vettier, R. Youngblood, G. Shirane, Phys. Rev.
B20, 4612 (1979)

22. K.A. McEwen, W.G. Stirling, C. Vettier, Phys. Rev. Lett. 41,
343 (1978).

POLARIZING MULTILAYER SPECTROMETER FOR NEUTRONS

C. F. Majkrzak, L. Passell, and A. M. Saxena
Brookhaven National Laboratory, Upton, NY   11973

ABSTRACT

Polarizing neutron monochromators were prepared by sputtering
thin-film multilayers with d-spacings from 40 to 85Å on large float-
glass substrates. Peak reflectivities as great as 90% and polarizing
efficiencies of 98% were measured. Increased angular acceptances
were obtained by fabricating multilayers with multiple d-
spacings. A planned polarized beam spectrometer which incorporates
the multilayers and which has a variable energy resolution independent of angular beam divergence is described.

INTRODUCTION

It has long been known that polarization analysis can enhance
the sensitivity and selectivity of many inelastic neutron scattering
experiments. Nevertheless, the technique is rarely employed because
those monochromating crystals which are capable of polarizing neutrons have relatively low reflecting efficiencies and consequently do
not yield adequate beam intensities.

Recently, a new type of neutron polarizing monochromator has
been developed. Lynn et al.[1] demonstrated that a composite of alternating thin films of Fe and Ge (with the Fe layers magnetized to
saturation) is not only an efficient polarizer but has a high reflectivity as well. However, the multilayers they made (by evaporation
on 1x6 inch substrates) had relatively large d-spacings, typically
of the order of 150Å. Thus for neutrons of wavelength λ = 1.5Å-a
wavelength commonly employed-the angle of reflection θ (given by the
Bragg condition $\lambda = 2d\sin\theta$) turns out to be $\sim$ 0.25°. This requires a
multilayer length in excess of 8 feet in order to reflect a beam
1/2 inch wide!

Using a specially constructed radio frequency sputtering apparatus, we have deposited Fe-Ge multilayers with d-spacings between
40 and 85Å uniformly on 2x18 inch float-glass substrates (up to 3x36
inch substrates can be accommodated). With these multilayers, peak
reflectivities as large as 90% and polarizing efficiencies of 98%
have been measured in applied magnetic fields of 100 Oersted. For
λ=1.5Å and d=40Å, the angle of reflection $\theta \sim 1.0°$ and 1/2 inch beam
widths can now be obtained with a multilayer only 2 feet long. We
will discuss the angular acceptance of these multilayers and a method of improving their energy resolution in the following sections.

MULTILAYER REFLECTIVITY

A mosaic crystal can be described in terms of an angular distribution of perfect microcrystallites which reflect a given wavelength
over a finite range of incident angles. The angular acceptance of a
multilayer (with parallel layers) depends, for a particular wavelength,

ISSN:0094-243X/82/890131-04$3.00 Copyright 1982 American Institute of Physics

132

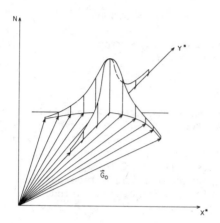

FIG. 1. Distributions of reciprocal lattice vector lengths and
directions.

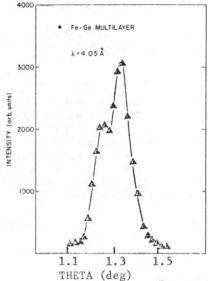

FIG. 2. θ:2θ scan of a multiple-d
multilayer.

not on an angular mosaic but on a
distribution of d-spacings.
Fig.1 shows a plane in reciprocal
space in which lie two distribu-
tions (taken to be Gaussian for the purpose of illustration) of reciprocal lattice vectors: one of the directions of the vector $\vec{G}_0$;the
other of reciprocal lattice vector magnitudes along a single direction. The former characterizes a mosaic crystal while the latter
corresponds to a multilayer with a distribution of d-spacings. Position along the vertical axis labeled N is proportional to the number
of reciprocal lattice vectors. Ideally, to measure directly the peak
reflectivity and angular acceptance first requires a perfectly parallel and monochromatic incident beam. To make such a measurement on a
mosaic crystal then requires that it be rotated about an axis perpendicular to $\vec{G}_0$ through the mean Bragg angle θ with a detector
fixed at a scattering angle of 2θ. This is the well-known "rocking
curve" scan which traverses the circular path of radius $|\vec{G}_0|$ in Fig.
1. For multilayers, on the other hand, a θ:2θ scan tracing a path
along the y* axis is appropriate. In practice, the peak reflectivity
and angular acceptance of our multilayers were measured in two ways:
first, using a conventional triple-axis spectrometer in the elastic
scattering mode; second, by placing two identical multilayers on a
double-axis spectrometer with one as monochromator and the other as
analyser. Peak reflectivity and angular acceptance were obtained
by deconvoluting the data with the spectrometer resolution functions.
Both methods yielded consistent results. Angular acceptances of
about 4 minutes of arc(full width at half maximum) in θ for a θ:2θ
scan were obtained for multilayers made with one nominal d-spacing;
the width is presumably due to a spread in d-spacings corresponding
to deposition rate fluctuations in the sputtering system . Smaller
θ widths for rocking curve scans were also detected which are
probably due to deviations of the substrate surface from perfect

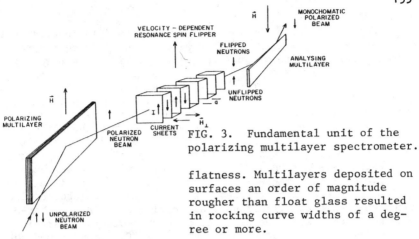

FIG. 3. Fundamental unit of the polarizing multilayer spectrometer.

flatness. Multilayers deposited on surfaces an order of magnitude rougher than float glass resulted in rocking curve widths of a degree or more.

As discussed above, the angular acceptance of a multilayer can be increased by additional d-spacings. Numerical calculations of the dynamical structure factor were performed in order to determine the interference effects to be expected for a particular sequence of d-spacings. These calculations are analogous to those encountered in the solution of thin-film optical interference problems.[2] One scheme is to deposit N_1 bilayers with d_1, N_2 with d_2 and so on. Another is to increment the d-spacing continuously as done by Mezei[3] in constructing "supermirrors," but without overlapping the region of total mirror reflection.

Fig. 2 shows a $\theta:2\theta$ scan for a multilayer with approximately 900 bilayers and 2 nominal d-spacings in an incident beam of 1' divergence and $0.006\Delta\lambda/\lambda$ wavelength resolution at $\lambda=4.05\text{Å}$. The peak reflectivity was measured to be greater than 90% at both this wavelength and at $\lambda=2.46\text{Å}$ (multilayer peak reflectivity is independent of wavelength, if true absorption effects are neglected). We are now preparing high reflectivity multiple d-spacing multilayers which are expected to have angular acceptances of 30'. Space limitations unfortunately prohibit any discussion of the technical problems associated with the actual fabrication of these multilayers here.

POLARIZING MULTILAYER SPECTROMETER

In a Bragg reflection process, the limiting wavelength resolution $\Delta\lambda/\lambda$ is given by $\cot\theta\Delta\theta$ where $\Delta\theta$ represents the angular divergence of a polychromatic incident beam. $\Delta\lambda/\lambda$ is therefore relatively poor for a typical multilayer since the angle θ is small. If the beam is polarized, however, the wavelength or energy resolution can be significantly improved by using a wavelength dependent flipper such as the one developed by Drabkin.[4,5] The fundamental unit of the polarized-neutron spectrometer we propose to build consists of a pair of polarizing multilayers between which a corrugated, current carrying Al foil is inserted as shown in Fig. 3. The foil produces a small, spatially oscillating magnetic field perpendicular to the beam direction. A larger, uniform magnetic field H_0 is superimposed

perpendicular to both the oscillatory field and the beam direction.
The resultant magnetic field acts as a velocity selective resonance
flipper; i.e., only those neutrons with velocities in the neighbor-
hood of $v_o = a \gamma H_o / \pi$ (where γ is the gyromagnetic ratio) undergo a spin
flip. Neutrons with other velocities are unaffected and are subse-
quently not reflected by the second multilayer in which the magnetization direction is opposite to that of the first. Together, the pair
of multilayers and flipper produce polarized monochromatic beams with
an energy or wavelength spread which can be varied by simply changing
the length of the current carrying section of the corrugated foil
(the energy resolution $\Delta E/E \propto 1/M$ where M is the number of reversals
in the oscillatory magnetic field direction). We believe such a
device to be practical for wavelengths $\gtrsim 1.5 \text{\AA}$. It should also be
noted that multilayers strongly discriminate against higher-order
beam components and thus will act as higher-order filters as well.[1]
The fundamental unit described above would function as monochromator
of a triple-axis spectrometer with a second identical unit serving as
analyser following the sample. Our expectation is that intense, polarized beams of high quality and adjustable energy resolution will be
obtained with this system.

ACKNOWLEDGMENTS

The authors would like to thank E. Caruso, F. Langdon, W. Lenz,
and P. Pyne for professional technical assistance. Work was supported by USDOE contract no. DE-AC02-76CH00016 and NSF grant no.
PCM77-01133.

REFERENCES

1. J. W. Lynn, J. K. Kjems, L. Passell, A. M. Saxena, and B. P.
Schoenborn, J. Appl. Cryst. 9, 454 (1976).
2. P. Croce and B. Pardo, Nouv. Rev. d'Optique Appliquée 1, 229(1970)
3. F. Mezei, Commun. on Physics 1, 81 (1979).
4. G. M. Drabkin, J. Exptl. Theoret. Phys. (U.S.S.R.) 43,1107(1962).
5. M. M. Aganalyan, G. M. Drabkin, and V. T. Lebedev, Zh. Eksp. Teor.
Fiz. 73,382 (1977).

THE ANALYSIS OF POWDER DIFFRACTION DATA

M.J. Cooper

Materials Physics Division, AERE Harwell, OX11 ORA, England

ABSTRACT

A comparison has been carried out between the results of analyses of several sets of neutron powder diffraction data using three different methods: the Rietveld method (H.M. Rietveld, Acta. Cryst. 22, 151-152 (1967): J. Appl. Cryst. 2, 65-71 (1969), a modification of the Rietveld method to include off-diagonal terms in the weight matrix (C.P. Clarke and J.S. Rollett, Acta. Cryst. In the press) and the SCRAP method, which involves the estimation of observed Bragg intensities (M.J. Cooper, K.D. Rouse and M. Sakata, Z. Krist. In the press). Two simulations have also been carried out to demonstrate the way in which the results can differ in more extreme cases. This study has confirmed that the values of the estimated standard deviations given by the Rietveld method are not reliable and that, of the methods considered, only the SCRAP method will in general give reliable values for the estimated standard deviations of the structural parameters.

INTRODUCTION

In an analysis of the Rietveld profile refinement method[1] we have shown that the method provides unreliable values for the e.s.d.s. (estimated standard deviations) of the refined parameters[2]. This has prompted the development of methods for the refinement of powder diffraction data which will give more reliable values using two quite different approaches to the problem.

In a recent paper Clarke and Rollett[3] have described a modification of the Rietveld method which allows for correlation between neighbouring residuals, whilst an alternative approach (SCRAP), which involves the estimation of Bragg intensities, has been developed by Cooper, Sakata and Rouse[4]. We have therefore carried out a detailed comparison of results obtained from the same experimental data using both these methods, as well as the Rietveld method, in order to compare their effectiveness in overcoming the limitations of the latter. Further details of this study have been reported elsewhere[5]. In the present paper we will simply attempt to clarify the reason why the Rietveld method is inadequate and summarise the results of the detailed comparison and the conclusions which can be drawn from them.

ISSN:0094-243X/82/890135-06$3.00 Copyright 1982 American Institute of Physics

THE RIETVELD METHOD

In an analysis of the Rietveld method we showed that correlation between neighbouring residuals can arise from differences between the observed and calculated values of the Bragg intensities and also that the residuals depend on the profile parameters and the structural parameters in two distinct and separable ways[2]. It is, however, important to note that correlation between neighbouring residuals can also arise from systematic differences between the observed and calculated peak shapes. For example, if the peak shape is assumed to be Gaussian but in practice contains additional intensity in the tails of the peak, perhaps due to thermal diffuse scattering, then there will be some correlation between neighbouring residuals even if the observed and calculated Bragg intensities are identical. Thus there are two different factors which can lead to correlation between neighbouring residuals.

Whilst this correlation is fairly easy to visualise, it is important to realise that the nature of the model which is used, from the point of view of the separability of the factors depending on the two different types of parameters, can cause errors in the values of the e.s.d.s. even when it causes no appreciable correlation. Such a situation could arise, for example, if the model provides an extremely good fit for the Bragg intensities.

The theoretical model used in the Rietveld method gives a calculated intensity which is derived from the product of two factors. One of these is the Bragg intensity, which is a function of the structural parameters (p_c) only, whilst the other is the peak shape function, which is a function of the profile parameters (p_p) only. In general, the calculated intensity thus has the form:

$$y_i' = K \sum_k I_k(k, p_c) \, G_{i,k}(i, k, p_p) \tag{1}$$

where K is a scale factor, I_k is the Bragg intensity and $G_{i,k}$ is the peak shape function for the kth peak.

As a consequence of this the residuals Δ_i, where $\Delta_i = y_i - y_i'$ and $y_i = (y_{obs})_i$, can be considered as two separate component terms:

$$\Delta_i = (\Delta_i)_S + (\Delta_i)_B \tag{2}$$

If Δ_k is the difference between the observed and calculated Bragg intensity these terms will have the form:

$$(\Delta_i)_S = \Delta_i - \sum_k G_{i,k} \Delta_k \tag{3}$$

$$(\Delta_i)_B = \sum_k G_{ik} \Delta_k \qquad\qquad (4)$$

Provided that the Bragg intensity is insensitive to the peak shape we can then assume that the first term depends only on the profile parameters and the second term only on the structural parameters. It therefore follows that the estimated standard deviations for the two types of parameters should be derived from the appropriate terms of the residuals and not from the total residuals.

ALTERNATIVE METHODS

The SCRAP method[4] separates the refinement of the structural parameters from that of the profile parameters. In the first refinement stage estimates of the Bragg intensities (or groups of unresolvable Bragg intensities) are refined with the profile parameters, so that the residuals are $(\Delta_i)_S$, as given in equation (3). The structural parameters are then refined separately from the estimates of the Bragg intensities and the residuals for this stage are $(\Delta_i)_B$, as given in equation (4). The SCRAP method thus determines the e.s.d. values from the residuals appropriate to the type of parameter concerned and is therefore statistically acceptable.

The method proposed by Rollett and implemented by Clarke and Rollett[3] permits the inclusion of off-diagonal terms in the weight matrix used in the Rietveld method in order to allow for the mean correlation between residuals separated by various numbers of steps on the profile. It will therefore take into account, to a certain extent, the correlation between neighbouring residuals resulting from the differences between observed and calculated values of the Bragg intensities, as well as any correlation resulting from systematic errors in the peak shape function. However, it does not allow for the separate dependence of the terms in the residuals on the two types of parameters and, although the e.s.d. values will be more reliable than those given by the unmodified Rietveld method, they will still be unreliable if one type of parameter is less well-determined than the other.

COMPARISON OF RESULTS

The three methods discussed above have been used to analyse a number of sets of neutron diffraction data in order to provide a comparison of the results which they give. Three experimental data-sets were used: UO_2, for which the peaks are essentially resolved; Al_2O_3, for which a certain amount of peak overlap occurs: and acetic acid, for which there is extensive peak overlap. Two simulations were also carried out, based on the UO_2 data, in order to investigate the results obtained in more extreme situations in which the two types of parameters are determined to widely differing degrees. The results of these analyses have been given elsewhere[5] and we shall therefore restrict the present discussion to a consideration of the main features which are of interest in structural studies and of their relevance to the differences between the three methods of analysis.

For the UO_2 data the structural parameters are relatively less well determined than the profile parameters and the standard deviations for the latter are therefore underestimated by the Rietveld method, in this case by a factor of about 1.4. The Rollett method, on the other hand, gives values which are in quite good agreement with those given by the SCRAP method, being about 10% larger.

For the Al_2O_3 data the profile parameters are very well determined, since an analyser was used and the peak shape is therefore quite well fitted by a Gaussian function. The underestimation of the standard deviations of the structural parameters by the Rietveld method is therefore larger, the factor being about 1.7, even though the structural parameters are also quite well determined. The Rollett method again gives values which are in good agreement with those given by the SCRAP method, in this case being smaller by about 7%.

The acetic acid data are more complex and therefore form a more realistic test of the various methods. They provide a poorer fit to the Bragg intensities and as a consequence the underestimation of the standard deviations of the structural parameters by the Rietveld method is by a factor of over 2. In this case the Rollett method gives values which are only 60% larger than the Rietveld values and both these methods therefore underestimate the values quite significantly.

In order to investigate the differences which might be found in more extreme cases we have also carried out two simulations based on analysis of the UO_2 data scaled in particular ways to give extremely good fitting of either the Bragg intensities (simulation 1) or the peak shapes (simulation 2). For these simulations only the SCRAP method, out of the methods considered, provides values for the e.s.d.s which are consistent with the model used. In the first case the standard deviations of the structural parameters are overestimated by a factor of about 6.5 by the Rietveld method and by a factor of nearly 8 by the Rollett method. In the second case they are underestimated by factors of about 4 and 3 by the two methods respectively.

DISCUSSION

The results of these analyses confirm that the values of the standard deviations given by the Rietveld method are unreliable. For a typical neutron diffraction dataset the standard deviations of the structural parameters are underestimated by a factor of about 2. However, the simulations demonstrate that these standard

deviations may also be overestimated if the structural parameters
are relatively better determined than the profile parameters and
that the factor by which they are in error can be much larger. It
is therefore not possible to apply a rule-of-thumb correction to
provide reliable values.

It would appear that the Rollett method always gives larger
values for the standard deviations of the structural parameters
than does the Rietveld method, even when the latter already over-
estimates them. In this respect it is therefore to be preferred to
the unmodified Rietveld method, since any underestimation of their
values will be reduced. However, the results for acetic acid imply
that the underestimation can still be significant for fairly complex
diffraction patterns.

It is clear that the main shortcoming of these methods is that
they do not distinguish between the two types of parameters in the
determination of their standard deviations, so that the values
obtained will necessarily be unreliable unless there is a comparable
goodness-of-fit for the peak shapes and the Bragg intensities. In
contrast, the SCRAP method separates the refinement of the two types
of parameters and thus is able to provide reliable values for their
standard deviations.

It should also be noted that correlation arising from deficienc-
ies in the peak shape function will not necessarily influence the
precision of the estimated Bragg intensities and hence the structural
parameters. This is illustrated by the results of simulation 1, in
which the SCRAP method is capable of giving very precise values
for the structural parameters, in spite of correlation between
residuals associated with the peak shapes. In these circumstances
similar precision would be obtained using an integrated intensity
method which did not require a knowledge of the peak shapes.

The main conclusions which can be drawn from this study are
therefore:

1. The values of the e.s.d.s. given by the Rietveld method are not
    reliable.

2. Introduction of off-diagonal terms in the weight matrix may
    give better values for the e.s.d.s. but will in general be
    insufficient to ensure their reliability.

3. Reliable values of the e.s.d.s. can only be determined if the
    form of the dependence of the calculated intensities on the two
    different types of parameters is taken into account.

4. Of the methods considered in this study only the SCRAP method
    will in general give reliable values for the e.s.d.s. of the
    structural parameters.

REFERENCES

1. H.M. Rietveld, Acta Cryst. <u>22</u>, 151 (1967); J. Appl. Cryst. <u>2</u>, 65 (1969).

2. M. Sakata and M.J. Cooper, J. Appl. Cryst. <u>12</u>, 554 (1979).

3. C.P. Clarke and J.S. Rollett, Acta Cryst. In the press (1981).

4. M.J. Cooper, M. Sakata and K.D. Rouse. Z. Krist. In the press (1981).

5. M.J. Cooper, Submitted to Acta Cryst. (1981).

THE POTENTIAL OF RESONANT NEUTRON SCATTERING AS A CONDENSED MATTER PROBE

R.E. WORD
UNIVERSITY OF MISSOURI, COLUMBIA, MO. 65211

G.T. TRAMMELL
RICE UNIVERSITY, HOUSTON, TX 77001

ABSTRACT

Resonant epithermal neutron scattering provides a new regime for the study of the properties of condensed matter. We have derived analogues to the Van Hove and Placzek developments for the resonant epithermal regime. Implications of the basic formulae for the study of anharmonic phonon interactions, the phase of phonon eigenmodes, non-Markovian diffusive processes, selective molecular diffusion, and the time dependent interatomic force density are summarized.

TEXT

The availability of high fluxes of epithermal neutrons, such as will be provided by the dedicated spallation sources, gives rise to the possibility of the utilization of resonant epithermal neutron scattering as a condensed matter probe. Hence, it becomes worthwhile to consider the unique features of this regime in comparison with those of thermal neutron scattering, in order to determine what new classes of information might be obtained.

We shall be concerned with the scattering of epithermal neutrons with energies less than a few electron volts. A number of heavy isotopes have resonances in this energy range[1] and their widths are in the range of several tens to several hundreds of millivolts. These widths are of the order of typical thermal vibrational energies of these nuclei in condensed matter or molecules, so that the prospect arises that they may be used to probe the character of these motions.

In the vicinity of a single epithermal neutron resonance, the scattering amplitude is given by the well known Breit-Wigner formula. Three characteristic features of this regime may be discerned: (1). At the center of resonance, the cross section for (n,n) scattering is three or four orders of magnitude larger than typical thermal neutron scattering cross sections, so that very small samples (or concentrations) of the resonantly scattering nucleus give rise to large scattering. (2). The magnitude of the resonant scattering length is highly energy dependent, in contrast to thermal neutron scattering lengths. (3). The phase of the resonant scattering length runs rapidly from zero to pi through the center of resonance as a function of incident energy, so that resonant neutron scattering may be used as a direct method to probe the phase of microscopic processes.

The amplitude for resonant neutron scattering from a condensed matter target system is given with an error less than about .1% in the single collision approximation[2], also referred to as Lamb's formula.[3] Applying the Van Hove development to the single collision ap-

142

proximation, one obtains the resonant scattering law in terms of a four point dynamical correlation function of target system variables:

$$W(\vec{k}_0, \vec{k}_f) = \sum_{i,i'} \int_{-\infty}^{\infty} dT \exp(-\frac{i}{\hbar}ET) \int_0^{\infty} dt \exp(\frac{i}{\hbar}\Delta Et) \int_0^{\infty} dt' \exp(-\frac{i}{\hbar}\Delta E^* t') \quad x$$

$$x <e^{-i\vec{k}_0 \cdot \vec{r}_i.(T)} \; e^{i\vec{k}_f \cdot \vec{r}_i.(T+t')} \; e^{-i\vec{k}_f \cdot \vec{r}_i(t)} \; e^{i\vec{k}_0 \cdot \vec{r}_i(0)}> \tag{1}$$

In (1), $\vec{k}_0$ and $\vec{k}_f$ are the incident and final state wave vectors, ΔE is $z + i\Gamma/2$ where z is the difference between the final state energy and the center the resonance and Γ is the total width of the resonance; the r's are the positions of the resonant scatterers in the Heisenberg representation, and the angular brackets refer to a thermal average. The variable E is the energy transfer to the neutron upon scattering. Thus, the resonant scattering law depends fully upon the variation of each of the six components of the scattering triangle. The Fourier-like integrals in (1) cannot be inverted in the usual way by virtue of the explicit k dependence of the energy variables; instead the space-space Fourier transform of (1) involves an integration of nine of the fifteen independent variables of the four point dynamical correlation function with a pair of neutron propagators entering as integral kernals. The space-space component of the resonant scattering law is a static quantity, but elsewhere we have shown how to introduce a time variation on it corresponding to the time variation in the more familiar space-time correlation functions of Van Hove.

A Placzek development for (1) may be obtained by writing the asymptotic development for the scattering law, and relating the coefficients of the formal series to moments of the scattering law by virtue of Watson's lemmas. (However, care must be taken to extract the explicit k dependence by appealing to symmetry arguments). One finds that in the case of chemical bonding of a resonant scatterer with a hard sphere (potential) scatterer that the time-dependent force density correlation function found in the second generalized z moment of the interference scattering law gives rise to a huge modification of the scattering law; we believe it would be useful to investigate this effect experimentally.

In the case of an amorphous target material, one may evaluate the space-space component of the scattering law in the static approximation; then the four point correlation function of resonant nuclear densities contains information on the modification of the nuclear probability density given the proximity of a second scatterer in the amorphous material.

One may consider the interpretation of the incoherent resonant scattering law for simple liquids (for which i=i' in (1)). The classical interpretation of the four point dynamical correlation function in (1) is the conditional probability that a partical will be found at $\vec{r}_4$ at time 0, then at r_3 at time t through diffusion, then again at $r_2 + r_3$ at time T+t', and finally at $r_1 + r_4$ at time T. Thus, the four point dynamical correlation function corresponds to a higher order diffusive propagator, giving information on the conditional diffusion over distinct intervals of time. In the few collision limit (obtaining for typical resonant parameters), the higher order diffu-

sive propagators contain information which cannot be obtained from
the usual two-time diffusive propagators measured in thermal neutron
scattering studies of simple liquids. When the conditional probabil-
ities may be ignored, the incoherent resonant scattering law may be
written[4,5]

$$W = \exp(-\tfrac{1}{2}\hbar\beta\omega)[2\mathrm{Re}\{P_w(\Delta\omega^*,k_f)P_w(-\omega-\Delta\omega,k_0)P_w(-\omega,\Delta k)\}$$

$$+ (2/\Gamma)\{\mathrm{Re}(P_w(-\omega-\Delta\omega,k_0))\mathrm{Re}(P_w(-\Delta\omega,k_f))\}] \qquad (2)$$

with the inclusion of the usual Schofield factor, and where P_w is the
two point diffusive propagator. The observation of the importance of
the higher order diffusive propagators would then consist in the measurement of deviations from the simple form (2). Elsewhere, we have
computed the form of the scattering law with the explicit inclusion
of the higher order diffusive propagators.

If the target is a molecular fluid, the incident neutron may ex-
cite virtual molecular vibrations during the scattering process in
accordance with the single collision approximation; provided that one
removes the higher order propagators via Plazcek corrections, (2) is
modified by the substitution $\Delta\omega \to \Delta\omega + R$, where R is the frequency of the
vibrational excitation; but the propagators P_w of this argument then
refer to the diffusion of the molecule while it is set in the state
of vibration of frequency R. Thus, it is seen that one may extract
from the variation of W information on the diffusion of those molecules which are set in definite states of vibrational excitation.
This question has been treated in detail elsewhere.[5]

We have computed the four point dynamical correlation functions
for a dynamical lattice via a diagrammatic technique: four points
are set down for the entry or exit of external phonon lines. Thus,
there are six categories (four things taken two at a time) of phonon
processes, corresponding to various mixtures of real and virtual processes. Resonant neutron scattering is found to be sensitive to the
details of the variation of the anharmonic phonon interaction of cubic order; a large effect arises when the sample material is brought
near its melting temperature. The limitations of space do not allow
us to treat this question in greater detail here.

The rapid phase variation of the Breit-Wigner resonant denomina-
tor allows the phase of microscopic processes to be probed. For example, one may obtain phase information about phonon eigenmodes in
large biological molecules.

The possibility of resonant neutron scattering opens up the possibility of the investigation of several categories of information
which are closed to thermal neutron scattering techniques.

REFERENCES

1. Brookhaven National Laboratory BNL-325.
2. G.T. Trammell and J.D. Chalk, Phys. Rev. __141__, 815 (1966).
3. W.E. Lamb, Phys. Rev. __55__, 190 (1939).
4. R.E. Word and G.T. Trammell, Phys. Rev. B (to be published).
5. R.E. Word, "The Use of Resonant Epithermal Neutron Scattering
   for the Study of Condensed Matter", (Ph.D. Thesis, Rice University, University Microfilm, 1979).

NEUTRON INTERFEROMETRIC MEASUREMENT OF THE REAL PART OF THE SCATTERING LENGTHS OF NATURAL AND ISOTOPICALLY PURE Sm-149*

R. E. Word and S. A. Werner
Physics Department and Research Reactor Facility
University of Missouri-Columbia, Columbia, MO 65211

ABSTRACT

We report here measurements of the real part of the scattering lengths of natural and of isotopically pure Sm-149 below and near the low-lying thermal resonance centered at E = 97.6 mev.

INTRODUCTION

Neutron transmission experiments provide a means of measuring the total neutron cross section, which by the optical theorem in quantum mechanics, is related to the imaginary part of the forward scattering amplitude. Neutron interferometry furnishes a means of directly obtaining the complementary information, namely the real part of the scattering amplitude (length). The phase shift $\Delta\phi$ suffered by the neutron wave function during passage through a sample of atomic density N, of uniform thickness t, and with a coherent neutron-nuclear scattering length b, is given by

$$\Delta\phi = -\lambda \ Nt \cdot Re(b), \qquad (1)$$

where λ is the neutron wavelength.[1] We report here data on the scattering length of natural samarium and of isotopically pure Sm-149 at a variety of incident wavelengths using neutron interferometer measurements of $\Delta\phi$.

EXPERIMENTAL

The experiments were carried out at the University of Missouri Research Reactor. Details of the experimental apparatus are given elsewhere.[1] We use a Bonse-Hart interferometer[2] constructed from a highly perfect silicon crystal shown schematically in Fig. 1. The monochromatic incident beam is produced by a double crystal pyrolytic graphite monochromator. This beam is coherently split in the first Si crystal slab at point A by Bragg reflection from the (220) lattice planes. The resulting two beams are coherently split again in the second Si slab near points B and C. Two of these beams are directed toward point D in the third Si slab, where they overlap and interfere. One of the outgoing beams is counted with a small high pressure He-3 detector. The Sm sample is inserted in the beam traversing the path ABD inducing a phase shift of this beam relative to the beam traversing the path ACD, as given by Eq. (1). This phase shift is measured directly by inserting a phase rotator (Si ar Al slab in our experiments) in the interferometer. Rotating this slab through an angle δ about an axis normal to the parallelogram ABCDCA results in a phase shift $\Delta\phi_{rot}$.[2] As we rotate this phase shifter, thus changing $\Delta\phi_{rot}$ continuously, the counting rate of the interfering beams oscillate sinusoidally as shown in the representative data displayed

*Work supported by the National Science Foundation through grant NSF-PHY 7920979

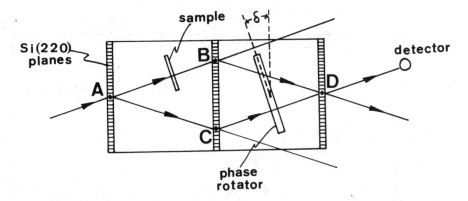

Fig. 1.   Schematic diagram of the Bonse-Hart interferometer.

in Fig. 2. Data is taken point by point, with the Sm sample in the beam path AB and then with it removed.   The two resulting sinusoids are phase shifted with respect to each other by the amount given in Eq. (1).   The numerical value for the phase shift is obtained by least squares fitting.

The Sm-149 sample is a thin foil of thickness t = 10.27 μm obtained from the Isotope Division, Oak Ridge National Laboratory. This thickness was chosen as a compromise between excessive absorption attenuation and a measurable phase shift.   At E = 82 meV, the transmission is only 6.6%.   The energy dependent scattering length obtained from this data is shown in fig. 3.   We are in the process of

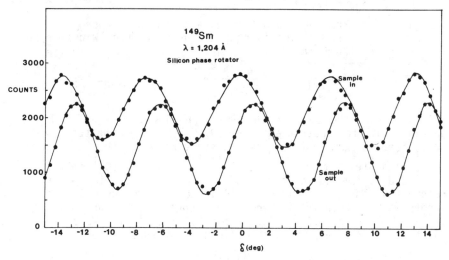

Fig. 2.   Counting rate as a function of phase rotator angle δ for Sm-149 sample in and out.   The counting times were 203 sec/point for sample in and 164 sec/point for sample out.

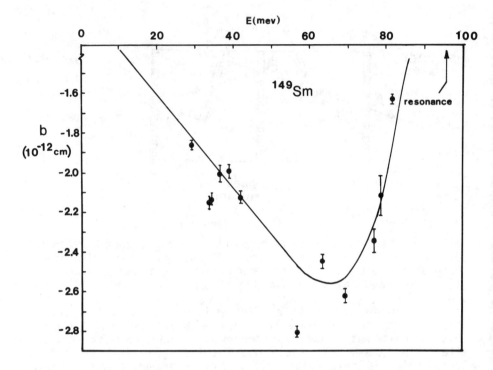

Fig. 3. Real part of the scattering length of Sm-149 as a function of energy E.

changing monochromators to obtain data on the high-energy side of the resonance. The solid line is obtained using a Breit-Wigner resonance line shape of the form

$$b = \frac{I+1}{2I+1} \; \lambdabar \; \frac{\Gamma_n}{2} \; \frac{E-E_0}{(E-E_0)^2 + \Gamma^2/4} \; + \; b_{pot}. \tag{2}$$

Here $I = 7/2$, the spin of the Sm-149 nucleus, $\lambdabar = \lambda/2\pi$, Γ_n is the partial width for neutron emission (0.50 meV), Γ is the total width (64 meV), $E_0 = 97.6$ meV, and b_{pot} is the scattering length due to potential scattering (taken to be $1.25 \times 10^{-12} A^{1/3} = 0.66 \times 10^{-12}$ cm).

Natural Sm is composed of about one part in seven of the resonantly absorbing isotope Sm-149. We have measured its scattering length at a number of different wavelengths and find it to be quite small. For example at $\lambda = 1.16$ A, $b = .04 \times 10^{-12}$ cm. We are currently in the process of refining the precision of these experiments.

REFERENCES

1. J.-L. Staudenmann, S. A. Werner and R. Colella, Phys. Rev. A21, 1419 (1980).
2. U. Bonse, in Neutron Interferometry (Clarendon Press, Oxford 1979).

EXPERIMENTAL DETERMINATION OF ANOMALOUS SCATTERING LENGTHS
OF SAMARIUM FOR THERMAL NEUTRONS

D. W. Engel
University of Durban-Westville, Durban, 4000, South Africa

T. F. Koetzle
Brookhaven National Laboratory, Upton, NY 11973

ABSTRACT

Anomalous scattering lengths of natural Sm for thermal neutrons
with wavelengths between 0.827 and 1.300 Å have been determined
using a single crystal of a Sm-complex of known structure. 140
selected reflections were measured at each wavelength and $b_o + b'$
and b'' refined in each case. The values obtained are in good
agreement with theoretical values obtained from a Breit-Wigner
calculation using tabulated resonance parameters for [149]Sm.
A value of $b_o = 4.3 \pm 0.2$ fm is deduced from the diffraction
experiment.

INTRODUCTION

Anomalous scattering of thermal neutrons by resonant nuclei
has been exploited for crystal structure determination[1-5]. For
successful application of phase determination techniques the anomalous scattering lengths must be known at all the wavelengths used.
We report here the determination of the scattering lengths of
natural Sm in a diffraction experiment. Anomalous scattering
lengths have previously been determined by diffraction[2, 6-9] for
Cd, Li, B and Sm.

BREIT WIGNER CALCULATION OF b

The scattering length of a nucleus is $b = b_o + b' + ib''$ where b'
and b'' are dispersion terms due to a resonance near the primary
neutron energy. The real and imaginary parts can be written as

$$b_{re} = b_o + b' = b_o + \frac{A(E-E_o)}{(E-E_o)^2 + B^2} \quad (1)$$

$$b'' = \frac{A B}{(E-E_o)^2 + B^2} \quad (2)$$

where $A = g\omega\lambda_o\Gamma_n/4\pi$ and $B = \Gamma/2$. g is a spin weighting factor
$\frac{1}{2}(2J+1)/(2I+1)$ where I and J are the spins of the nucleus and
compound nucleus respectively, ω is the abundance of the relevant
isotope, λ_o the resonance wavelength, E_o the resonance energy, E
the primary neutron energy, Γ_n the resonance width for re-emission
of the neutron, Γ_a that for absorption of the neutron and $\Gamma = \Gamma_n + \Gamma_a$
the total width.

The wavelength-independent term is given by

$$b_o = R + \Sigma \; g\omega\lambda_o \Gamma_n / (-4\pi E_o) \tag{3}$$

where the sum is taken over all higher energy resonances of all isotopes present. R is the scattering length for potential scattering (nuclear radius).

Values of A, B, E_O and b_O were calculated for natural Sm from tables of absorption data[10] and are shown in Table 1.

EXPERIMENTAL

A single crystal of a sodium samarium edta complex, $Na^+[Sm(C_{10}H_{12}N_2O_8).3H_2O]^-.5H_2O$ was used for the diffraction experiment. The crystal was mounted in a helium atmosphere at 37 K on an automated four-circle diffractometer at the Brookhaven High Flux Beam Reactor. The neutron beam was monochromated using the 002 face of a Be crystal. 1800 independent reflections were measured at a wavelength of 1.300 Å and the crystal structure refined to a final R-value of 0.058. 450 parameters including the real and imaginary Sm scattering lengths were varied in the final refinement.

140 reflections with strong Sm-contribution to the structure factor were now selected and measured at a number of different wavelengths. Using the structural parameters obtained in the full refinement at 1.300 Å the real and imaginary scattering lengths of Sm were refined together with the scale factor for each limited data set.

DISCUSSION

In Figure 1 the refined scattering lengths of natural Sm (points) can be compared with the curves calculated from the absorption data. (The curve of b_{re} was drawn using $b_O = 4.3$ fm to provide a slightly better fit.) The parameters A, B E_O and b_O were also calculated by a least squares fit of the values of b_{re} and b" from the diffraction experiment to equations 1 and 2. The refined values are compared to those deduced from absorption data in Table 1.

Table 1 Parameters for scattering lengths of natural Sm in equations 1 and 2. Comparison of values deduced from tables of absorption data and from the diffraction experiment.

		Absorption	Diffraction from b_{re}	from b"
A	(fm.eV)	0.302(6)	0.299(32)	0.310(23)
B	(eV)	0.0305(3)	0.0310(24)	0.0310(12)
E_O	(eV)	0.0976(3)	0.0966(11)	0.0972(8)
b_O	(fm)	≲4.02	4.32(21)	–
Γ_n	(meV)	0.533(8)	0.526(60)	0.546(41)
Γ	(eV)	0.0611(6)	0.0619(49)	0.0620(24)

The agreement is good. The e.s.d.'s of A, B and E_O deduced
from absorption data are lower than those obtained in the diffraction experiment. However, the value of b_O from equation 3 may be
in error due to omission from the sum of resonances not tabulated
and due to uncertainty in the value of R, taken here as 7.98 fm for
natural Sm. The diffraction value of b_O = 4.3 ± 0.2 fm is thus
probably the more reliable one.

Fig. 1   Scattering lengths of natural Sm plotted against neutron
wavelength. Full curve: Breit-Wigner calculation from tables of
absorption resonance data. Experimental points are values refined
from diffraction data. Triangles are values from Sikka[8].

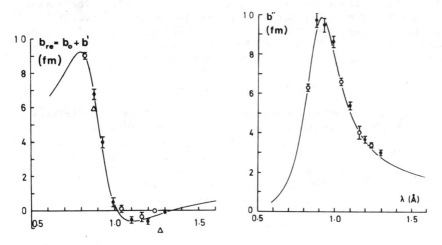

DWE   wishes to thank the University of the Orange Free State
and the South African Council for Scientific and Industrial Research and the Brookhaven National Laboratory for financial assistance.

REFERENCES

1.   B. P. Schoenborn, In "Anomalous Scattering", Ed. S. Ramaseshan
     & S. C. Abrahams (Munksgaard, Copenhagen, 1975) p 407.
2.   T. F. Koetzle & W. C. Hamilton, In "Anomalous Scattering",
     p 489.
3.   S. K. Sikka & H. Rajagopal, In "Anomalous Scattering", p 503.
4.   R. J. Flook, H. C. Freeman & M. L. Scudder, Acta Cryst. B33,
     801 (1977).
5.   H.D. Bartunik, Acta Cryst. A34, 747 (1978).
6.   H. G. Smith and S. W. Peterson, J. Phys. Radium 25, 615 (1964).
7.   A. C. McDonald & S. K. Sikka, Acta Cryst. B25, 1804 (1969).
8.   S. K. Sikka, Acta Cryst. A25, 621 (1969).
9.   H. Fuess & H. D. Bartunik, Acta Cryst. B32, 2803 (1976).
10. S. F. Mughabghab & D. I. Garber, Neutron Cross Sections, BNL 325
     3rd ed. Vol. I (Brookhaven National Laboratory, Upton NY, 1973).

COMPARATIVE STUDY OF MOSAIC STRUCTURE IN CRYSTALS BY REFINEMENT OF
EXTINCTION PARAMETERS FROM NEUTRON DATA AND BY X-RAY TOPOGRAPHY

A. Sequeira, H. Rajagopal & R. Chidambaram
Bhabha Atomic Research Centre, Trombay, Bombay 400 085, India

Krishan Lal & B.P. Singh
National Physical Laboratory, New Delhi 110 012, India

ABSTRACT

The domain size and mosacity in two crystals have been studied
using high precision neutron diffraction data to refine the extinction parameters and these have been compared with the values
obtained by x-ray topography and multicrystal diffractometry. The
combination of these techniques help in obtaining physically
meaningful mosaic width and particle size parameters.

INTRODUCTION

It is now possible to correct extinction effects in diffraction
data precisely and routinely, using the Zachariasen's theory[1] or
some of its modifications[2,3]. However, the physical reasonableness
of the derived extinction parameters has often been in doubt. We
have compared such parameters obtained in two crystals with direct
estimates of the domain sizes obtained from x-ray topography and
the mosaic spreads from multicrystal diffractometry.

RESULTS AND DISCUSSION

1. KCl Crystal: The neutron data set consisting of 54 reflections
in an octant of reciporocal space was recorded in the θ-2θ mode from
a single crystal $(2.2 \times 2.6 \times 4.2 \text{ mm}^3)$ using the computer controlled 4-
circle neutron diffractometer at Trombay[4]. The data were moderately
affected by extinction (minimum $Y = F_O^2/F_C^2 = 0.7$). Refinement of the
isotropic extinction parameter along with a scale factor and two
isotropic thermal parameters yielded an R-value of 0.0154. As anisotropic extinction effects were apparent in the F_O^2 values, the
refinement was continued using various anisotropic extinction models
and the results are summarised in Table I. It is clear from the
results that one cannot choose between the primary and secondary
extinction models either at isotropic level (models 1 and 3) or at
the anisotropic level (models 4 and 7). Hence, the parent crystal
was examined by x-ray topography and by multicrystal diffractometry.
Large width (14 min.) of the resulting diffraction curve (Fig.1a)
seems to rule out the secondary extinction models which indicate
an equivalent mosaic spread of < 2 min, while the average particle
size of 8 microns indicated by topography (Fig.1b) seems to agree
fairly well with the primary extinction model.

Table I Results of KCl refinement from various extinction models

Model	R-Factor(F)	$B(K)\overset{\circ}{A}{}^2$	$B(Cl)\overset{\circ}{A}{}^2$	Avg-particle radius (rms) (microns)
1. Isotropic primary (Zachariasen)	0.0152	1.82	1.88	13.5
2. Isotropic primary (Becker and Coppens)	0.0168	1.57	1.76	21.3
3. Isotropic secondary	0.0154	1.83	1.87	0.12
4. Anisotropic primary (Zachariasen)	0.0067	1.82(3)	1.89(3)	10.5
5. Anisotropic primary (Becker and Coppens)	0.0097	1.83	1.94	-
6. Anisotropic Secondary (Zach. Type I)	0.0098	1.83	1.87	-
7. Anisotropic secondary (Zach. Type II)	0.0066	1.80	1.86	0.10
8. Anisotropic secondary (Becker and Coppens)	0.0118	1.70	1.81	-

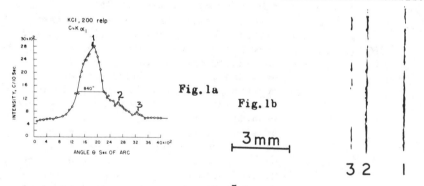

Fig.1a

Fig.1b

3mm

3 2

2. L-glutamic acid.HCl: The data set[5] consisting of 639 independent
reflections (λ =1.406A) was severely affected by extinction (Y_{min} =
0.06). The positional and anisotropic thermal parameters were
refined along with a scale factor and extinction parameters. The
Zachariasen's secondary extinction model was able to correct reflections with $Y > 0.25$ fairly satisfactorily but was consistently
undercorrecting the stronger reflections and leading to unrealistic
thermal parameters. However, when both primary and secondary
extinctions were accounted for simultaneously, corrections were
quite satisfactory. Refinements were carried out using Zachariasen,
Becker and Coppens[2] and the modified[3] Zachariasen formalisms, all
of which yielded consistent structural parameters. The resulting

R-values and extinction parameters are listed in Table II.

Table II Results of L-glutamic acid. HCl refinement

Isotropic Extinction Model	R-Factor(F) (NO=639)	Extinction Parameters*	
		g x 10^{-4}	r(microns)
1. Zachariasen			
a) Secondary	0.0530	17.1(1.3)	-
b) Primary	0.0542	-	178(8)
c) Secondary + Primary	0.0478	9.5(1.0)	24(1)
d) -do-	0.0489	0.29$^+$	133(4)
2. Becker & Coppens (Type I Lorentzian)			
a) Secondary	0.0493	10.1(0.6)	-
b) Secondary + Primary	0.0475	8.5(0.6)	29(2)
3. Modified Zachariasen $Y = \left[1 + 2x + ax^2 \right]^{-\frac{1}{2}}$	0.0475	9.9(0.8)	a = 0.12(0.01)

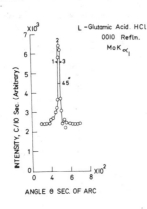

Fig. 2

* The notation follows Zachariasen[1]
+ Constrained at this value

A large crystal of the same crop when examined by multicrystal diffractometry gave a mosaic width of 45 sec. (see Fig.2) The equivalent mosaic spread (~1 sec.) obtained by a combined primary and secondary extinction refinement is in comparison much too small and so is the resulting particle size. However, a constrained refinement using a g-value of 0.29 (corresponding to the experimentally measured mosaic width of 45 sec.) gives a particle radius of 133 microns which is not inconsistent with the x-ray topographs.

REFERENCES

1. W.H. Zachariasen, Acta Cryst. 23, 558 (1967)
2. P.J. Becker and P. Coppens, Acta Cryst. A30, 129 (1974)
3. A.Sequeira,H.Rajagopal,& R.Chidambaram,Acta Cryst. A28,S193(1972)
4. A.Sequeira, S.N.Momin, H. Rajagopal, J.N.Soni, R. Chidambaram, Pramana, 10, 289(1978)
5. A.Sequeira,H.Rajagopal,& R.Chidambaram,Acta Cryst.B28,2514(1972).

ECONOMICAL DESIGN OF A 4-CIRCLE NEUTRON DIFFRACTOMETER

Hans Bartl

Universität Frankfurt, Institut für Kristallographie
D 6000 Frankfurt/M., Germany

ABSTRACT

By extensive use of commercial X-ray diffractometer components considerable reductions of cost and time, necessary for the design of single crystal neutron diffractometers, can be achieved. This example shows, that the
Nicolet Instr. diffractometer control system can be
adapted to an instrument of different origin though larger and heavier in its dimensions.

MECHANICAL PART

Neutron diffractometers must be in the position to carry
heavy detector shieldings as well as refrigeration cham-
bers: figure. A spectrometer arm, well supported by high
precision bearings, is guided around a large stationary
toothwheel, which forms the goniometer table. This construction minimizes the total weight of the moving parts.
All spectrometer elements are made of steel.[1]

The crystal goniometer, made of aluminium, is commercially designed by R. Huber [2]. This 3-circle orienter is
well sized for the accommodation of Air Products' Displex closed cycle refrigerators. The applied resolving
and driving aggregates are identical with those being
used in the Nicolet diffractometers.

ELECTRONICS

The electronic hardware is based on Data General's Nova
series of computers, as they are employed by Nicolet and
furnished with special interface boards for scaling and
driving. The present system is controlled by a Nova 1200,
familiar to many X-ray diffractionists from the Syntex
$P2_1$ systems.[3]

There are no major modifications necessary, exept, that
there must be a second scaler board for the monitor detector and that there should be some simple ratemeter.
Both BF_3 as well as ^3He detectors are in use. The minimum configuration of periphery comprises a video terminal and a magnetic tape. Comfort is improved by a printer and a disk system. Full compatibility to the commercial Nicolet crystallographic systems is maintained,
even servicing may be obtained.

ISSN:0094-243X/82/890153-03$3.00 Copyright 1982 American Institute of Physics

SOFTWARE

Backbone of the diffractometer control and data processing system is the software package originally supplied by Syntex Analytical Instr.(B. Sparks[3], S. Byram[4]). Whereas the Fortran versions provide for more flexibility, it seems, that single load assembler programs guarantee for the most reliable performance.

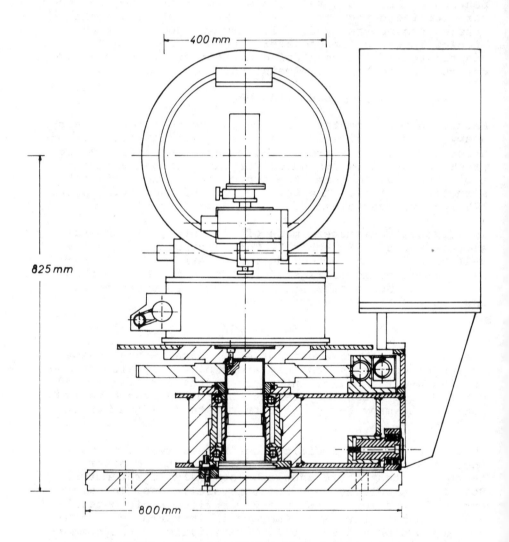

4-circle neutron diffractometer.

Some minor modifications had to be applied to the driving routines, since the Huber goniometer moves the χ-circle independently from φ.

It is recommended to use X-ray rotation photographs for the initial centering procedure. Additional advantage results from an X-ray tube and scintillation detector directly attached to the diffractometer, since the Nicolet centering algorithm can be very time consuming with low counting rates.

A new software contribution is the correction for background by profile analysis.

CONCLUSION

A prototype of the described instrument has been in use for 4 years[5]. There were no major faults, neither mechanical nor electrical. For everyone who wants to build up a classical neutron 4-circle diffractometer the given example outlines an economic way, which needs just little personal contribution.

REFERENCES

1. Ing. Krisch, drawing: F 400-9-0-2, Ges. f. Kernforschung m. b. H., D 7500 Karlsruhe, Postfach 3640, (1978).

2. R. Huber, Diffraktionstechnik G. m. b. H., D 8219 Rimsting, Germany.

3. Syntex P2$_1$ Operations Manual, 10040 Bubb Road, Cupertino, Calif., USA.

4. S. Byram, priv. comm. (1979).

5. H. Bartl, KfK-report 2719, Kernforschungszentrum, Postfach 3640, D 7500 Karlsruhe (1978).

A CONSTANT-ENERGY WEISSENBERG SPECTROMETER

D. Hohlwein

Institut für Kristallographie der Universität Tübingen, Charlotten-
strasse 33, D-7400 Tübingen, West Germany

ABSTRACT

A new crystal spectrometer is proposed which allows the registration of scattered neutrons with a chosen energy transfer in a
large and continuous range of scattering angles simultaneously.
This is achieved by flat analyser crystals which reflect the elastically or inelastically scattered neutrons out of the horizontal plane and a linear position sensitive or banana-like multidetector. The
properties of such an instrument are discussed. Some of the calculations are compared with test measurements.

INTRODUCTION

The combination of a crystal spectrometer with a multidetector
has been described with a flat analyser crystal[1] (MARX-spectrometer)
and with curved ones[2]. In both cases the reflected beams stay in the
horizontal plane. This limits the angular range which can be simultaneously covered by the analyser crystal(s).

In this paper we propose a vertical reflection at the analyser
crystals. In this way large and continuous ranges of scattering angles can be analysed simultaneously for a nearly constant energy
transfer as shown in the following.

OUT-OF-PLANE DIFFRACTION AT
A FLAT ANALYSER CRYSTAL

The scattering geometry is shown in Fig. 1. One analyser crystal
accepts the scattered neutrons in a continuous angular range of about
$10°$. The central beam ($\phi_s = \phi_o$; $\phi = 0$) is reflected with a polar angle
(relative to the horizontal plane) $\mu = 2\Theta_A$ and an azimuthal angle
(relative to ϕ_o) $\phi_A = 0$. The values of ϕ_A, μ and the wavelength λ as a
function of ϕ are:

$$\tan\phi_A = \tan\phi / \cos2\Theta_A \qquad (1)$$

$$\tan\mu = \sin2\Theta_A/(\tan^2\phi + \cos^2 2\Theta_A)^{1/2} \qquad (2)$$

$$\Delta\lambda/\lambda = (1 - \cos\phi) / \cos\phi \qquad (3)$$

From equation (1) follows that the angular range of the azimuth
is expanded for $2\Theta_A < 90°$ and contracted for $2\Theta_A > 90°$. At $2\Theta_A = 90°$,
ϕ_A is $90°$ and the neutrons are reflected in a vertical plane with
$\mu = 90° - \phi$. The equation has been experimentally verified for the

ISSN:0094-243X/82/890156-03$3.00 Copyright 1982 American Institute of Physi

reflection of 2.4 Å neutrons at a graphite analyser crystal and a photographic plate as detector, Fig. 2.

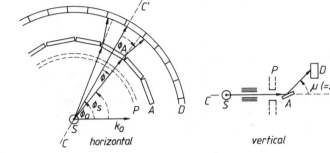

Fig. 1.
Spectrometer
design; sample(S)
protection (P),
analyser crystals
(A), multidetec-
tor (D).

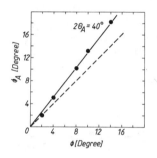

Fig. 2. Experimental verification of equation (1).

Equation (2) has also been proved with the same experimental conditions. For this particular case μ varies in an angular range of $\phi = \pm 5^{\circ}$ only from 40.0 to 40.4 degrees.

The wavelength variation in a range of ϕ from -5 to $+5$ degrees is only 0.4 percent, equation (3). Therefore the scattered neutrons will have the same energy (for a moderate resolution). For a monochromatic beam of 2.4 Å neutrons we have measured the reflectivity as a function of ϕ with a graphite analyser. We found indeed a constant reflectivity in a range of about 12°.

SPECTROMETER DESIGN

The analyser crystals are arranged tangential to a circle around the sample as is schematically drawn in Fig. 1. For $2\theta_A < 90^{\circ}$ there will be an expansion of the azimuthal angle after the analyser. This will cause an overlap of the angular regions of neighbouring analysers if they are too close together. The overlap region can be kept small if the distance between analyser and detector is made as small as possible. There will be no problem in back-scattering geometry.

A cylindrical shielding between sample and analyser crystals will catch the neutrons which are not scattered in the horizontal plane.

A cylindrical shaped multidetector ('banana') can be placed concentric to the analyser circle if $2\theta_A$ is not too near 90°. Then only a device for a vertical movement of the detector has to be installed. A small version of the instrument for a limited angular range

can consist of one or two analyser crystals and a linear position
sensitive detector mounted on the third axis of a triple axis spectrometer.

The vertical sample size can be made large if vertical soller
collimators are installed between sample and analyser crystals.

RESOLUTION

Only the resolution of the analyser part will shortly be discussed. The horizontal dimensions of the sample limit the q-resolu-
tion vertical to the scattering angle ϕ_s for a given sample to analyser distance. The energy resolution can be controlled by a vertical soller collimator, the mosaic angle and d-spacing of the analyser crystal in a well-known way. The energy change with ϕ is given
by equation (3).

SCAN IN RECIPROCAL SPACE

Because of the small energy variation with scattering angle
(equations (3)) we can collect simultaneously all scattering events
in the horizontal plane for a given energy transfer ($2\theta_A$) and a
given sample orientation. In reciprocal space this corresponds to a
circular scan which is concentric to the elastic Ewald circle.

APPLICATIONS

By turning the sample crystal around a vertical axis all possible scattering events in the horizontal plane can be recorded systematically as with a Weissenberg camera, but here also for a given energy transfer. In this way, e.g. the elastic diffuse scattering of disordered crystals can be measured. Also inelastic or elastic
scattering from non-crystalline samples and incoherent scattering
can be analysed efficiently.

The test measurements mentioned have been done at the photographic instruments P111 at the FR2 reactor in Karlsruhe and D12 at
the ILL in Grenoble. We are indepted to the Kernforschungszentrum
Karlsruhe and the Institute Laue-Langevin Grenoble for providing
the facilities. This work was supported by the Bundesministerium
für Forschung und Technologie (project KNF 03-41E03P; 04-45E03I).

REFERENCES

1. J. K. Kjems, P. A. Reynolds, Neutron Inelastic Scattering 1972
    (IAEA, Vienna 1972), p. 733.
2. R. Scherm, V. Wagner, Neutron Inelastic Scattering 1977
    (IAEA, Vienna 1978), p. 149.

VARIABLE HIGH RESOLUTION SPECTROMETER

P.S. Goyal, B.A. Dasannacharya and N.S. Satya Murthy
Nuclear Physics Division
Bhabha Atomic Research Centre
Bombay-400085.

ABSTRACT

The paper describes a spectrometer design which can cover a Q range from 0.15 to 3.15 Å^{-1} and an energy detection range from 3.2 to $\sim$9 meV with a resolution of 20 to 60 μeV at $\sim$5.2 meV. The design is based on a new type of analyser system.

INTRODUCTION

Most conventional neutron spectrometers provide an energy resolution of $\sim$100 μeV at $\sim$5000 μeV. The back scattering and the spin echo spectrometers, on the other hand, have a resolution typically about 1 μeV. In the present paper we propose a spectrometer design based on a new type of analyser system, which, with a properly tailored incident beam, is capable of bridging the gap in resolution by continuously varying it from about 20 to 60 μeV at $\sim$5220 μeV. Compared to conventional spectrometers, it will have a better resolution at a comparable energy and a better momentum transfer, $\hbar Q$, range at a comparable resolution. It has a poorer resolution than a back scattering spectrometer but can cover a larger energy range.

The spectrometer is well suited for a pulsed neutron source as it is possible to match the resolution of the analysing spectrometer with the resolution of the incident beam easily. We, therefore, discuss the spectrometer in conjunction with a pulsed neutron source though it is straight forward to convert the design to suit a steady state reactor.

SPECTROMETER DESIGN

The schematic drawing of the spectrometer is shown in Fig.1. Neutrons from a reactor or a pulsed neutron source are tailored by conventional time-of-flight techniques to reach the sample at a 40-meter distance from the moderator down a curved guide tube with a resolution of about 15 μeV (or more) at about 5220 μeV (which is the Bragg-cut-off energy of Be). This, incidentally is also the inherent resolution of the neutrons from SNS at Rutherford laboratory, since the pulse width from the 20°K moderator is $\sim$60 μsec for neutron of this energy[1]. To avoid frame overlap one employs a conventional chopper restricting the incident energy from about 3.2 to 9 meV.

ISSN:0094-243X/82/890159-03$3.00 Copyright 1982 American Institute of Physics

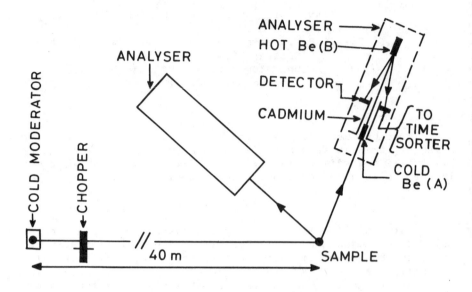

Fig.1. Schematic drawing of the spectrometer.

The main new feature of the spectrometer consists in the use of a window-filter analyser[2] (WFA) whose window width can be continuously varied. In a WFA the neutron beam first passes through a filter with a Bragg cut-off wavelength λ_1, and then falls on another polycrystalline material with a cut-off λ_2 slightly larger than λ_1. All the neutrons with wavelengths λ such that $\lambda_2 > \lambda > \lambda_1$ get nearly back-reflected and are detected in a suitable detector (Fig.1). The feasibility of such a system was demonstrated with Be and BeO combination long ago[2]. In the present setup it is proposed to use a combination of two Be blocks, A and B, at different temperatures. The first Be, acting as filter, is at 100°K, whereas the temperature of the second Be block can be varied from room temperature to, say, 500°K. This allows a variation in the window width from 13 to 42 μeV with the mean energy varying from 5212 to 5227 μeV, without any change in the geometry of the analyser system. In order to make full use of the reflected neutrons an annular ring counter would be ideal. For A and B at 100° and 500°K respectively it is sufficient to have a detector of 11 cm diameter at a distance of 50 cm behind B. The inner diameter of the detector is restricted by the size of the filter. The detector, of course, can be used in a multi-detector mode to cover larger range of scattering angles at a time. A range of scattering angles from 5 to 165° gives a q-range from 0.15 to 3.15 Å^{-1}.

It is possible to estimate the expected intensity for such a system. Assuming numbers relevant to SNS at Rutherford Laboratory and a scattering probability per unit solid angle per unit energy interval to be 10^{-3}, one obtains an intensity of about 10 counts per minute in an energy width $\Delta E = 20$ μeV for a sample size of 20 cm^2. In a quasielastic scattering experiment the intensity would be better.

Compared to a back scattering spectrometer with a temperature gradient analyser,[3] the present system avoids the problem of the detection of directly scattered neutrons without cutting the time period of the incoming neutrons. The range of Q, however, is limited since higher mean energies cannot be used.

1. G. Manning, Contemporary Physics, 19, 505 (1978).
2. P.K. Iyengar, Nucl. Inst. Methods 25, 367 (1964).
3. B. Alefeld Z. Physik 228, 464 (1969).

USE OF THE PEARSON TYPE VII DISTRIBUTION
IN THE NEUTRON PROFILE REFINEMENT OF THE STRUCTURES
OF $LiReO_3$ AND Li_2ReO_3

A. Santoro,* R. J. Cava,** D. W. Murphy,** and R. S. Roth*
*National Bureau of Standards, Washington, DC 20234
**Bell Laboratories, Murray Hill, NJ 07974

ABSTRACT

The crystal structures of the compounds $LiReO_3$ and Li_2ReO_3 have been refined with the Rietveld method. [1] Neutron powder diffraction data collected at room temperature were used in these calculations. Since the shapes of the diffraction lines for both materials could not be approximated by Gaussians with sufficient accuracy, the Pearson type VII function was used in all refinements. [2,3] The value of m was assumed to be 2θ-independent in these calculations. The best fits to the experimental observations were obtained with m = 1.5 for $LiReO_3$ and m = 3 for Li_2ReO_3. These values indicate that the lines of the first compound are almost Lorentzians, while those of the second are close to the so-called "modified Lorentzians." The final R-factors are R_I = 5.66, R_p = 5.55, R_W = 7.04, and χ = 1.41 for $LiReO_3$ and R_I = 8.05, R_p = 7.58, R_W = 9.77, and χ = 1.27 for Li_2ReO_3. The non-Gaussian peak shape cannot be attributed to instrumental factors since the instrumental function of the diffractometer used to collect the data is Gaussian to a very good approximation. Crystallite size effects could explain the observed intensity distributions, but other factors could also be present. Both compounds crystallize with the symmetry of space group R3c, and the lattice parameters (hexagonal axes) are a = 5.0918(3), c = 13.403(1) Å for $LiReO_3$ and a = 4.9711(4), c = 14.788(1) Å for Li_2ReO_3.

INTRODUCTION

In the course of a powder neutron diffraction study of $LiReO_3$ and Li_2ReO_3, it became apparent that the peaks produced by samples of the two compounds had tails too broad to conform to a Gaussian distribution. In fact, an attempt to fit some isolated reflections with Gaussians gave unreasonably high values of the goodness of fit χ (Fig. 1). The agreement between observed and calculated values, however, improved dramatically when the Pearson type VII function [2,3] was used to fit the same experimental data (Fig. 2). This function was taken into consideration because it can be varied gradually from a Lorentzian to a Gaussian by changing the value of one of its parameters. On the basis of these preliminary results, it was decided to use the Pearson distribution in an attempt to refine the structures of $LiReO_3$ and Li_2ReO_3 with the profile analysis method.

ISSN:0094-243X/82/890162-04$3.00 Copyright 1982 American Institute of Physics

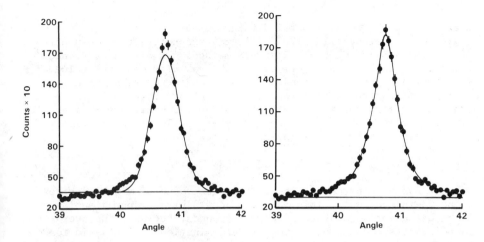

Fig. 1. Least-squares fit of a Gaussian to the intensity data of a single peak of LiReO₃. (χ = 2.21).

Fig. 2. Least-squares fit of a Pearson distribution to the same data of Fig. 1 (m = 1.2, χ = 1.16).

PEARSON TYPE VII DISTRIBUTION

The Pearson function is represented by the equation

$$y_i = 2I\left(\frac{2^{1/m} - 1}{\pi}\right)^{1/2} \frac{\Gamma(m)}{\Gamma(m - 1/2)} \frac{1}{H}\left[1 + 4\left(\frac{2\theta_i - 2\theta_o}{H}\right)^2 (2^{1/m} - 1)\right]^{-m} \quad (1)$$

where I is the integrated intensity; H, the full width at half maximum; $2\theta_o$, the Bragg angle of the reflection; and m, a parameter defining the peak shape. For m = 1, Eq. (1) becomes:

$$y_i = (4I)/(H\pi)[1 + 4(2\theta_i - 2\theta_o)^2/H^2]^{-1} \quad (2)$$

which is a Lorentzian. For m = 2, we obtain:

$$y_i = [4I(2^{1/2} - 1)^{1/2}/(H\pi)][1 + \{4(2\theta_i - 2\theta_o)^2/H^2\}(2^{1/2} - 1)]^{-2} \quad (3)$$

which is the so called "modified Lorentzian."[4,5,6] Finally it can be shown that, for m → ∞, Eq. (1) becomes a Gaussian:

$$y_i = 2I[(\ln2)/\pi]^{1/2}(1/H) \exp[-(4\ln2)(2\theta_i - 2\theta_o)^2/H^2] \quad (4)$$

For values of m as low as 20 or 25, the Pearson function is practically undistinguishable from a Gaussian.

EXPERIMENTAL

Neutron diffraction data were collected with a five-detector diffractometer at the National Bureau of Standards Reactor using a wavelength λ = 1.5416(3) Å produced by the reflection 220 of a Cu monochromator. The horizontal divergences of the first, second, and third collimator were 10', 20', 10' arc, respectively. The neutron data were analyzed with the Rietveld method, [1] modified by Prince [7] to include the Pearson function. In the calculations, the value of m was assumed to be 2θ-independent. The background was assumed to be a straight line with finite slope and was refined together with the profile and structural parameters. Both structures were refined with different values of the Pearson parameter m. The final results for the refined structures are given in Tables I and II.

DISCUSSION

It has been noted previously that the value of m was considered to be 2θ-independent. This is almost certainly an approximation which was adopted for both rhenium compounds, because line overlapping prevented the experimental determination of the variation of m with the diffraction angle. In those cases in which well separated peaks are available over the entire angular range, the value of m, in

Table I Results of the profile refinement of the structure of $LiReO_3$

Atom	Position	x	y	z	B
Re	6a	0	0	0	0.18(6)
O	18b	0.3801(7)	-0.012(1)	0.2540(9)	0.27(6)
Li	6a	0	0	0.227(1)	1.6(3)

$m = 1.5$, $R = 5.66$, $R_p = 5.55$, $R_w = 7.04$, $R_E = 5.00$
Space group: R3c, $Z = 6$; Number of observations: 2464
Number of independent Bragg reflections: 49

Table II Results of the profile refinement
of the structure of Li_2ReO_3

Atom	Position	x	y	z	$B^{(*)}$
Re	6a	0	0	0	0.21
O	18b	0.3580(9)	0.008(2)	0.2449(8)	0.21
Li(1)	6a	0	0	0.188(1)	1.4
Li(2)	6a	0	0	0.831(1)	1.4

$(*)$Thermal parameters were not refined for this compound.

$m = 3$, $R = 8.05$, $R_p = 7.58$, $R_w = 9.77$, $R_E = 7.71$
Space group: R3c, $Z = 6$; Number of observations: 2505
Number of independent Bragg reflections: 65

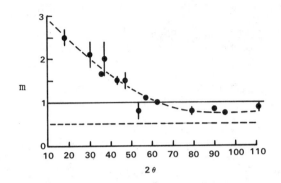

Fig. 3. Plot of m versus 2θ for a
sample of Fe_3O_4.

general, decreases as 2θ
increases. For example,
in the case of a sample
of Fe_3O_4, the peaks have
the shape of a modified
Lorentzian at low dif-
fraction angles and
become close to a
Lorentzian at high
angles (Fig. 3). It is
difficult to find a
unique explanation for
the departure of the
powder diffraction lines
from Gaussian shape.
Instrumental factors can
be excluded by the fact that the diffractometer, used with identical
experimental conditions, gives Gaussian peaks with standard materials
such as Al_2O_3 (χ between 1.1 and 1.2). Broadening caused by particular distributions of crystallite sizes may be a possible reason for
the observed shapes. However, other factors, such as structural
distortions or thermal diffuse scattering, could play an important
role in determining the observed intensity distributions.

The results obtained for the two rhenium compounds show conclusively that the Pearson function can be applied successfully to
neutron profile refinements. However, a detailed understanding of
the factors which may cause departures from the Gaussian shape is
still needed, especially to predict the variation of m with 2θ.

REFERENCES

1. H. M. Rietveld, J. Appl. Cryst. 2, 65-71 (1969).
2. W. D. Elderton and N. L. Johnson, Systems of Frequency Curves
   (Cambridge Press, N. Y., 1969), pp. 77-78.
3. M. M. Hall, V. G. Veeraraghavan, H. Rubin, and P. G. Winchell,
   J. Appl. Cryst. 10, 66-68 (1977).
4. E. J. Sonneveld and J. W. Visser, J. Appl. Cryst. 8, 1-7 (1975).
5. G. Malmros and J. O. Thomas, J. Appl. Cryst. 10, 7-11 (1977).
6. C. P. Khattak and D. E. Cox, J. Appl. Cryst. 10, 405-411 (1977).
7. E. Prince, Natl. Bur. Stand. US Tech. Note 1117 (edited by F. J.
   Shorten, 1980), pp. 8-9.

ASSESSMENT OF ACCURACY IN NEUTRON TIME-OF-FLIGHT DIFFRACTOMETRY

K.J. Tilli and H. Pöyry
Reactor Laboratory, Technical Research Centre of Finland, Espoo,
Finland, SF-02150

ABSTRACT

The assessment of the overall accuracy in a time-of-flight experiment requires the estimation of the reliability with which the
inherent wavelength dependent corrections have been accounted for.
In the present work a new method is described for estimating the
errors arising from the inaccuracies in the corrections. The application of the method to specific examples showed that the bias in the
refined parameters and the corrections to the standard deviations
can be considerable, if the counting precision of the diffraction
pattern and the accuracy of the corrected incident neutron flux are
of the same order of magnitude.

INTRODUCTION

The aim of a diffraction experiment is to gain information on
some physical quantities, such as the atomic positions and amplitudes
of vibration. The accuracy of these parameters depends not only on
the counting precision of the observations, but also on the reliability with which the possible sources of error can be recognized and
accounted for. In white-beam diffractometry, most factors capable of
producing errors in a measurement are the same as those in a conventional experiment, but the wavelength dependence of the relevant corrections has to be taken into consideration, as well. Typical of
these factors are the incident white-beam spectrum, beam attenuation
in air and in the sample, extinction and thermal diffuse scattering.
It is clear that the refined parameters will be biased if these factors are ignored, but the accuracy of the corrections is also of vital
importance in the estimation of the standard deviations. This will
be true especially as the counting precision and the data collection
rate will increase via the advent of advanced techniques, such as
the use of synchrotron radiation or pulsed neutron sources. The
present work aims at giving an account of the significance of the
wavelength dependent corrections and their accuracies in the analysis
of a neutron time-of-flight (TOF) diffraction experiment.

INCIDENT NEUTRON SPECTRUM IN THE PROFILE REFINEMENT

The analysis of a diffraction pattern is usually carried out by
means of the profile refinement method[1]. The inaccuracies of the
wavelength dependent factors can be included into the analysis by
treating the corrected incident neutron flux as a set of observations
with a known variance-covariance matrix V_ϕ. Then the observations
$\{Y_j\}$ are linearly related not only to the usual profile and structural parameters $\{\beta_j\}$, but also to the neutron flux values $\{\gamma_j\}$,

as given by the following matrix equation:

$$Y = X\beta + Z\gamma + \varepsilon, \tag{1}$$

where X and Z are known design matrices. ε corresponds to the errors
in the observations and it is taken to be normally distributed with
a zero mean and with a known variance-covariance matrix V_y. In eq.
(1) the linearization is performed using the experimental values for
the incident spectrum, so the matrix γ denotes corrections in the
neutron flux parameters.

With the aid of the ordinary least-squares method we obtain for
β the estimator

$$b = b^o - (X^T V_y^{-1} X)^{-1} X^T V_y^{-1} Zc, \tag{2}$$

where b^o is the estimator obtained by omitting the inaccuracy of the
incident flux, while c denotes the least-squares estimator for γ.
If the ordinary model with $\gamma=0$ is assumed to be reasonable, we will
obtain for c the expression

$$c = V_\phi Z^T V_y^{-1} \{Y - Xb^o\}. \tag{3}$$

The variance-covariance matrix for the estimator b is given by

$$V(b) = (X^T V_y^{-1} X)^{-1} + (X^T V_y^{-1} X)^{-1} X^T V_y^{-1} Z V_\phi Z^T V_y^{-1} X (X^T V_y^{-1} X)^{-1}. \tag{4}$$

Eq. (2) shows the bias in b^o to depend essentially on the corrections c of the neutron flux values. It can be shown that the values
of c are calculated as weighted means of the measured flux values and
the residuals $Y-Xb^o$ of the ordinary refinement. Therefore, the bias
in the estimator b^o depends not only on the relative accuracies of
the diffraction pattern and the corrected incident spectrum, but also
on the goodness-of-fit of the ordinary refinement.

APPLICATION TO SPECIFIC EXAMPLES

To illustrate the above theory by concrete examples, we analyzed
first a hypothetical model containing two peaks without a background.
Then the same postulated value $\phi\beta_1$, where ϕ is the incident neutron
flux, was fitted to the experimental intensities $I(1-\delta)$ and $I(1+\delta)$.
The bias in the ordinary estimator b_1^o was found to depend strongly
both on the residual $Y-Xb_1^o$ (i.e. on δ) and on the inaccuracy of ϕ.
However, the bias will be considerable only if δ and the e.s.d. of ϕ
are unexpectedly large. Moreover, we found that the error in the
variance can be great, if the inaccuracies of the incident flux and
the peak intensities are equal. If the correction is then neglected,
the variance will be underestimated by a factor of two.

The influence of various kinds of error was investigated by
propagating simulated inaccuracies into the analysis of a nickel
powder pattern measured with the reverse Fourier TOF method[2]. The
theoretical profile was chosen to contain not only the scale factor

Table. Effects of the inaccuracy of the incident neutron flux (ϕ) upon the bias and error in the variance.

$\dfrac{\sigma\phi}{\phi}$	$\dfrac{\Delta SF}{\sigma SF}$	$\dfrac{\Delta V(SF)}{V(SF)}$	$\dfrac{\Delta B}{\sigma B}$	$\dfrac{\Delta V(B)}{V(B)}$
0.2 %	0.038	0.066	0.028	0.045
0.6	0.28	0.60	0.23	0.39
2.0	1.5	6.6	1.3	4.5

and the isotropic Debye-Waller parameter, but also parameters for the background, peak shape and peak position.

The most important question in all white-beam measurements is to find a satisfactory solution for describing the incident spectrum. The effect of the accuracy of the neutron flux is shown in the table, in which results are given only for the scale factor (SF) and the Debye-Waller parameter (B), since the effect upon the other parameters was minor. Examination of the table indicates that both the bias and the errors in the variances will be great, if the average inaccuracy of the incident flux is as large as the weighted R-factor of the experiment (R_w = 1.84 %).

In the error of the second kind we analyzed the effect of a wavelength-dependent correction by using the equation $A(\lambda)=1+C(\lambda/\lambda_m)$, where C is a constant and λ_m the maximum value for the wavelength λ. At the accuracy level of the experiment, the bias was negligible (< 1.2 %), but the relative errors in the variances were 2.7 and 0.34 for the scale factor and for the Debye-Waller parameter, respectively.

DISCUSSION AND CONCLUDING REMARKS

By means of the new method proposed for analyzing the effect of the precision of the wavelength-dependent factors, we have shown the inaccuracy of the corrections to greatly affect the refined para- meters and their variances. Although the extent of bias depends on the investigated model, the effects can be considerable if the in- accuracies of the corrected neutron flux and the diffraction pattern are of the same order of magnitude. Since the bias and the errors in the variances will be small only if the precision of the corrected incident flux is high in comparison with the experiment, the analysis of an accurate diffraction pattern will lead to considerable require- ments not only in the determination of the incident spectrum, but also in the accuracies of all factors having a large wavelength variation. In the present work we have not examined errors in the coordinate parameters. Abrahams has, however, shown that errors increasing linearly with the scattering angle are absorbed entirely into the scale factor and the temperature factors[3]. Since the in- accuracy of the wavelength-dependent corrections leads to this error type, it is apparent that this is true also in TOF diffractometry.

1. H.M. Rietveld, J. Appl. Cryst. 2, 65 (1969).
2. K.J. Tilli, A. Tiitta and H. Pöyry, Acta Cryst. A36, 253 (1980).
3. S.C. Abrahams, Acta Cryst. A25, 165 (1969).

THE EQUIVALENCE BETWEEN PROFILE REFINEMENT
AND INTEGRATED INTENSITY REFINEMENT

T. M. Sabine,
N.S.W. Institute of Technology, N.S.W.   2007   Australia

ABSTRACT

It is shown that the results of least squares refinement for crystallographic parameters are identical when the ordinates of the profile are refined or when integrated intensities are extracted as an intermediate step.

INTRODUCTION

The basis of the Rietveld method (Rietveld, 1969) is the expression of the ith ordinate of the kth reflection as

$$y_{ik} = G_{ik} \, I_k \qquad (1)$$

where the profile function G_{ik} is normalised so that $\sum_i G_{ik} = 1$. I_k is the integrated intensity. The quantity minimised is

$$M = \sum_i \sum_k w_{ik} \, (y_{ik}^{obs} - y_{ik}^{calc})^2 \, .$$

The weight, w_{ik}, associated with the observation y_{ik}^{obs} is $\text{var}^{-1}(y_{ik}^{obs})$.

An integrated intensity refinement minimises

$$M = \sum_k W_k \, (I_k^{obs} - I_k^{calc})^2 \, .$$

The weight, W_k, associated with the observed integrated intensity is $\text{var}^{-1}(I_k^{obs})$.

To prove equivalence between the two methods it is sufficient to show that the normal equations of least squares are the same (Cramer, 1955).

THEORY

For profile refinement the components of the matrices in the normal equations are

$$\sum_i \sum_k w_{ik} \, \frac{\partial y_{ik}^{calc}}{\partial U_j} \, \frac{\partial y_{ik}^{calc}}{\partial U_\ell} \qquad (2)$$

and

$$\sum_i \sum_k w_{ik} \frac{\partial y_i^{calc}}{\partial U_j} (y_{ik}^{obs} - Y_{ik}^{calc}) . \tag{3}$$

For integrated intensity refinement these become

$$\sum_k W_k \frac{\partial I_k^{calc}}{\partial U_j} \frac{\partial I_k^{calc}}{\partial U_\ell} \tag{4}$$

and

$$\sum_k W_k \frac{\partial I_k^{calc}}{\partial U_j} (I_k^{obs} - I_k^{calc}) . \tag{5}$$

On using the relation $y_{ik} = G_{ik} I_k$ (2) and (3) become

$$\sum_i \sum_k w_{ik} G_{ik}^2 \frac{\partial I_k^{calc}}{\partial U_j} \frac{\partial I_k^{calc}}{\partial U_\ell} \tag{6}$$

and

$$\sum_i \sum_k w_{ik} G_{ik}^2 (I_k^{obs} - I_k^{calc}) . \tag{7}$$

Inspection of (4), (5), (6) and (7) shows that the normal equations are the same if, and only if,

$$W_k = \sum_i G_{ik}^2 w_{ik} . \tag{8}$$

It will now be shown that this is so (dropping the superscript obs.)

Each ordinate y_{ik} provides an independent estimate of I_k from the relation $I_k^{(i)} = G_{ik}^{-1} y_{ik}$ (the superscript (i) means "evaluated at i"). The variance of $I_k^{(i)}$ is given by

$$var (I_k^{(i)}) = G_{ik}^{-2} var (y_{ik}).$$

It can be shown (Aitken, 1949) that the optimal value of I_k is the weighted mean of $I_k^{(i)}$ with weights, $w_{(i)}$, equal to $var^{-1}(I_k^{(i)})$.

The variance of the weighted mean is

$$var (I_k) = \frac{\sum_i w_{(i)}^2 \ var \ I_k^{(i)}}{(\sum_i w_{(i)})^2} ,$$

now
$$W_k = \text{var}^{-1}(I_k)$$
$$= \sum_i G_{ik}^2 \, \text{var}^{-1}(y_{ik})$$
$$= \sum_i G_{ik}^2 \, w_{ik} \qquad .$$

Hence equation (8) holds in the general case. G_{ik} can have any form.

CONCLUSION

It has been shown that profile refinement and integrated intensity refinement using the same data lead to the same results.

A recent claim (Sakata and Cooper, 1979) that the results of the two methods of refinement are different is incorrect.

ACKNOWLEDGEMENTS

I thank Dr. J. K. Mackenzie for helpful discussion.

REFERENCES

Cramer, H. (1958) The Elements of Probability Theory, John Wiley and Sons, New York

Rietveld, H. M. (1969) J. Appl. Cryst. 2, 65–71

Sakata, M. and Cooper, M. J. (1969) J. Appl. Cryst. 12, 554–563

Aitken, A.C.L. (1949) Statistical Mathematics, Oliver and Boyd, London

172

NEUTRON SCATTERING FACILITIES AT THE McMASTER NUCLEAR REACTOR

J. R. D. Copley, M. F. Collins, K. J. Lushington and C. V. Stager
McMaster University, Hamilton, Ontario, L8S 4K1, Canada

P. A. Egelstaff
University of Guelph, Guelph, Ontario, N1G 2W1, Canada

ABSTRACT

The thermal neutron scattering facilities at the McMaster Nuclear Reactor (MNR) are described. A brief progress report on the new Small Angle Neutron Scattering (SANS) facility is included.

INTRODUCTION

The MNR first went critical in 1959. It is an enriched uranium light water moderated reactor licensed to operate at 5 MW, but currently operated at 2 MW, 120 hours/week. Three of the six radial beam ports are used for thermal neutron scattering experiments.[1]

DIFFRACTOMETERS

Wide angle diffractometers are located at beam ports 5 and 6.
The BP5 instrument, known as GWELFNEUD I, is operated by the Guelph group. A large Cu(200) or (111) monochromator reflects a 0.94 or 1.08 Å beam to the sample position, at a monochromator scattering angle $2\theta_M$ of 30°. The ~ 21 cm diameter source area is stopped down at present in order to reduce background. Four detectors are placed in a common shield, 120 cm from the sample position, ~ 3.7 cm (1.75°) apart. The detectors are automatically stepped in 0.5° steps up to 90°, and the data is written on standard cassette tapes. The instrument is mainly used as a "testbed" for the larger GWELFNEUD II spectrometer at Chalk River, but experiments on N_2 and HCl have been published.[2]

In 1981 the collimation will be relaxed allowing the whole source area to be viewed, and a 15 cm long single crystal Al_2O_3 fast neutron filter[3] will be installed. Also planned is a reconstruction of the counter assembly and track to include six detectors at 150 cm, and scattering angles up to 120°.

At BP6 there are two diffraction instruments.[4] One operates at $2\theta_M$ = 42°, giving wavelengths of 1.0, 1.4$_5$ and 2.4 Å using Al(220), Al(200) and pyrolytic graphite (PG) monochromators respectively. The detector can be lifted ~ 25° out of the scattering plane. The $2\theta_M$ angle at the other instrument can only be scanned with difficulty: it is normally operated at 1.40 Å using a Cu(200) monochromator. The flux at the sample position at 2 MW is

ISSN:0094-243X/82/890172-03$3.00 Copyright 1982 American Institute of Phys

1×10^5 n cm^{-2} s^{-1}. A position-sensitive detector (PSD) was recently installed at this instrument, but the original single detector (OSD) can still be used if so desired. (Indeed the instrument may still be operated as a three-axis spectrometer.)

The PSD consists of two 2.54 cm diameter 61 cm long ^{3}He detectors mounted one above the other with axes horizontal, normal to the central scattered neutron direction. Pulses from each end of a detector are fed through matched preamplifiers, amplifiers and 9-bit ADC's to a Motorola 6809 microprocessor which performs the necessary charge division using double precision table-lookup of the inverse followed by hardware multiplication.

Spectra taken using copper powder samples contained in thin-walled vanadium cans 4, 8 and 12 mm in diameter, demonstrate that the angular resolution of the PSD does not change significantly over this range of sample diameters. Comparing the PSD and the OSD preceded by an appropriate Soller collimator such that the angular resolutions of the two systems are comparable (0.9 ± 0.1° FWHM), we find that to measure the integrated intensity of a single peak, scanning over 3° in scattering angle, the OSD takes 4-5 times as long as the PSD. The useful angular acceptance of the PSD is roughly 30° so that if a continuous scan is required the OSD takes 40-50 times longer than the PSD. If the sample is a very weak scatterer the above statement must be modified, because the signal-to-background ratio is approximately 2.5 times higher for the OSD.

In the near future we intend to replace the punched paper tape control units at the BP6 diffractometers with individual Synertek SYM-1 microcomputer systems, each of which will include a video terminal, a printer, and a cassette tape-recorder.

SMALL ANGLE NEUTRON SCATTERING AND LONG WAVELENGTH TOTAL CROSS-SECTION SPECTROMETERS

The SANS spectrometer located at BP3 is presently under construction. An in-pile cooled beryllium filter will be followed by a single crystal of PG, or possibly Si, at $2\theta_M = 90°$, which directs a vertical monochromatic beam (mean wavelength 4.74 or 4.43 Å) to the sample position. In this respect the design follows closely that of the SANS facility at the Missouri University Research Reactor.[5] A Borkowski-Kopp type detector,[6] with an active area 65 cm square and a wire separation of 5 mm, is presently being fabricated. The computer system will be based on a DEC LSI 11/23 CPU, which will also be interfaced to the BP6 diffractometers. At 2 MW the thermal flux at the monochromator position in the absence of any filter is $\sim 2 \times 10^9$ n cm^{-2} s^{-1}, in good agreement with calculation. We hope to have the SANS facility in operation by late 1982.

The beam transmitted by the SANS monochromator will pass through ~ 100 cm of room-temperature bismuth shot. The in-pile beryllium and the bismuth will together provide a clean subthermal

174

$(\lambda > 6.7\text{Å})$ neutron beam with an estimated flux 200 cm from the reactor face of $1\text{-}5 \times 10^4$ n cm^{-2} s^{-1}. This beam, chopped and time-analyzed, will be used for measurements of the temperature (and wavelength) dependent total cross-sections of a variety of hydrogen-containing molecular solids at temperatures down to ~ 0.3 K. The results will be combined with the results of calorimetric and neutron inelastic scattering measurements to provide a clearer picture of the low-lying energy states associated with the tunneling of molecular reorientations in these materials.[7]

CONCLUSION

The neutron scattering facilities of the MNR are well used at the present time[2,8] and we anticipate considerable growth in this respect as the BP3 instruments come on-line.

We thank Dr. J. Mizuki and Messrs. J. D. Couper and G. R. Mulligan at McMaster, and Mr. L. Hahn at Guelph, for their invaluable assistance with the instruments described in this paper. We also thank colleagues, particularly at the University of Missouri and at NBS, who have offered advice and assistance with regard to the PSD and SANS projects. This work was supported by the Natural Sciences and Engineering Research Council Canada.

REFERENCES

1. McMaster Nuclear Reactor Research Report 1978-79-80, unpublished (1980).
2. See, for example, A. K. Soper and P. A. Egelstaff, Mol. Phys. 39, 1201 (1980).
3. H. F. Nieman, D. C. Tennant and G. Dolling, Rev. Sci. Instrum. 51, 1299 (1980).
4. B. N. Brockhouse, G. A. de Wit, E. D. Hallman and J. M. Rowe, Neutron Inelastic Scattering (IAEA, Vienna, 1968), Vol. II, p. 259.
5. R. Berliner et al., A Small Angle Neutron Scattering Spectrometer at MURR, Final Report, unpublished (1979).
6. C. J. Borkowski and M. K. Kopp, Rev. Sci. Instrum. 46, 95 (1975).
7. K. J. Lushington and J. A. Morrison, Can. J. Phys. 55, 1580 (1977).
8. See, for example, C. W. Turner, M. F. Collins and J. E. Greedan, J. Magn. Magn. Mat. 23 (1981) (in press).

DIFFUSE SCATTERING WITH SPIN ANALYSIS USING A SUPERMIRROR POLARISER
AND 5 SUPERMIRROR ANALYSERS: RESULTS ON PARAMAGNETIC SCATTERING,
CRYSTAL FIELD TRANSITIONS, SEPARATION OF COHERENT AND INCOHERENT
SCATTERING IN LIQUID SODIUM USING TIME OF FLIGHT ANALYSIS

O. Schärpf
Institut Laue Langevin, 156X, 38042 Grenoble Cedex, France

ABSTRACT

A description of the alterations of the existing diffuse scattering machine D7 for use with multidetector polarisation analysis,
including the possibility of time of flight measurements, is given.
Results of three test experiments are reported. The first test experiment was with paramagnetic scattering by a CrFe(10%) alloy to see
whether it can be achieved simultaneously in many directions. One can
identify by such a measurement the form factor, inelasticity and multiple paramagnetic scattering, and other depolarisation sources. The
second test experiment was with $TmAl_2$ showing crystal field transitions. The third test was to investigate the possibility to separate
coherent and incoherent linewidth in the time of flight spectrum of
liquid sodium simultaneously in different directions about $Q=2$ $Å^{-1}$.

INTRODUCTION

It is common wisdom that polarised neutron work means precious
extra information often at too high a price in loss of intensity.
Diffuse and quasielastic neutron scattering with polarisation analysis and time of flight analysis suffers especially from the low
intensity and from the mechanical difficulty of the use of multidetectors[1]. In its usual version polarisation analysis uses polarising
crystals as analysers[2], which forbids an arrangement using simultaneously 64 detectors distributed in the whole angular range as in
the diffuse scattering set-up D7 in the ILL[3]. Another approach to
the problem is the use of iron filters as polarisers and analysers,
which requires heavy magnets for achieving saturation of the filters
and suffers from a low polarisation of only 47%[4]. The newly developed supermirror multislit curved guide polarisers eliminate of all
these difficulties and allow a simple set-up of a diffuse scattering
machine that can be immediately integrated in the existing diffuse
scattering machine D7 at the ILL. These polarisers can handle convenient beam divergences and cross sections for wavelengths above
2 Å with very good transmission figures[5].

ALTERATIONS OF THE DIFFUSE SCATTERING APPARATUS D7

To show the realisability of this project, time was foreseen
to make some test measurments with a preliminary set-up. The polarisation of the neutrons was achieved by a polariser described in[5],
that simultaneously filtered the second harmonic of the pyrolythic

ISSN:0094-243X/82/890175-07$3.00 Copyright 1982 American Institute of Physics

graphite monochromator and thereby replaced the Be-filter. That was
also the best place to mount the polariser because it didn't use
more space than the Be-filter. This set-up already brought a gain
factor of 30 of intensity in comparison with the previously used version for polarised neutrons: a badly matched combination of graphite
crystal and a Heusler crystal[3,7]. This was necessary because the
magnet for the Heusler crystal could not be mounted in the neutron
beam of D7. For this combination[7,9], the intensity of the polarised
neutrons was always 0.01 of that for the unpolarised neutrons
(2×10^7 ncm^{-2}s^{-1} in [7] or 4.5×10^6 in [9] which will be increased by
a factor of 5 with the now installed double focussing monochromator).
Using the new polariser the intensity was 30% of the intensity without polarisation and didn't change the collimation at all.

As analysers I installed 5 curved guide multislit supermirrors
with a height of 100 mm and a width of 2 cm. They could be mounted
directly in front of the existing detectors on the same table that
carried the large collimators of D7, as they replace these collimators. Also guide fields of 50 cm length were fixed on these tables
so that the tables with the detectors and the adjusted analysers
and guide fields could be moved under computer control. In this way
the whole program facilities of D7 could be used for the following
measurements. Between the polariser and the sample I installed a
Mezei flipper. The sample was situated in an arrangement of three
orthogonal Helmholtz coil pairs to allow the field at the scatterer
to be brought to every wanted direction. This instrument extends the
existing polarisation analysis machines in three points: 1. it enlarges the possible intensity for polarisation analysis enormously;
2. it extends the possibility of analysis in now 5, and for the
future, 32 directions simultaneously; 3. in principle it enlarges the
usable wavelength range to long wavelenghts by the use of mirrors.

APPLICATION TO MEASURE PARAMAGNETIC SCATTERING

I used the described set-up to look at the alloy CrFe (10%) at
12 K. S.K. Burke and P.W. Mitchell lent me their sample and introduced me to the problems of this material. I don't want to solve
the metallurgical problems of this material but my aim is rather to
show that with the new possibility of diffuse scattering with spin
analysis at high intensity one can find and resolve interesting
problems. I used for the investigation a method to look at paramagnets that I have learned from F. Mezei[8], and which he used in the
spin echo machine also for spin analysis. I could even use his coils
for the sample field. For paramagnetic scattering in the spin echo
set-up IN11 the polarisation direction can be brought in the direction of the scattering vector. For the 5 different directions that
were needed, this was not possible. But the method can be generalised by measuring with P (=polarisation) in three orthogonal directions. And one gets simultaneously information about depolarisation,
multiple scattering and even inelasticity for as many directions as
one has analysers.

PRINCIPLE OF THE X,Y,Z-POLARISATION ANALYSIS OF PARAMAGNTIC
SCATTERING

Moon, Riste, Koehler[2] use the spin flip behaviour to separate paramagnetic scattering from all other magnetic scattering. For $\underline{P}||\hat{\varkappa}$ ($\varkappa$ = scattering vector) with flipper off you get no paramagnetic scattering, with flipper on you get paramagnetic scattering. With $\underline{P}\perp\hat{\varkappa}$, flipper on and flipper off give the same paramagnetic scattering and this is half of that before. Nuclear spin incoherent scattering independent of the direction of $\underline{P}$ relative to $\varkappa$ doubles the intensity if the flipper is on as one can see in the time of flight spectra in fig. 4 from the scattering by the vanadium container. If one works with flipper on and measures the scattering with $\underline{P}||\hat{\varkappa}$ and with $\underline{P}\perp\hat{\varkappa}$ and takes the difference of the two measurements, nuclear spin incoherent scattering drops out and one obtains half of the paramagnetic scattering alone. This method suffers from the necessity to bring $\underline{P}||\hat{\varkappa}$. If one wants to measure simultaneously in many directions one can apply the behaviour[10], that $\underline{P}'_{para} = -\hat{\varkappa}(\hat{\varkappa}.\underline{P})$ i.e. in a direction where one has an angle α between $\underline{P}$ and $\varkappa$ one measures only $P\cos^2\alpha$ and with $\underline{P}$ normal to that (in the scattering plane) $P\sin^2\alpha$. Adding up these two gives then the wanted half of the paramgnetic scattering[8]. ($\hat{\varkappa} = \varkappa/|\varkappa|$, $\underline{P}'_{para}$=polarisation of scattered beam).

We choose z normal to the scattering plane, x in the scattering plane and normal to the incident beam, y in the scattering plane and in the direction of the incident beam. Using large Helmholtz coil pairs of 30 cm diameter, three pairs orthogonal to each other at the sample position outside of the cryostat, one can adiabatically rotate the polarisation P_z of the polarised beam at the scattering plane in succession in the z,x,y direction. After the scattering the resulting P_z,P_x,P_y spin direction is brought back into the original z-direction and analysed at the different analysers with and without spin flip. The resulting curves are given in fig. 1 for the z-direction of $\underline{P}$. Fig. 1a shows the scattered intensity and 1b the polarisation of the scattered beam. The direct beam remains highly polarised. Directly beside the direct beam yet still well separated from it, there appears a scattered beam of high intensity at $5.7°$ that is nearly totally depolarised and which I found accidentally. As I used a polychrystalline material, one should expect at the corresponding negative angle a similar peak. The measured points don't show an indication for this, but may be it is so sharp that one has really to look for it. It corresponds to a Bragg reflection for a lattice constant of 24 Å - 30 Å for the used wavelength of 4.75 Å. At higher angles the scattered intensity goes down but remains 40% to 50% polarised. This remaining nearly isotropic scattering can be nuclear, nuclear incoherent, paramagnetic or other magnetic scattering. If one flips the spin, only nuclear spin incoherent, paramagnetic and other magnetic scattering can be present. Paramagnetic scattering is independent of flipping if $\underline{P}\perp\hat{\varkappa}$. To see whether it is paramagnetic one rotates the polarisation in the x and y direction and measures the scattering again. If you then take the difference between the P_z and the P_x measurement you can see whether for large angles you have a paramagnetic part in the scattered intensity, because if flipped, this part should be doubled and subtracting then the part from the measurement with P_z should leave half of the paramagnetic scattering. In our case for large angles this difference is zero, i.e. for the flipped intensity here remains only the possibility of

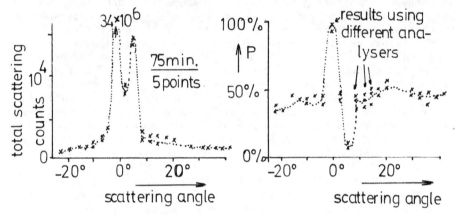

Fig. 1. Scattered intensity (a) and polarisation (b) of the scattered beam with spin not flipped as a function of the scattering angle. $\lambda=4.75$ Å, CrFe (10%), polarisation of the incident beam in z-direction i.e. normal to scattering plane.

nuclear spin incoherent scattering or another depolarising scatte-ring, perhaps some isotropic remnant of that which causes the totally depolarised Bragg reflection in the direction 5.7° of fig. 1.

For smaller angles you get really a paramagnetic scattering from the CrFe(10%) alloy given in fig. 2. As the polarisation is not in the direction of the scattering vector one has to measure also with the polarisation in the y-direction, to obtain the remaining compo-nent of the paramagnetic scattering. Again one obtains this as a difference between two measurements: one with P_z and one with P_y by

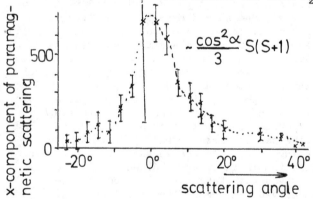

Fig. 2. x-component of the paramagentic scattering as a function of the scattering angle $\lambda=4.75$ A, CrFe(10%) measured with spin parallel (x) and antiparallel (o)

adiabatic spin rotation as in the case of P_x. The results are given
in fig. 3. There was no significant contribution to this part that

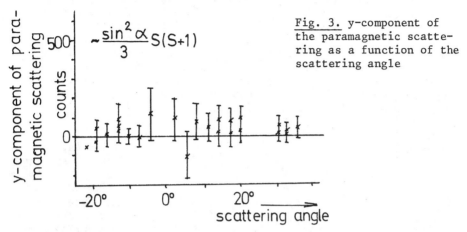

Fig. 3. y-component of
the paramagnetic scatte-
ring as a function of the
scattering angle

would be certainly other than zero. This component would also give
information about an inelastic contribution to the scattering which
at 12 K was not significantly high, as one sees in fig. 3. We measured also at higher temperature(220 K) where one has much lower
scattering intensity and the x-component went down to the same magnitude as the y-component in fig. 3. so that one would need more
time to measure exactly.

OTHER TEST MEASUREMENTS

Other test measurements were done (with Loewenhaupt) on the
system $TmAl_2$ which has very different ranges of behaviour. It is
ferromagnetic at low temperature(below 4.2 K as roughly measured
by the depolarisation) where it totally depolarised even in the
direct beam. It shows crystal field transitions which can be found
by time of flight with unpolarised neutrons. These above 4.2 K gave
paramagnetic scattering in a wider angular range than the C̲r̲Fe(10%)
alloy and showed increasing contributions coming from the y̲-compo-
nent. To observe the crystal field transitions we needed TOF measurements. There were at this stage difficulties with the chopper wheel,
which depolarised the beam totally.

For the measurements of the linewidth of the coherent and incoherent scattering of liquid sodium (together with Gläser) I had to
mount the chopper in front of the polariser. The time of flight curves
resulting from these measurements are given in fig. 4a-c. Fig. 4a
shows the spectrum at $Q = -2$ Å^{-1} without analyser and fig. 4b shows
the same spectrum for $Q = +2$ Å^{-1} with the analyser of 2 cm width.
The detectors are 5 cm wide. The countrate difference is nearly
exactly a factor 10. For an analyser of 5 cm width this factor will
be less than 4. The loss by the polariser was a factor 3 so that one

looses by spin analysis with the curved multislit super mirror analyser a factor of 12 in intensity. Fig. 4c shows the spectrum of
fig. 4b with spinflip. The narrow line on top of this spectrum is in
this case higher than in fig. 4b. It is the incoherent scattering
of the vanadium container, that is expected to show this behaviour.
The best of these measurements is with a fast statistical flipper,
that is nearly finished now by Badurek in Vienna and which, for the
final version of the instrument, will be at our disposal as well.

The result of these tests is that a version of the polarisation
analysis set-up is now under construction,that part of which using
8 analysers will be available for use at the end of 1981.

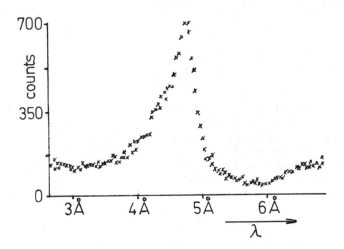

Fig. 4a. Time of
flight spectrum
without analyser
simultaneously
measured with
that of fig. 4b,
measuring time
11 hours, scat-
tering by liquid
sodium with
$Q=-2\text{Å}^{-1}$,$T=130°C$,
$\lambda=4,75\text{Å}$; elastic
line on top is
from vanadium
container.

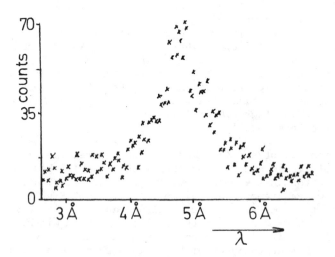

Fig. 4b. Same as
4a but using an
analyser with
spin not flipped

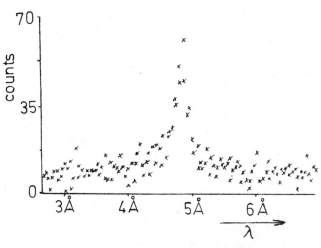

Fig. 4c. Same as 4b but spin flipped; height of the narrow spin incoherent line from vanadium container is doubled by the spin flip.

REFERENCES

1. F. Mezei: 'Diffuse Elastic and Quasielastic Neutron Scattering from Defects in Solids', St. Pierre de Chartreuse, October 1978 ILL Internal Report 78ME214
2. R.M. Moon, T. Riste, W.C. Koehler, Phys. Rev. 181 (1969) 920
3. G. Bauer, E. Seitz, W. Just, J. Appl. Cryst. 8 (1975) 162
4. T.J. Hicks, Nukleonika 24(1979) 795
5. O. Schärpf, this conference proceedings
6. M. Löwenhaupt, JMMM 2 (1976) 99-105
7. W. Just, W. Schmatz, G. Bauer, Proceedings of the Neutron Diffraction Confernece Petten 1975 (RCN234) 588-597
8. F. Mezei and A.P. Murani, JMMM 14 (1979) 211-213
9. W. Just, St.Pierre de Chartreuse 1978 Internal Report
10. O. Halpern and H.R. Johnson, Phys. Rev. 55 (1939) 898

RECENT ADVANCES WITH SUPERMIRROR POLARISERS

O. Schärpf

Institut Laue-Langevin, 156X, 38042 Grenoble Cedex,France

ABSTRACT

A combination of supermirrors, antireflecting layers, and curved multislit guides with thin glass sheets as substrates uses nearly all of the improvements of simple polarising mirrors develop - ed at different places. The realisation requirements and the measur- ed properties of such superbender polarisers with cut-off wave length of 2 Å and 4 Å are described.

INTRODUCTION

The use of polarised neutrons in parity violation experiments,[21] in diffuse scattering with polarisation analysis,[6] in time of flight investigations and for the measurements of incoherent inelastic scattering in fluids by TOF are always in want of higher intensity polarised neutron beams. To achieve such beams there were developed different improvements of simple polarising mirrors: 1.curved neu- tron guides [1,2,3,4] using sheets of iron cobalt alloys 2.focusing mirror systems [5,6] 3. Soller collimator using thin plastic foils [7]. 4.to be able tu use good quality glass surfaces for the mirrors, Drabkin[8], developed the antireflecting layer of gadolinium and ti- tanium 5.to reduce the necessary mirror surface, Mezei and Dagleish developed the supermirrors[9,10]. In some of these improvements, com- binations of two refinements were already used.But all of these single improvements could not be combined, because they had properties that excluded eachother by disadvantages connected with them.

The use of the glass substrate needs the antireflecting layer to avoid the total reflection of unpolarised neutrons on the glass itself. The antireflecting layer of Drabkin was so thick and of such a bad quality and roughness that supermirrors evaporated on top of it did not work. The super mirrors out of iron and silver on a glass substrate had to be very carefully aligned to avoid the reflection of unpolarised neutrons at the smaller angles. The latter excluded the utilisation of supermirrors of this sort in curved guide polarisers and soller collimator polarisers. It was necessary to evaporate the super mirror on solid glass plates with well de- fined flat surfaces, which led to heavy and big arrangements for po- larisers, especially together with the required huge magnet systems. By inventing remedies against all these disadvantages I was able to combine all of the above improvements and to arrive at a small size, short (320 mm), easy to handle and light polariser with high polari- sing efficiency (97% and more) and a transmission of 60 to 80 % for the correct spin at wavelengths larger than 4 Å, and of 30 to 40 % for wavelengths around 2 Å. The beam cross section I achieved was 25 mm x 86 mm and 30 mm x 50 mm. The polarisers need only a magnetic field of 100-200 Oe, which can easily be achieved by permanent magnets of the ferrite type in a compact geometry.

It is not possible to give even an indication of a detailed description of the theory of the single improvements. They can be found in the literature in more or less complete form (polarising mirros:[11,5,1,2,3,4,] curved guide:[12,13,14] antireflecting layer:[8], supermirros:[9,10,15,16,17,18] soller polariser:[19,4,7,21]. Only their application toward the desired result will be given here.

CURVED SOLLER GUIDE

A curved soller guide is mainly characterized by a cut-off wavelength λ^* that gives the wavelength for which the transmission in phase space is 66%[13]. It depends on d=spacing between mirros, R=radius of curvature of the curved mirros and Δ k=wave vector at cut-off at normal incidence and is given by $\lambda^* = (2\pi/\Delta k)(2d/R)^{1/2}$. The super mirror gives here a higher Δk and thus a smaller characteristic wave length. The Δk determines also the limits of the transmitted divergence of an incident beam. This is also higher for curved guides using supermirros and as well, gives a very effective improvement for a beam that does not use the whole divergence. Thus a polariser with 1 mm spacing and R = 10m, using the supermirrors of cobalt and titanium with a $\Delta k = 2.2 \times 10^{-2}$ Å^{-1} (roughly twice that of Ni) have a $\lambda^* = 4$ Å. The cold neutron spectrum of PN7 is very flat in the range of 3 to 4 Å and has at the leading edge its half height at 2.57 Å with a divergence that corresponds to that of a Ni-guide. The superbender polariser transmitted a spectrum with the leading edge with half height at 3.38 Å, i.e. it cuts off ~0.8 Å of the cold spectrum. Because the incident spectrum in this range is very flat, on can evaluate directly an effective $\lambda^*_{eff} = 3.4$ Å, if used behind a nickel guide. The reason of this is that with the small divergence of the incident beam behind a nickel guide the part that is transmitted at λ^* is much higher than 66 %, so that the wavelength with 66 % transmission is at a smaller wavelength giving a smaller λ^*_{eff}. For the described superbender the measurement gives a gain of 0,6 Å. The transmission for the total integral cold spectrum of PN7 measured with gold foil activation is 53 % for the correct spin i.e. 6 x 10^8 ncm^{-2} sec^{-1} directly behind the polariser, at 3m distance behind the guide 36 % are measured (compared with half of the free beam intensity without polariser). This is the highest polarised neutron flux ever achieved.(Oakridge 1,2 Å 10^7 n /cm^{2}sec.)

For the thermal spectrum a polariser was constructed with a mirror spacing of 0.5 mm. For this the transmitted spectrum was also measured. The incident spectrum began at 0.6 Å, had its half height at the wavelength of 0.8 Å, its maximum at 1.28 Å, and went down to half height again at 2.2 Å. The transmitted spectrum began at 0.9 Å, goes to half height at 1.5 Å, and has its maximum at 2.2 Å. The characteristic wavelength of this guide is $\lambda = 2$ Å. Also in this case the measured point of 66 % transmission behind a nickel guide is shifted. But the spectrum is not flat so that one can not directly evaluate the result. The transmission of this guide for the cold spectrum, measured with gold foil activation was 40 %, and for the

184

thermal spectrum it was measured to be 17 % for the neutrons with
the correct spin. The angular range, where the superbender transmits
neutrons of the wave length λ,is (0.2λ)degrees, i.e. for 7 Å neutrons it transmits in a range of 1.4 deg.

The transmission and polarisation of the superbenders on thin
glass substrates are both very homogeneous over the whole cross
section. The polarisers using plastic foils as substrates show
parts where the transmission and/or the polarisation is bad, i.e.
both are very inhomogeneous over the cross section.

The construction of the super mirror bender is given in fig.1.
As it is only elastic deformation by which the curvature is adjusted,
the curvature can be easily changed if one wants to exclude certain
shorter wavelength. By this one can achieve simultaneously the
effect of polarisation and of a Be-filter to suppress higher order
monochromator reflections. Thereby I suppressed the second harmonic
of 20% intensity at 2.41 Å of a pyrolatic graphite monochramotor at
the desired wavelength of 4.8 Å down to less than 1% with simultaneously 60 % transmission of 4.8 Å neutrons polarised to 98 %.Fig.
1 shows the simple construction of the bender. As it is so easy to
change mechanically you should not be concerned about the different
sizes that were used. The width of the polariser, the spacers and
even the curvature can be adapted to very different needs in less
than half an hour.

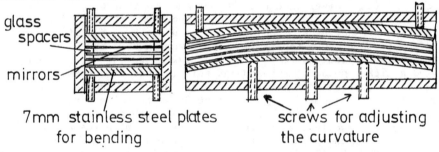

Fig. 1 Mechanical construction of the curved
guide supermirror multislit polariser

ANTIREFLECTING LAYER

To be able tu use thin glass plates in the curved soller guide
polariser the substrate must be coated first with an antireflecting
layer to avoid reflection of neutrons with the wrong spin direction.
One can calculate a theoretical alloy of Ti and Gd that would have
the index of refraction n=1 and the correct thickness to attenuate
the reflection to 1 %. But to evaporate such an alloy with just this
compositiondoes not give layers with this same composition. Good
antireflecting layer properties were achieved by the following method. The desired alloy is obtained by evaporating 140 thin layers

of 10 Å Gd and 17 Å Ti. This composition was calculated by taking
into account the change of the scattering length of Gd by the absorption i.e. by the imaginary part of the scattering length. I obtained a layer structure that gave Bragg reflections at 8.5 degrees
corresponding to a 27 Å lattice constant at 8 Å wavelength, if I
evaporated the layers on a cold substrate. The antireflecting efficiency in the range of small angles was the same as for an evaporation on the hot substrate. In the latter case I could not observe
Braggreflections i.e. the layers are well alloyed by diffusion. The
reflectivity measurements of such layers are as theoretically expected.

To test the efficieny of such an antireflecting layer for a
polariser, I first evaporated on top of this layer a simple polarising alloy of 50 % iron and 50 % cobalt, where the alloy was produced in the same manner by evaporation of 1000 Å in portions of
10 Å Co and 10 Å Fe. The reflectivity curve of such a mirror is given
in fig. 2 showing the reflectivity for spin parallel and of spin antiparallel. For the spin antiparallel one can see an interference
effect in the layer, that was already mentioned by Hamelin[20], who
found it in his calculations. It becomes less pronounced with
thinner layers. Using such mirrors on thin glasses of 0.08 mm thickness and the construction of fig.1 it was already possible to set up
an efficient polariser that avoided the unevenness of the plastic
foils and their detoriation in a high neutron flux, as it was observed after one years use in the high flux beam PN7 in the ILL for
the Rutherford polariser.[21]

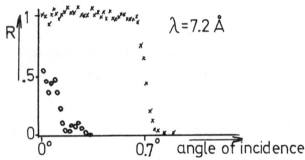

Fig. 2 Reflectivity curve for spin up[(x)] and down[(0)]
of iron-cobalt mirror on top of an antireflecting
layer, prepared as given in the text.

SUPERMIRROR

To use the possibility of the enlargement of Δk by the super
mirror action without losing the possibility of the curved guide
soller I had to develop another sort of supermirror: one that polarises in the whole range where it reflects. I tried first a
super mirror with iron cobalt alloy and as contrast material vanadium. Here, as well, it was not possible to evaporate the FeCo
alloy directly. When I tried this first, I got a strange super

mirror angular dependence of reflectivity with the behaviour of the iron silver super mirror at small angles resulting from evaporation of an iron rich alloy first. At higher angles I got a bad quality polarising super mirror because in this region of layer thickness the real alloy was badly adapted to the contrast material V, not being 50% Fe   50 %Co.

The alloy of Fe and Co with 50 % had to be evaporated in portions of 10 Å Fe and 10 Å Co. By this procedure I produced a good quality polarising super mirror, polarising over the whole range of reflectivity from angles of incidence $\theta = 0°$ to $\theta = (0.23\lambda(Å))°$. But this sort of super mirrors is very laborious and not well suited to produce thousands of mirrors. For this I tried a super mirror consisting of layers of Co and Ti as contrast material. This proved to be the best solution because the evaporation of those materials is simple. This super mirror needed also a better adaptation of the antireflecting layer to the Ti and a simplification of the above given method of producing the correct alloy of Ti and Gd, that needed only 40 layers instead of 140 layers. It begins with Gd and ends with Ti and uses a smooth transition of the concentration of the one to that of the other with a sort of antisuper mirror action, that I observed experimentally. It is also apparent that all of the so produced antireflecting layers are flat enough to allow super mirror effect on top of them. This comes perhaps from a flattening   effect by the evaporation of the alloy in small portions of very thin layers. The above reported properties of curved super mirror multislit polarisers were all obtained with such mirrors.

Fig. 3 compares reflectivity curves for iron/silver super-mirrors, cobalt/titanium super mirrors and the cobalt/iron alloy on top of the antireflecting layer, all measured for $\lambda = 7.2$ Å neutrons.

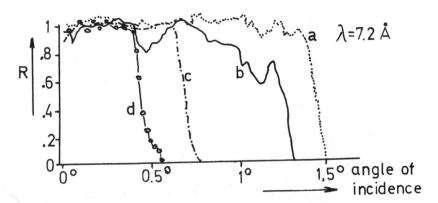

Fig.3 Reflectivity curves for iron/silver (a), Co/Ti super-mirror (b) and CoFe alloy mirror (c)for spin up and for iron/silver supermirror for spin down (d)

Fig. 4 gives another reflectivity curve for nickel/titanium
super mirrors produced by the method described by the Japanese[16,17].
They give their results only in logarithmic scale. It was easier for
me to produce mirrors using their method and measuring their behaviour than to use their curves for comparison. The reflectivity goes
down at small angles even in the range where one should have total
reflection for nickel mirrors. Only the total reflectivity of the
glass substrate remains. Also the curves in their paper show, as do
ours, in fig.4, for their layer sequence a reflectivity of only 60 %
even in the range of total reflection. The reason for this behaviour
is one of the main differences between our method and the method of
the Japanese. We begin with the thinnest layers and use two thickness corrections for the different layers,[9] which is not used by the
Japanese. The effect of omitting these corrections is that the phase
relation in the Japanese super mirrors is lost, so that they get a
partial deletion of intensity even at small angles in the range of
total reflection. The cut-off at the higher angles is also not as
well pronounced as with our mirrors, for apparently the same reasons.
For a rough view of the layer thickness sequences of the two methods
one would not expect such a large difference in the reflectivity behaviour.

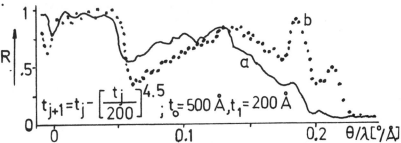

Fig. 4 Reflectivity curves for super mirrors made by the
method given by the Japanese 16,17 (a) 90 layers
(b) 200 layers.

Another observation I made is that the super mirror cut off seems
to be always at the same angular region of 2-2.3 times the cut off
angle of nickel mirrors. I tried super mirrors with 160 and even
with 600 layers and could never overcome this limit. That may have
two different reasons: the coherence of the layers begins to become
insufficient because of island formation during evaporation or: the
phase relations become such that the reflectivity at different portions of the layers combine with destructive interference and so destroy in part the super mirror action. I could observe indications
of such a behaviour in fig.4 and also in the on-line investigation of
the reflectivity curves with neutrons during evaporation. Immediately at the beginning of the thinnest layers the reflectivity curve
shows two maxima that grow simultaneously, one in the range of total
reflection and one in the range corresponding to Bragg reflections
by the thinnest layers. They grow until they reach nearly the same

height with a wide valley between them. With the last 10 layers one sees a more than proportional disappearance of the depth of the valley and the last five layers replenish it then totally, but only by lowering sometimes the reflectivity at the higher angles. The latter behaviour shows that an already achieved reflectivity can be lost again by interference with a wrong phase. One should do more investigations to get a better insight of the reasons for this, but here in the ILL one is more interested in an immediate use of the mirrors for different experiments such as parity violation,[21] in beam nuclear magnetic resonance and diffuse scattering with polarisation analysis using 32 detectors with analysers simultaneously.[6] For this sort of experiments the achieved properties of the mirrors are already sufficient and these experiments are already working with supermirror polarisers or are in construction. The improvement of the cut off angle would be interesting for the possibility to use the mirrors for very· short wavelengths and for position sensitive detectors of very large size. By the use of the above described polarisers one can now handle convenient beam divergencies and cross sections for wavelengths above 2 Å with very good transmission figures.

I wish to thank R. Gähler and the Marburg group for their help to get the results for PN7 and S6 respectively.

189

REFERENCES

1. K. Berndorfer, Z. Physik 243 (1971) 188-200
2. M. Hetzelt and A. Heidemann, Nucl. Instr.and Meth.133(1976)51-55
3. O. Schärpf, J.Phys.E:Sci.Instr.8(1975)268
   O. Schärpf and W. Vorbrugg, Proceedings of the Neutron Diffraction Conference, Petten 1975, RCN 234, 156-162
4. W.H. Kraan, On designing and testing a set-up with magnetised mirrors, Thesis 1974, Interuniversitair Reactor Instituut te Delft
5. K. Abrahams, W. Ratynski, F. Stecher-Rasmussen and E. Warming, Nucl. Instr. and Meth. 45 (1966) 293
6. F. Mezei: 'Diffuse Elastic and Quasielastic Neutron Scattering from Defects in Solids', St. Pierre de Chartreuse, October 1978 ILL Internal Report 78 ME 214
   W.G. Williams, Nukleionika 25 (1980) 775
7. J.B. Hayter, J. Penfold and W.G. Williams, J. Phys.E.Sci. Instr.11 (1978) 454
8. G.M. Drabkin, A.I. Okorokov, A.F. Schebetov, N.V. Borovikova, A.G. Gokasov, V.A. Kurdriashov, V.V. Runov, Nucl. Instr. and Meth. 133 (1976) 453-456
9. F. Mezei, Comm. Phys. 1, (1976)81-85
   F. Mezei, Inst. Phys. Conf. Ser. No.42 (1978) 162
10. F. Mezei and P.A. Dagleish, Comm. Phys. 2 (1977) 41-43
11. D.J. Hughes and M.T. Burgy, Phys.Rev.81 (1951)498-506
12. H. Maier-Leibnitz and T. Springer, J. Nucl. Energy A/B 17 (1963) 217-25
13. B. Jacrot, Instrumentation for Neutron Inelastic scattering Research, IAEA, Wien 1970, p.225-247
14. O. Schärpf and D. Eichler, J. Phys.E:Sci.Instrum.6(1973)774-80
15. J. Schelten and K. Mika, Nucl.Instr. and Meth.160(1979)287-94
16. Sh. Yamada, T. Ebisawa, N. Achiwa, T. Akiyoshi and S. Okamoto, Annu.Rep.Res.Reactor Inst. Kyoto Univ.11 (1978)8-27
17. T. Ebisawa, N. Achiwa, Sh. Yamada, T. Akiyoshi and S. Okamoto J. of Nucl.Science and Tech., 16(1979)647-679
18. A.G. Gukasov, V.A. Ruban and M.N. Bedrizova, Sov.Tech.Phys.Lett. 3(1977)52
19. H.B. Möller, L. Passelll and F. Stecher-Rasmussen, Reactor Science and Technology (1963) 17 p.227-231
   W. Fiala and H. Rauch Nucl. Instr. and Meth.52(1967) 15-24
20. B. Hamelin, Nucl. Instr. and Meth. 135 (1976)299-306
21. B.R. Heckel, Parity non-conserving neutron spin rotation: the tin isotopes, thesis 1981,Harward University, Cambridge Massachusetts ILL Internal Report 81 HE
    M. Forte, B.R. Heckel, NF. Ramsey, K. Green, G.L. Greene, J. Byrne, and M. Pendlebury, Phys.Rev.Lett.45(1980)2088

PRELIMINARY NEUTRON STUDY OF GRAPHITE INTERCALATION COMPOUNDS IN
VIEW OF APPLICATION AS MONOCHROMATORS

A. Boeuf, R. Caciuffo
J.R.C., Ispra, Italy and Institut Laue-Langevin, Grenoble, France

A. Freund
Institut Laue-Langevin, Grenoble, France

A. Hamwi, P. Touzain
Lab. Abs. et Réac. de gaz sur solides, St. Martin d'Hères, France

F. Rustichelli
J.R.C., Ispra, Italy and Università di Ancona, Italy

ABSTRACT

A neutron diffraction study of KC_8 (d_{001}=5.35 Å), KC_{48} (d_{001}=
15.44 Å) and $KC_{24}(C_6D_6)_2$ (d_{001}=9.2 Å) intercalation graphite compounds
has been performed in view of their use as large wavelength neutron
monochromator. The peak and integrated reflectivity for (00ℓ) type
reflections, the width of the rocking curves and homogeneity of the
samples were investigated.

INTRODUCTION

The intercalation compounds of graphite present a lattice structure with a periodicity greater than that of graphite crystals. If
their reflectivity were sufficiently great they could be suitable for
long wavelength neutron monochromatization.
In order to investigate this opportunity, we have performed a
neutron diffraction study of KC_8, KC_{48} and KC_{24} $(C_6D_6)_2$ compounds.
The stage 1 potassium intercalation compound KC_8 is characterized by a
fully ordered structure[1,2], with an interplanar distance d=5.35 Å for
(001) planes. KC_{48} which corresponds to stage 4 has no stacking correlation between different K layers. The interplanar distance is
d=15.44 Å for (001) planes. For $KC_{24}(C_6D_6)_2$ the interplanar distance[3]
is d=9.2 Å for (001) planes.

EXPERIMENTAL RESULTS

The measurements were performed at the two axis spectrometer D13
of I.L.L. Grenoble. The polychromatic neutron beam emerging from a
neutron guide was monochromatized by a Cu(111) crystal (λ=1.766 Å) or
by a Ge(111) crystal (λ=2.220 Å and λ=2.897 Å). The dimensions of the
samples were 1x1x0.15 cm. The peak (r_{max}) and integrated (R_s^θ) reflectivity for (00ℓ) reflections, the full width at half maximum ($L_{1/2}$)
of the rocking curves and the homogeneity of the samples were investigated for different values of incident neutron wavelengths. The
high order neutron contamination was eliminated using an oriented
graphite filter. Fig. 1 shows a typical rocking curve obtained for
KC_{48}(005) reflection (λ=2.897 Å).

In Gaussian approximation[4], $L_{1/2}$ is given by

$$L_{1/2}^2 = \beta^2 + \alpha^2 \tag{1}$$

where β is the effective mosaic width, $\alpha = (tg\theta_2/tg\theta_1 - 1)^2\alpha_1$ is the instrumental resolution, α_1 being the primary neutron beam divergence, θ_1 and θ_2 the monochromator and sample Bragg angles. The β values were determined by extrapolation as shown in Fig. 2 in the case of KC_{48} for $\lambda=1.766$ Å. The obtained values ($\beta=358'$ for KC_8, $\beta=122'$ for KC_{48} and $\beta=191'$ for $KC_{24}(C_6D_6)_2$) are large in comparison with the mosaic width of the pristine graphite ($\beta \simeq 20'$, $r_{max}=60\%$). However preliminary γ-ray diffraction studies performed on KC_{24} compound have shown that the mosaicity can be reduced to the mosaicity of the pristine graphite by applying a pressure of $150 \times 10^6 N/m^2$.

Table 1 reports the measured (R_s^θ) integral reflecting powers of the different compounds for different neutron wavelengths and different reflections. The theoretical values were evaluated by the kinematical approximation $R_t^\theta = Qt_0/\sin\theta_B$ where t_0 is the crystal thickness, θ_B the Bragg angle and Q the crystallographic quantity

$$Q = \frac{\lambda^3}{\sin 2\theta_B} \frac{F^2}{V_c^2} \tag{2}$$

λ being the neutron wavelengths, F the structure factor and V_c the volume of the elementary cell ($V_c=112$ Å^3 for KC_8 and $V_c=485.3$ Å^3 for KC_{48}). The elementary cell volume of $KC_{24}(C_6D_6)_2$ being not yet known, no theoretical evaluation has been done for this compound. The theoretical values are reported only for reflections corresponding to Bragg angle sufficiently large that a pure Bragg geometry was obtained. The peak reflectivity (r_{max}) reported in table 1 was obtained by the equation $r_{max} = R_s^\theta/\beta$ which give values corresponding to ideal experimental conditions.

In general the experimental reflectivities are lower than the theoretical ones. The fact that the Debye-Waller factor was neglected in the calculation, that no corrections for parasitic reflections, absorption and diffuse scattering was applied, is not sufficient to explain this discrepancies. The crystals seem not to behave

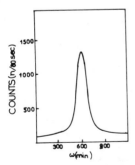

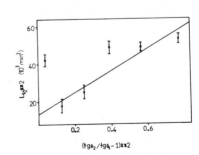

Fig. 1. Rocking curve for (005) KC_{48} reflection at $\lambda=2.897$ Å

Fig. 2. Effective mosaic width determination for KC_{48} compound.

as ideal mosaic crystals. Furthermore the compounds could not be absolutely pure and could contain fraction of other stages. This could explain why in few cases the experimental reflectivity is higher than the theoretical. Another explanation could be the non exact position of the K atoms inside the lattice cell. Finally, integral reflecting powers R_s^θ increase with the neutron wavelength λ in qualitative agreement with eq. (2).

However, much work should be done in the sample preparation technique to increase the neutron reflectivities of these compounds in view of their application as neutron monochromators. The large mosaic spread which could be reduced by pressure and the small thickness of the sample are also partially responsible of this low reflectivity.

Table I   Value of R_t^θ in radian $(\times 10^3)$
R_s^θ in radian $(\times 10^3)$ and r_{max} (%)

Reflections ($\lambda=1.76$Å)	KC_{48}			KC_8			$KC_{24}(C_6D_6)_2$	
	R_t^θ	R_s^θ	r_{max}	R_t^θ	R_s^θ	r_{max}	R_s^θ	r_{max}
001	–	1.50	4.26	–	6.83	6.56	1.32	2.40
002	–	1.30	3.69	10.32	5.26	5.05	2.50	4.60
003	–	1.33	3.77	3.60	2.96	2.84	3.0	6.00
004	1.99	1.05	2.99	3.25	1.55	1.48	–	–
005	23.37	4.61	13.10	–	–	–	–	–
($\lambda=2.22$Å)								
001	–	2.25	6.39	–	9.42	9.04	3.60	6.60
002	–	1.91	5.42	13.45	5.9	5.66	4.90	8.90
004	–	1.97	5.59	–	–	–	6.60	11.94
003	–	3.89	11.05	–	–	–	–	–
005	–	7.45	21.11	–	–	–	–	12.12
($\lambda=2.897$Å)								
001	–	5.73	16.27	48.60	5.36	5.14	–	–
002	–	2.66	7.75	–	–	–	–	–
003	5.84	4.39	12.47	–	–	–	–	–
004	3.30	5.62	–	–	–	–	–	–
005	41.54	4.91	13.95	–	–	–	–	–

REFERENCES

1.   D.E. Nixon, G.S. Parry, Brit.J.Appl.Phys.ser.2,1,291-8 (1968).
2.   W. Rudorff, E. Schultze, Z.anorg.Chem.,277,156-71 (1954).
3.   A. Hamwi, P. Touzain, C. Riekel, to be published in Synth.Met.
4.   H. Dachs, Neutron Diffraction (Springer-Verlag, Berlin Heidelberg, 1978) p. 29.

THIN FILM MULTILAYER NEUTRON MONOCHROMATORS

A. M. Saxena and C. F. Majkrzak
Brookhaven National Laboratory, Upton, NY   11973

ABSTRACT

Thin film multilayer monochromators have been prepared in a specially designed RF sputtering chamber by translating a glass substrate back and forth under cathode targets made of selected materials.
These monochromators have high reflectivities, and their angular
acceptance and bandwidth can be selected according to experimental
requirements to produce a neutron beam of desired characteristics.
Multilayers are especially useful for monochromatizing long wavelength neutrons from a cold source, and for small angle scattering
experiments in which $\Delta\lambda/\lambda$ of about 10% is acceptable.

INTRODUCTION

Preparation of multilayer monochromators, made by depositing
thin films of two materials in an alternating sequence on a glass
substrate, has been reported earlier[1,2,3]. These multilayers were
made by resistive heating of the materials in a vacuum chamber. This
method of preparation had the following limitations- (1) a film of
uniform thickness cannot be deposited over a large area due to geometrical constraints. The length of a multilayer, ℓ , required to reflect a beam of width w is given by ℓ = w/sinθ = 2dw/λ , where θ is
the Bragg angle for the multilayer, d is the thickness of one bilayer,
and λ is the wavelength of neutrons. A useful multilayer should have
a length of 12" to 30", (2) total number of bilayers in a multilayer
prepared by evaporation cannot exceed a few hundred, (3) due to fluctuations in deposition rates, the multilayers have a substantial
aperiodicity built into them. In order to overcome these limitations,
a system has been built in which the films were deposited by RF sputtering. The deposition rates in this system were stable over long
periods of time. The thickness of a bilayer was controlled by a stepping motor placed outside the vacuum system, hence it could be regulated to deposit any sequence of d-spacings on the substrate.

SPUTTERING SYSTEM

Two water-cooled cathode assemblies containing targets of the
selected materials were placed in the middle of a 104" long stainless
steel cylindrical chamber. The substrate was mounted on a carriage
which was moved back and forth under the targets with a stepping
motor placed outside the vacuum system, and coupled to the carriage
by a rotary motion feedthrough. The chamber was evacuated to a pressure of 1x10^{-6} Torr by a turbomolecular pump and then argon gas was
introduced in it. RF power was fed from a 2 KW generator to the two
matching networks in parallel. Tuning of matching networks was
adjusted to produce the desired voltages at the targets. Plasma was
fired at an argon pressure of 40 μ , and then the pressure was

lowered to about 5 μ for actual deposition. Both the sputtering guns were left on for the entire deposition period so that each material was deposited twice in one complete cycle. The targets were ¼" thick and 5" in diameter. A typical substrate was 2"x18"x¼", but a longer substrate could be used. The deposition rates were very stable and deposition could be continued without interruption for many days.

REFLECTIVITY OF MULTILAYERS

According to kinematical theory,[3] the reflectivity of a perfect multilayer for odd orders is given by

$$|q|^2 = 4N^2d^4(f_1-f_2)^2/\pi^2n^4 \qquad (1)$$

where N is the total number of bilayers, n is the order of reflection, and f_1 and f_2 are the neutron scattering amplitude densities for the two materials. The values of scattering amplitude densities for Fe and Mn are $0.81x10^{11}$ and $-0.30x10^{11}$ cm^{-2} respectively. Most of the multilayers were made of these two materials because of the good contrast in their scattering densities. The choice, however, is not unique and multilayers with good reflectivities have been made of other materials. For any pair of materials, the quality of films deposited depends on a number of parameters such as deposition rate, argon pressure, temperature of the substrate, partial pressure of oxygen in the system. If the films are granular, or, if there is too much interdiffusion, then the reflectivity may saturate at a low value and may not increase significantly by increasing the number of bilayers.

Multilayers with d-spacings ranging from 45 to 100 angstroms, and containing upto 3000 bilayers have been made by this method. A single set of bilayers usually gives rise to a Bragg reflection of width (FWHM) 0.07 degrees. Since the acceptable value of beam divergence for a typical experiment is significantly greater, the use of single d-spacing multilayer will result in a loss of intensity. However, a multilayer can be fabricated with a number of closely spaced d-spacings so that the acceptance angle of the multilayer is equal to the divergence of the neutron beam. Such multilayers were made in an uninterrupted process, in which enough bilayers of one d-spacing were deposited to obtain good reflectivity, and then the speed of the carriage was changed to deposit the second set of bilayers. When the total number of bilayers is large, the adhesion of the multilayers to the substrate may be poor. This phenomenon is quite significant in Fe-Mn multilayers. Multilayers of other materials may have to be prepared for stacking a number of different d-spacings.

Since the angular width of Bragg reflection from a multilayer is very small, the measured reflectivity (R = I(n=1)/I(INC.)) will, in general, be smaller than the true reflectivity. A well collimated beam is required for a measurement of reflectivity. The number of parameters that affect the quality of films in a multilayer is large, and it is not always possible to predict the reflectivity of a multilayer. For d = 70 Å, a reflectivity of 85% has been obtained with

195

400 bilayers. Multilayers with reflectivities greater than 80% have been made for d-spacings within the range 45Å - 100Å.

BANDWIDTH

An important parameter for a monochromator is $\Delta\lambda/\lambda$, the bandwidth of the reflected beam. Since the angle of reflection from multilayers is small, divergence of the incident and reflected beams will make a large contribution to the bandwidth. Another contribution to the bandwidth will be from Δd of the multilayer. Data collected with different beam divergences show that for multilayers made in the present setup, the former contribution is much greater than the latter. Typical values of bandwidth for d=70Å are 5% for divergence (FWHM) of $0.07°$, and 15% for a divergence of $0.5°$. Because the contribution of beam divergence to bandwidth decreases rapidly with increasing Bragg angle, bandwidth may be decreased by using a multilayer with small d. The smallest d-spacing for which good multilayers can be made has not been established as yet, but it appears that it would be difficult to decreases $\Delta\lambda/\lambda$ much below 5%.

CONCLUSIONS

With the advent of cold neutron sources, multilayer neutron monochromators have become extremely useful. Such a monochromator has high reflectivity, adjustable angular acceptance, and a bandwidth that can be selected within certain limits. Because the multilayers have good reflectivities, it is also possible to use two similar multilayers in a parallel geometry to eliminate the aberrations introduced by a single multilayer.

ACKNOWLEDGEMENTS

The authors are indebted to Drs. B. P. Schoenborn and L. Passell for useful discussions, and to E. Caruso and P. Pyne for technical assistance. This work was performed under NSF grant no. PCM77-01133 and USDOE contract no. DE-AC02-76CH00016.

REFERENCES

1. B. P. Schoenborn, D. L. D. Casper, and O. F. Kammerer, J. Appl. Cryst. 7, 508(1974).
2. A. M. Saxena and B. P. Schoenborn, Brookhaven Symposia in Biology, No. 27,VII-30 - VII-48(1975).
3. A. M. Saxena and B. P. Schoenborn, Acta Cryst. A33, 805(1977).

MULTIDETECTOR DEVELOPMENT:
TESTS WITH A PHTHALOCYANINE CRYSTAL

by R.F.D. Stansfield

Institut Laue-Langevin, 156X, 38042 Grenoble Cédex, France

ABSTRACT

As part of a program to develop methods of using multidetectors
for neutron crystallography, a total of 1277 diffracted intensities
from a crystal of metal-free phthalocyanine were collected using a
prototype 2-dimensional detector. Two different algorithms for peak
integration are under consideration. These are described briefly and
the results are compared in detail.

INTRODUCTION

D19A is a test diffractometer situated on a thermal neutron
guide at the ILL. For this experiment the multidetector simply replaced a point detector on the 2θ-arm of a conventional 4-circle
diffractometer. Thus single reflections were directed to the centre
of the detector in the equatorial plane. The prototype detector is
a one-eighth subunit of a large curved "banana" multidetector (64 x
4°) designed for routine use with crystals of unit cell volumes
between ~10^3 and ~10^5 Å^3.

This experiment was the first attempt at collecting a data-set
using the prototype detector. Phthalocyanine was chosen as a sample
because it has been studied[1] by neutron diffraction at Lucas Heights,
Australia ; Harwell, England ; and on a conventional diffractometer
(D8) at the ILL. In the event, only structure factors from D8 obtained at low temperature (77 K), were immediately available for comparison with the room temperature D19A results. This crystal had also
already been used to determine some sample and instrument characteristics[2] on D19A.

The entire set of raw counts was saved on magnetic tape, which
will not normally be practicable. However, the measurement prompted
several modifications to both data reduction methods described here,
and structure factors were therefore derived from the raw data after
the measurement, from two improved programs. Our aim in this first
comparison was to find a satisfactorily fast and accurate on-line
data reduction (peak integration) method that can be used (with
appropriate modifications) when the large multidetector receives
simultaneous reflections in the normal-beam setting.

THE DETECTOR

The detector is one of a new generation of multiwire proportional counters[3] at the ILL. The "subunit" on D19A comprises 1024 cells
made up of 64 anode wires 2.5 mm apart and 16 cathode wires 5.0 mm
apart. At 1.15 m from the sample this corresponds to a possible

angular resolution of 0.125 and 0.250°, and a total angular range of 8 x 4°. (The complete detector will subtend 64 x 4° like an upright banana). The multiwire chamber contains 10 atmospheres of ^{3}He and 1 atmosphere of Argon and is about 80 % efficient for 1 Å neutrons.

EXPERIMENTAL

Crystal data : $C_{32}H_{18}N_8$, $P2_1/C$, $Z=2$, $a=14.7866(10)$, $b=4.7293(3)$, $c=19.3871(19)$ Å, $\beta=121.985(4)°$, $V=1177$ Å^3. A 5 mm^3 needle-like sample was used in a flux of about 5.10^6 ncm^{-2}s^{-1}, $\lambda=1.572$ Å.

Reflection data : 258 pairs of equivalents (h0ℓ) and ($\overline{h}$0ℓ) to $\theta_{max} = 65°$ plus 761 reflections (K > 0) to $\theta_{max} = 45°$.

Variable width ω-scans of 31 steps were used, counting for about 20 s per step. Reflections were therefore characterised by 31 x 1024 = 31744 numbers, which were transferred to magnetic tape after filling intermediate RK05 cartridge disks 56 times. No corrections were applied for (small) systematic effects such as extinction and absorption. 1275 usable reflections, 989 independent, were treated with the two programs BRAGV3 and PEAKINT to obtain sets of $|F|^2$ and $\sigma(|F|^2)$.

THE PROGRAMS

BRAGV3[2,4] works sequentially on the frames of data at successive ω steps, first determining a background level for the frame based on Poisson's statistics, and then contouring that frame to include in the peak points significantly above background that are not isolated occurrences. A rudimentary form of 3-dimensional contouring is effected by checking forward and backward one frame for "significant" neighbours to a significant point. The integrated intensity is accumulated in the pass through all the frames for a reflection.

For a single strong peak PEAKINT[5] treats as a whole the 3-D data array extended in ν(vertical on the detector), $\gamma(=2\theta$ since the detector centre is at $\nu=0$) and ω. An intensity contour is taken to be ellipsoidal in shape and can be characterised by the inertia tensor for equally weighted points contained within that contour. By choosing a suitable contour level (typically 5 % of peak height is chosen) and expanding and contracting the volume of the ellipsoid while keeping the shape fixed, the fractional intensity contained within volumes of that shape can be calculated. A library of strong peaks with their shape functions is kept updated by the program. Weak peaks can be processed periodically in batches, by interpolation between nearby strong peaks in the library, for greater accuracy (see ref. 5, these proceedings, for more detail). The background for any peak is measured in a sufficiently distant ellipsoidal shell enclosing the peak.

The following analyses showed the results from these two programs to be very similar. (i) a merging R factor $\Sigma|I_1-I_2|/\Sigma I$ (writing I for $|F_o|^2$) was 1.6 %. (ii) the data were on exactly the same scale, i.e. for $\delta m_i = (I_{1i}-KI_{2i})/\{\sigma^2(I_{1i})+K^2\sigma^2(I_{2i})\}^{1/2}$, when $\Sigma\delta m_i^2$=a minimum, K=1.000. (iii) crystallographic refinement gave conventional R factors not significantly different at 5.2 % (BRAGV3) and 5.4 % (PEAKINT) for 222 independent parameters, with no significant difference in the final parameters or their precision.

Merging R factors defined as above but summed over the symmetry equivalents within each set of results were 3.5 % (BRAGV3) and 2.9 % (PEAKINT). These reflect a real systematic difference (of ~3 %) observed between equivalent pairs that may be due to beam inhomogeneity or extinction or absorption effects, but they also indicate that PEAKINT views the equivalents as more similar than does BRAGV3. Further subtle differences are as follows.

A simple plot of I against $\sigma(I)$ showed distributions of similar form, but with uniformly higher σ values attributed by PEAKINT. This may be because PEAKINT uses fewer points to determine the background for a peak, and the then greater uncertainty in background is incorporated as greater uncertainty in the integrated peak intensity. However, the normal probability plot comparing the two sets of results may be interpreted, if the intensity results are reasonable, as indicating that one or both sets of σ values are underestimated by up to (a combined total of) 30 %. The same plot indicated that systematic discrepancies between the results existed, mostly for strong reflections, to a lesser extent for weak reflections, and hardly at all for the bulk of intermediate reflections.

For strong reflections BRAGV3 intensities were larger, by up to 10 %. For many more weak to intermediate reflections the reverse held true but by only a few per cent. This may be because the background determined by PEAKINT is weighted towards the value on the central frames of the ω-scan, where M. Thomas[2] has observed a diffuse background peak from scattering from the Al window of the detector.

CONCLUSION

Both algorithms have been shown to give good quality results at about the same speed, and the small variations between them are partly understood in terms of the different techniques employed. BRAGV3 is thought to be a good immediate tool, its sequential action is easy to include in the control software. For the future, PEAKINT may be more appropriate to incorporate a multi-dimensional resolution function for the large detector, and possibly to process overlapping or "edge-affected" reflections.

Grateful acknowledgements are made to M. Thomas, A. Filhol, C. Wilkinson, H. Khamis, S.A. Mason, G. Greenwood, M. Berneron, B.E.F. Fender and J.B. Forsyth.

REFERENCES

1. Hoskins, B.F., Mason, S.A. and White, J.C.B., J.C.S. Chem. Commun, 554 (1969) ; Mason, S.A. Ph.D Thesis, University of Melbourne (1971), and unpublished work.
2. Thomas, M. and Filhol, A., ILL report 81THO8T (1981).
3. Jacobe, J. et al., ILL report 80AJ34T (1980).
4. Filhol, A. and Greenwood, G., BRAGV3 (1981).
5. Wilkinson, C. and Khamis, H., PEAKINT (see these proceedings).

EVALUATION OF A LINEAR POSITION SENSITIVE NEUTRON DETECTOR FOR POWDER SPECTROSCOPY*

I-Ping Chang and A.C. Nunes
University of Rhode Island, Kingston, RI 02881

ABSTRACT

A one meter long position sensitive detector was acquired from Harshaw Corp. for a small angle neutron camera operated by the University of Rhode Island at the 2 MW reactor of the Rhode Island Nuclear Science Center. In the course of testing the detector system, to determine linearity, resolution, stability and efficiency, it was employed in a powder spectrometer. Once geometrical effects are properly taken care of, it appears that the present detector, though not optimized for powder diffraction, permits collection of data essentially equivalent to that taken with the original step scanned detector though 15 times faster.

INTRODUCTION

Neutron linear position sensitive detectors have been commercially available for several years. They are potentially of great value in speeding data collection in scattering experiments in which the scattering pattern is not a strong function of sample orientation such as for example, small angle solution scattering, and scattering from fluids, amorphous solids, and powders. Properties of the detector system, such as efficiency, resolution, linearity and stability dictate the extent of application of these detector systems. In this note we consider a long straight linear PSD spanning an angular range of 55° as a possible powder diffractometer.

THE DETECTOR SYSTEM

The U.R.I. PSD is a stainless steel tube 1 meter long by 2.5 cm diameter filled with 3 atmospheres of ^{3}He, 5 atmospheres of Ar, and 5% CO_2, obtained from Harshaw, Inc. The anode, a resistance wire of 6000 Ω is accessible by HV connectors at each end of the detector. Position encoding employs the charge divider method of Alberi, et al.[1] The crucial preamps and divider circuit were obtained from Brookhaven National Laboratory. Other electronics are standard NIM modules. A PDP 11/40 with fast ADC is used as the multichannel analyser.

Detector counting efficiency was designed to be approximately 85% for 4 angstrom neutrons traversing a diameter (about 38% at one angstrom). Transmission measurements indicate a four angstrom capture efficiency of 83% at 4 angstroms (36% at 1 angstrom).

System spatial resolution is limited by the ion path lengths in the detector gas, but there is also an electronic component as

*Work supported by the National Science Foundation and the Research Corporation.

illustrated in Fig. 1. The total system resolution is about 5.5 cm FWHM, which is adequate for present work.

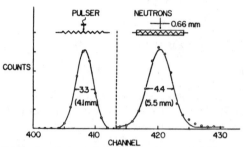

Fig. 1. Spatial resolution of PSD electronics (left) and total system (right).

Position vs. channel number was found to be very slightly non linear, but adequately described by a quadratic function of channel number over 95% of the active length of the detector. Assuming that the detector lies in the scattering plane, the scattering angle 2θ vs. channel number x is thus described by:

$$2\theta = 2\theta_o + \tan^{-1}(Ax^2 + Bx + C) \tag{1}$$

where $2\theta_o$, A, B, and C are calibration parameters determined by a calibrating program employing a standard powder. $2\theta_o$ is the angle between the incident beam and a perpendicular drawn from detector to sample (see inset, Fig. 2). The precision of eqn. 1 in describing 2θ vs. x was tested using $CoFe_2O_4$ powder sample and comparing true powder peak positions (2θ) with those calculated from (1). The differences were attributable to counting statistics.

The divider circuit is temperature sensitive. Structures near the center of the tube appear to move approximately .5 channel per degree Celsius change in ambient air temperature. This problem is eliminated by using a proportional temperature controller to stabilize the nimbin ambient temperature to within 0.3°C.

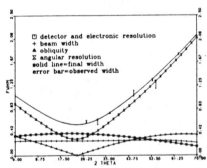

Fig. 2. Angular resolution of PSD powder spectrometer.

The angular resolution of the system can be expressed as the sum of three parts: 1) the detector spatial resolution for a normally incident beam (both detector resolution and beam width enter), 2) an additional spatial width introduced when the beam strikes the detector obliquely, and 3) the angular resolution introduced by collimators and monochromator mosaic. These are plotted in Fig. 2, and the square root of the sum of the squares of these contributions (solid line) is compared with observed widths of Ni and Al powder peaks. Agreement is quite good especially considering that the solid line is not a fit but a prediction.

The widths are also closely fit by the $W = \tan^2\theta + Y\tan\theta + Z$, suggesting that PSD data may be used in profile refinements[2] without serious modifications to the refinement programs.

Powder patterns of Fe_3O_4 were taken on one spectrometer of the

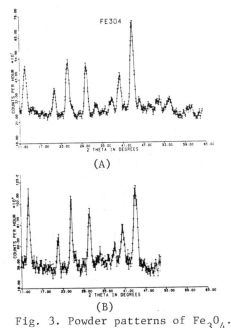

(A)

(B)

Fig. 3. Powder patterns of Fe_3O_4.
(A) PSD 4.5 hour exposure,
(B) Step scan 17.5 hours.

RINSC with the PSD and step scan detector. The same wavelength (1.1Å) sample, monochromator and collimators were used in each case. The PSD data was transformed to counts/angle and corrected for sample attenuation. The step scan data was corrected for sample attenuation. The results are displayed in Fig. 3. Signal to noise ratios are the same for each curve. Counting rate per angular increment is lower for the PSD than for the step scan detector due to its tighter vertical collimation and reduced efficiency. Overall data collection time for the two patterns was, however, almost a factor of four lower with the PSD. Moreover, the absolute counting statistics are not equal for the two curves, being worse for the step scan. To take data of similar statistical precision over the same range of angles would have required 70 hours of step scan time, or more than 15 times longer!

CONCLUSION

Use of a long linear PSD subtending a large range of (2θ) as a neutron powder diffractometer seems feasible. Stability, angular definition and resolution, and intensities (after appropriate corrections for geometric factors) are comparable with those obtained with a step scanned detector suggesting use of PSD data in profile refinement. Great improvement in data collection rates are realized.

REFERENCES

1. J. Alberi, et. al. TEEE Trans. Nucl. Sci. (1975) NS-22, 255-269.
2. H.M. Rietveld J. Appl. Cryst. (1969) 2, 65-71.

ANHARMONIC CONTRIBUTIONS IN ZnS POWDER DIAGRAMS[*]

H. Boysen, G. Steger
Institut f. Kristallographie d. Universität
8 München 2, Theresienstr.41, FRG

A.W. Hewat, J.L. Buevoz
Institut Laue-Langevin, 38042 Grenoble Cedex, France

ABSTRACT

In ZnS contributions from third order anharmonic thermal vibrations are important at high temperatures being proportional to T^2. Neutron powder measurements at different temperatures confirm this behaviour. The magnitude of the temperature independent anharmonicity parameter is $-5.5(1.4)*10^{-12} erg/Å^3$ similar to that from a single crystal determination at room temperature[1].

INTRODUCTION

For atoms in non-centrosymmetric positions third order terms in the one-particle potential may become important. In zincblende (space group F$\bar{4}$3m) for both Zn and S atoms lying in tetrahedral site symmetry the appropriate potential is

$$V_j(\vec{u}) = V_{oj} + \frac{1}{2}\alpha_j(x^2+y^2+z^2) + \beta_j xyz \qquad (1)$$

leading to anharmonic contributions to the temperature factor. It can be shown[1], that only the reflections with all indices odd have a considerable contribution due to anharmonicity. For h+k+l=4n+1 there is an alternating positive and negative change of intensity

$$(\Delta I)_a = \mp 32 b_{Zn} b_S T_{Zn} T_S hkl\beta, \qquad (2)$$

where T_j is the normal harmonic temperature factor

$$T_j = \exp(-Q^2 kT/2\alpha_j) = \exp(-B_j \sin^2\Theta/\lambda^2) \qquad (3)$$

and the anharmonicity is described by the parameter

$$\beta = (kT)^2 (2\pi/a)^3 (\beta_S/\alpha_S^3 - \beta_{Zn}/\alpha_{Zn}^3). \qquad (4)$$

$Q=4\pi\sin\Theta/\lambda$ is the scattering vector, Θ the scattering angle, λ the wavelength, b_j the scattering length,

*work supported by funds of BMFT

a the lattice constant, k Boltzmann's constant and T
the absolute temperature. The anharmonic contribution
is not correlated with the harmonic component, since
the latter decreases monotonically with Q. As ß is
proportional to T^2 its relative importance grows at
higher temperatures. E.g. at T=1000 K intensity changes
of about 6% can occur.

RESULTS

Neutron powder diagrams of ZnS have been recorded
at different temperatures on instrument D1A of the ILL
in Grenoble with λ=1.388 Å in the range $0° \leq 2\Theta \leq 160°$.
After correction for the background from the furnace
and the sample can the peaks were first fitted with
separate peak heights to give integrated intensities,
which were afterwards fitted to the modified intensity-
equation (2). The refined parameters are listed in
table I. For comparison the results for the simple
harmonic model (ß=0) are also included. Only the data
at the two highest temperatures had sufficient statistical accuracy (and a large enough anharmonic component) to yield a clear ß≠0 as shown by the e.s.d.'s,
the relative R-factors and the corresponding significance levels estimated from the significance tables of
Hamilton[2]. As anticipated before the calculated P-
values are independent of ß. They also show the expected linear increase with T (comp. eq.(3)).

Table I. Temperature parameters in ZnS.

T (K)	$-ß$	B_S ($Å^3$)	B_{Zn} ($Å^3$)	R (%)	sign. level
293	.0005(8) 0	.86(6) .86(6)	.88(3) .89(3)	2.01 2.05	.5
523	.0006(8) 0	1.64(9) 1.65(9)	1.95(5) 1.95(4)	2.29 2.35	.5
773	.0006(12) 0	2.39(13) 2.40(13)	2.65(6) 2.65(6)	2.45 2.49	.5
973	.0022(9) 0	3.33(9) 3.35(12)	3.58(5) 3.60(6)	2.60 3.19	.025
1223	.0025(14) 0	4.00(12) 3.99(14)	4.65(7) 4.61(8)	2.18 2.54	.05

DISCUSSION

Although the errors for the individual β's are relatively large, their variation is not inconsistent with the predicted T^2-dependence of eq.(4) (see fig.1) and from the slope a temperature independent parameter

$$\beta' = \beta_S - (\alpha_S/\alpha_{Zn})^3 \beta_{Zn} \qquad (5)$$

can be calculated: $\beta'=-5.5(1.4)*10^{-12}erg/\AA^3$. This value is in good agreement with the result from a thorough single crystal determination at room temperature: $\beta'=-4.2(1.1)*10^{-12}erg/\AA^3$ [1].

In the evaluation of the powder data corrections to the measured intensities were not necessary unlike in the single crystal case, where extinction has been appreciable. TDS corrections were estimated to amount up to about 4%. However, TDS again varies monotonically with Q and should thus only influence the magnitude of the β's, but not that of β. The same argument holds for absorption.

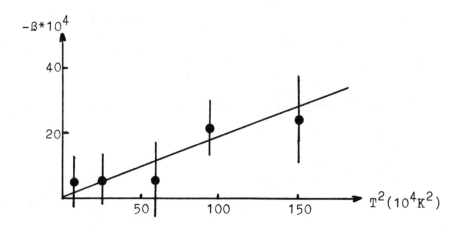

Fig.1. Temperature dependence of anharmonicity parameter β (plotted as function of T^2).

REFERENCES

1. M.J. Cooper, K.D. Rouse and H. Fuess, Acta Cryst. A29, 49 (1973).
2. W.C. Hamilton, Acta Cryst. 18, 502 (1965).

CRYSTAL STRUCTURE ANALYSIS AND REFINEMENT FROM POWDER DIFFRACTION DATA BY PROFILE ANALYSIS AND POWLS

G. Will, E. Jansen & W. Schäfer

Mineralogisches Institut der Universität, Bonn, West Germany

ABSTRACT

The refinement of crystal structures from powder diffraction data is separated into two independant steps: Profile Fitting and Least Squares Analysis. The method has been applied to the refinement of several crystal structures and has given low R-values, for example R = 0.37 %, 1.57 % and 1.76 % for silicon, corundum and quartz resp.

INTRODUCTION

The refinement and analysis of crystal structures from powder diffraction data, using either neutrons or X-rays, was first sucessfully applied in 1963 [1]. Integrated intensities of overlaping peak multiplets were used as the observed quantities. This is in contrast to the total profile fitting procedures, as it is used for example in the *Rietveld* program [2]. We have chosen to perform the analysis in two separate steps thereby avoiding possible bias between the profile model function and the actual refinement of structural parameters.

STEP 1: PFA = PROFILE FITTING ANALYSIS
         The Experiment gives G * S. With G known, the contribution of S = integrated intensity = I(hkl) is determined.

STEP 2: POWLS = POWDER LEAST SQUARES REFINEMENT
         Adjustment of structural parameters x_j, y_j, z_j, B_j, etc. to the observed intensities.

METHOD

Like in single crystal structure analysis we like to determine intensity values for each individual reflection. For this purpose the diffraction diagram is first analyzed by a profile fitting procedure yielding precise and unbiased integrated intensities of most of the individual peaks. In neutron diffraction the profile of the recorded diffraction maxima can be described by *Gaussian* functions. We have tested for a possible asymmetry in the *Gaussian* line shape, but found it to be negligible. The one variable which strongly depends on the experimental setup is the halfwidth of the diffraction maxima, FWHM. This variable in dependence of the scattering angle 2θ can be determined experimentally from a diffraction diagram of a standard sample. We have used Si from the NBS and we have fitted each peak individually with a *Gaussian* profile. According to *Cagliotti* [3] the FWHM vs. θ can be described by equ. (1):

$$FWHM = U \cdot tg\theta + V \cdot tg\theta + W \tag{1}$$

A least squares curve through the FWHM datapoints gave the constants U = 166, V = - 101, W = 38, which were then used in the following diagram analysis.

The goodnes of the profile-fit is judged in the conventional way by calculating a profile R-value R_{PF} for each peak or peak multiplett:(Table 1)

$$R_{PF} = \sqrt{\frac{\sum w \cdot (Y_i(obs) - Y_i(calc))^2}{\sum w \cdot Y_i^2(obs)}} \tag{2}$$

INTEGRATED INTENSITIES

The diffraction pattern contains through its integrated intensities the complete crystal structure information. The observed pattern, $Y(2\theta)$, is the true diffraction effect S of the specimen, modified by the geometrical and instrumental aberrations G. Mathematically Y is the convolution of G with S, plus the background BG

$$Y = G * S + BG \tag{3}$$

Parrish & Huang [4] have described an approach for the case of X-ray diffraction, by which they obtain the integrated intensities by deconvoluting $Y(2\theta)$ with the aid of measured values of the profiles with standard samples. This technique has recently been successfully appiied to several structure refinements using X-ray data [5]. It is also a good approach to analyze neutron diffraction data. Since G is known from the analysis of the standard sample, S can be calculated in a straightforward way by fitting *Gaussian* curves of known FWHM to the observations, STEP 1. Fig. 1 shows as an example the diffraction diagram of quartz with the resulting profile fitted curves inserted.

STRUCTURE REFINEMENT

The ensuing crystal structure refinement, STEP 2, is done using the program POWLS, POWder Least Squares [6]. The program is designed to refine on the unbiased set of observations determined by the profile fitting procedure from the overlapping powder diffraction maxima the atomic positional parameters, temperature factors and other parameters as needed (like magnetic moments). It is the special feature of POWLS that it can handle the cases of overlapping peaks which either cannot be separated intrinsically, like cubic 333/551, or in quartz h0l/0kl, or peaks which we wish to treat simultaneously.

The refinement in POWLS is carried out on a general quantity G(obs), which in general is the intensity I(hkl). POWLS forms and solves the matrix equation

$$(A^T PA)X = A^T PR \tag{4}$$

X is the (m x 1) dimensional matrix of the m parameter adjustments to be determined. R is the (n x 1) matrix of the n residuals (n > m)

$$r_i = g_i(obs) - g_i(calc) \tag{5}$$

P is the weight matrix, the inverse to the variance-covariance matrix of R, which in our case is determined by the profile fitting least squares adjustment. A is the matrix of the partial derivatives with respect to the parameter adjustments. A ^{T} is the transposed of A. For details and a write-up of the program see ref. 7.

EXAMPLES

POWLS has been used to analyze numerous samples in the past 15 years. In a new, modified version we now deconvolute the diffraction diagram into its contributions G and S by fitting *Gaussian* profiles to the observed data by least squares methods. This is similar to the approach which we have develloped for X-ray data [5]. Results on silicon Si, corundum Al_2O_3, and quartz SiO_2, are given below.

Silicon	R = 0.37 %
Corundum	R = 1.57 %
Quartz	R = 1.76 %

Table 1. Observed and calculated intensities of Si

HKL	I_{obs}	I_{calc}	I_{obs}-I_{calc}(%)	R_{PF}
111	78080	78766	0.88	0.022
220	88254	87936	0.36	0.015
311	63675	63730	0.08	0.026
400	21875	21825	0.23	0.015
331	36860	36724	0.37	0.022
422	58099	58212	0.37	0.011

Table 2. Atomic parameters of Al_2O_3

Al	x = 0.0	y = 0.0	z = 0.3526(3)	B = 0.06(6)
O	0.3065(4)	0.0	0.25	0.28(5)

Table 3. Atomic parameters of SiO_2 in comparision with single crystal measurements

Atom	this study	ref. 8	ref. 9	ref. 10	ref. 11
x(Si)	0.4686(10)	0.4697(1)	0.4699(1)	0.4697(2)	0.4698(3)
x(O)	0.4131(7)	0.4135(3)	0.4141(2)	0.4125(4)	0.4145(8)
y(O)	0.2661(8)	0.2669(2)	0.2681(2)	0.2662(4)	0.2662(7)
z(O)	0.1188(6)	0.1191(2)	0.1188(1)	0.1188(2)	0.1189(4)
B(Si)	0.44(13)	0.62(2)	0.55(2)	0.55(4)	0.41(6)
B(O)	0.80(6)	1.05(2)	0.98(3)	1.08(7)	0.98(10)

There is good agreement among various sets of positional parameters, including values from single crystal work [8,9,10,11,12]. Noteworthy is the good agreement between the temperature factors. The esd's derived from the total pattern methods are generally reported smaller than the ones calculated in POWLS, and other programs, by a factor of 2 - 3, but as has been discussed by *Cooper et al.* [13] the esd's in the total pattern routine are highly overestimated.

208

CONCLUSION

We have also used the Rietveld program routinely in the past, and we have come across many cases where the procedure of a two step analysis as presented here has proven to be supirior or at least more convinient. We consider our method true alternative. Since it avoids possible bias between the profile fitting and the actual refinement a two step analysis is liable to give more reliable data.

REFERENCES

1. G. Will, B.C. Frazer & D.E. Cox, Acta Cryst. 19, 854 (1965).
2. H.M. Rietveld, J.Appl.Cryst. 2, 65 (1969).
3. G. Cagliotti, A. Paoletti & F.P. Ricci, Nucl. Instr. Methods 3, 223 (1958).
4. W. Parrish & T.C. Huang, Proc."Symposium on Accuracy in Powder Diffraction", Natl. Bur. Stds., Gaithersburg, Maryland, USA, June 11 - 15, 1979.
5. G. Will, W. Parrish & T.C. Huang, ACA Spring Meeting 1981, College Station Texas, USA.
6. G. Will, J.Appl. Cryst. 12, 483 (1979).
7. G. Will, E. Jansen & W. Schäfer, REPORT JÜL - 1646, 1980, KFA Jülich.
8. L. Levien, Ch. Prewitt & D.J. Weidner, Am. Mineral. 65, 920 (1980).
9. Y. Le Page & G. Donnay, Acta Cryst. B32, 2456 (1976).
10. W.H. Zachariasen & H.A. Plettinger, Acta Cryst. 18, 710 (1965).
11. G.S. Smith & L. Alexander, Acta Cryst. 16, 462 (1963).
12. R.A. Young & B. Post, Acta Cryst. 15, 337 (1962).
13. M.J. Cooper, M. Sabate & K.O. Rouse, Proc."Symposium on Accuracy in Powder Diffraction", Natl. Bur. Stds., Gaithersburg, Maryland, USA, June 11 - 15, 1979.

ACKNOWLEDGMENT

This work was supported by the Bundesministerium für Forschung und Technologie, which is gratefully acknowleged.

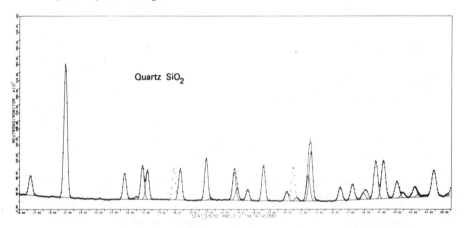

Fig. 1.   Diffraction diagram of Quartz. + are the experimental points, lines are the profile fitted curves. The remaining peaks come from the Al sample holder.

ANION DISORDER IN THE FAST-ION PHASE OF FLUORITES

M. T. Hutchings
Materials Physics Division, AERE Harwell,
Didcot, Oxon, OX11 0RA, UK.

ABSTRACT

Several neutron scattering techniques have been used to
investigate the nature of the dynamic disorder and diffusion processes which occur in four compounds with the fluorite structure at
high temperatures. The results of the experiments and the principal
conclusions are reviewed, and a model for the defect structure in
terms of fluctuating clusters of anions is discussed.

1. INTRODUCTION

Fast-ion conductors, also known as solid electrolytes or
superionic conductors, have been the object of considerable
experimental and theoretical research over recent years because of
their interesting physical properties and their technological use in
energy storage and other devices[1]. These materials are good conductors of electricity in which the charge carriers are ions rather than
electrons, and the conduction mechanisms are therefore particularly
amenable to study by neutron scattering techniques. Several classi-
fications of types of fast-ion conductor have been made by different
authors[2], but undoubtedly one of the simplest systems which exhibits
the phenomenon is provided by those compounds with the fluorite, or
antifluorite, crystal structure. The fluorite structure is shown in
figure 1, and may be most easily viewed as a simple cubic lattice of

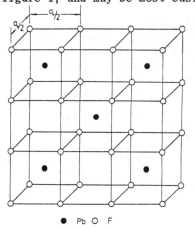

● Pb ○ F

Fig.1. The fluorite structure
of PbF_2.

anions with alternate cube centres
occupied by cations. We shall
refer to the f.c.c. unit cell of
side a_0 with a basis of a cation at
the origin and anions at $\pm$ ($\frac{1}{4},\frac{1}{4},\frac{1}{4}$)
positions.

Most fluorites exhibit a broad
Schottky-type anomaly in their
specific heat at a temperature T_c a
few hundred degrees below their
melting temperature T_m, which marks
the onset of a relatively high
dynamic-disorder of the anion sublattice[3]. In the region of T_c their
anion conductivity rises rapidly to
a value close to that of the molten
salt[4]. Typical values of T_c are
given in table 1.

In this paper the results of a
series of experiments using a
variety of neutron scattering

ISSN:0094-243X/82/890209-12$3.00 Copyright 1982 American Institute of Physics

techniques to investigate four fluorites, PbF_2, $SrCl_2$, BaF_2 and CaF_2, are reviewed. The experiments form part of a continuing collaborative programme of research, aimed at understanding the detailed mechanism of the conduction process in fluorites, involving Harwell, the Clarendon Laboratory, Oxford, Risö National Laboratory, Denmark, and the Institut Laue-Langevin (ILL) Grenoble. The simple nature of the fluorite lattice makes it especially suitable for theoretical work, and both static defect-energy calculations[5] and molecular dynamics computer simulations[6,7] have been made at Harwell, University College London, and elsewhere.

2.DIFFRACTION

Elastic Bragg diffraction gives a time-averaged probability density for the nuclear distribution within the unit cell, and for a well-ordered crystal this is usually represented by a model of ionic positions, with associated harmonic and anharmonic temperature factors to describe the mean effects of ionic vibrations. In the case of highly defective structures, such as fast-ion conductors, this form of analysis becomes difficult as various different models of the averaged ionic positions must be developed and tested, and high-order anharmonicity parameters may be needed[8,9]. Nevertheless this is the approach which has been adopted in the present work to give a first approximation to a representation of the nuclear probability density. Lack of a priori knowledge of extinction corrections makes the usual alternative approach, that of a direct Fourier transform of the data, inaccurate. However a third alternative has recently been suggested involving interpretation of data on fast-ion conductors directly in terms of atomic potential functions, and this may in future prove to be the most appropriate way in which to treat such data[10,11].

At low temperatures it is well established that the intrinsic defect is a Frenkel-pair of an anion interstitial in the cube-centre position and an associated vacancy at least more distant than on the next-nearest neighbour anion site[12]. Many calculations have been carried out on the assumption that such defects contribute to the main disorder in fluorites at high temperatures, and early analysis of diffraction data was based on these Frenkel pairs[13]. However both molecular dynamics calculations[6] and recent analysis of X-ray[14] and neutron scattering data[15] now indicate that the actual cube-centre site is not occupied for any length of time at high temperatures when the defect density becomes relatively large. The actual numbers of Frenkel pairs created at high-temperatures in the fluorites has been one of the main questions attracting attention over recent years. Early suggestions of a 'melting' of the anion sublattice with the number of defective anions approaching 100% above T_c are now discounted.

Both single-crystal and powder diffraction experiments have been carried out on the four compounds, but the single crystal data has proved more reliable since some powders exhibited marked preferred orientation at high temperatures. The measurements were made at several temperatures below and above T_c on small crystals of a few m.m. size encapsulated under vacuum. The Mark VI diffractometers at Harwell and the D8 or D15 diffractometers at ILL were used, with neutrons of wavelength $\sim 1.1\text{Å}$. The data were corrected for extinction and T.D.S. effects, and the intensities of approximately 25 symmetry-unrelated Bragg peaks were used in the least-squares fitting procedure except in the case of BaF_2 for which the data were less extensive[16].

Three basic models have been applied to fit the diffraction data[17], and their main features will be described here. Although all of these have not yet been fitted to the data for each compound the main conclusions are becoming clear. Each model allows for a fraction D of the anions to leave their regular sites and move to occupy sites within the unit cell, the average occupancy of which is determined. The thermal vibrations are allowed for by an isotropic temperature factor for each site, and in the case of the regular anion site an anharmonic temperature factor favouring motion towards the empty cube-centre[18]. We can describe the three models in terms of the occupancy, by the displaced anions, of two types of sites in the empty cube. These are the eight 'R' sites at $\pm(\frac{1}{2}-x), \pm(\frac{1}{2}-x), \pm(\frac{1}{2}-x)$, and the twelve 'I' sites at $\pm(\frac{1}{4}-y), \pm(\frac{1}{4}-y), 0$, etc., relative to the empty cube centre. The units are in a_o. The R sites are at positions $\pm(x,x,x)$ relative to the cations, and the 'I' sites are displaced from the mid-anion positions towards the empty cube centre by $\pm(y,y,0)$ etc. In model 'A' the interstitial anions occupy the R sites only, with D/4 at each site. In model 'B' one third of the displaced anions occupy the I sites and two thirds occupy the R sites, so that there are D/18 and D/6 anions in each I and R site respectively. In model 'C', equal numbers of displaced anions occupy each type of site giving D/12 and D/8 anions in each I and R site respectively.

At 295K the data for each compound are fitted well by the regular fluorite structure D=0. But as the temperature is increased all the models show D rises rapidly in the region of T_c, shown for PbF_2 in figure 2, as do the anion and to a lesser extent the cation temperature factors. Model A is found to fit the data quite well for all compounds, and gives x in the region $0.33-0.37$[15]. It should be noted that in the case of $SrCl_2$ early fits were made with the displaced anions at the cube-centre site, x=0.5, and D determined to be $\sim 3.5\%$[13]. However if x is allowed to vary a second minimum in the weighted R-factor[19] is found, giving an equally good fit with $x=0.37\pm0.1$ and D rising to 27% at the highest temperature[16]. Such a high correlation between x and D is not found for the other data.

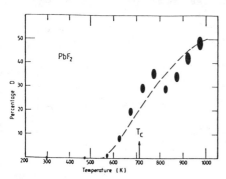

Fig.2. Fraction of anions, D%, leaving their regular site in PbF_2 as determined by models B and C. The values from Model A lie within the errors (ref.17).

Analysis in terms of model A indicates values of D at high temperatures in the range 23-49%, and for this model all these anions form Frenkel pairs. However at such concentrations, vacancies and interstitials would tend to be on adjacent sites, an unstable situation[12]. Furthermore estimates from specific heat data[3] suggest a maximum number of Frenkel defects of $\sim$ 10%, and Catlow has argued on static energy calculations, albeit based on cube-centre interstitials, that the number of defect pairs is unlikely to be larger than 10-20%[5]. Model B and C are based on a cluster picture of the defects in which a distinction is made between the true-interstitial anions of a Frenkel pair and neighbouring anions relaxed from their regular sites due to the presence of these interstitials. These models are suggested by the diffuse quasielastic scattering discussed in Section 4, which cannot be accounted for by model A with anions on the R sites alone. As described in Section 5 the clusters are centred on the mid-position of the regular-anion sites, with true interstitials located at I sites causing the nearest-neighbour anions to relax towards the empty cube-centres, to R sites. Model B has one true interstitial, and model C two true interstitials with two relaxed anions in each cluster (see figure 7).

In general models B and C give a significantly[19] better fit to the diffraction data than model A,[17] with similar values of D and x. The parameter y is found for PbF_2 to lie in the range 0.08 ± 0.01, somewhat smaller than the value deduced from the diffuse scattering (0.18). The regular-site temperature factors deduced from Models A-C differ only slightly. The values of D at the highest temperature of measurement, T_e, are summarised in table 1, the number of Frenkel pairs, or true interstitials, being D/3 or D/2 for models B or C respectively. At the moment it is not possible to distinguish

Table 1 Values of T_c and T_m for the four fluorites investigated together with values of D determined at T_e, and the maximum observed value of the coherent quasielastic scattering linewidths.

	T_c(K)	T_m(K)	Model	T_e(K)	D%	MAX $2\Gamma_{coh}$ (meV)
PbF_2	711	1128	B,C	973	49 ± 3	2.3
$SrCl_2$	1001	1146	B	1073	21 ± 6	0.65
BaF_2	1275	1628	A	1373	34 ± 6	3.2
CaF_2	1430	1696	B	1403	23 ± 3	2.6

between models B or C from the diffraction data, as the R-factors are very close, but model C gives the better account of the diffuse quasielastic scattering from CaF_2, as shown in Section 5. We can however conclude that the maximum number of Frenkel pairs occuring is likely to be $\sim$ 24%, in PbF_2 above T_c. It is interesting to note that simulated diffraction data from molecular dynamics calculations have recently been analysed in terms of the above models, and the best fits are given by model B with x = 0.35 and D $\sim$ 27%[10]. This work illustrates how important the definition of the defect is in making comparisons between different studies, since analysis of the same simulated data on a dynamic criterion indicates only 2-3% defective anions.

3.COHERENT SCATTERING FROM PHONONS

Measurements of the temperature variation of phonon excitations using inelastic coherent neutron scattering have been made on single crystals of PbF_2[20-23], $SrCl_2$[24], and BaF_2[24], and a weighted phonon frequency distribution has been measured using a powder sample of PbF_2[25]. These measurements were carried out on samples of a few cc in volume using the 3-axis spectrometers in PLUTO at Harwell, and IN2, 3, 4 and 8 at I.L.L. Grenoble.

In each case the optic and high-energy acoustic modes are found to broaden rapidly even below T_c[20,22,25]. In the case of PbF_2 for example the phonon dispersion relation was measured at 10K because the neutron groups were too broad at 295K to measure accurately[21]. All the optic modes are found to broaden in the same manner, and there appears to be no distinction in behaviour between modes involving only anion motion and those involving only cation motion in the harmonic approximation. No soft mode behaviour of the type predicted by Boyer[26] was observed. although the broadening of the modes tended to an overdamped situation making it difficult to determine the mode frequency[24].

A detailed examination of the temperature dependence of the energies and line-widths of long-wavelength acoustic modes has been made in PbF_2[23] and $SrCl_2$[24]. In the case of PbF_2 the elastic constants C_{12} and C_{44} fall steadily as the temperature is raised to above T_c; but C_{11} falls more rapidly and the anisotropy of the modes increases at T_c. This behaviour is attributed to the anion defects changing the long-and short-range inter-ionic forces in an additive manner for C_{11}, but in a manner which nearly cancels for C_{12} and C_{44}. The transverse acoustic mode line-widths, 2Γ(FWHM), in PbF_2 increase with the square of their wavevector q at all temperatures in the range 573-898K. To a good approximation $2\Gamma = dq^2$ where q is in $Å^{-1}$, for all three principal directions. This q^2 dependence indicates all the data were taken in the hydrodynamic regime. The

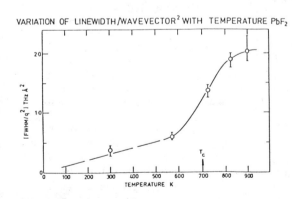

VARIATION OF LINEWIDTH/WAVEVECTOR^2 WITH TEMPERATURE PbF$_2$

Fig.3. Temperature variations of $2\Gamma/q^2$ for transverse acoustic phonons in PbF_2 (ref.23).

variation of the proportionality constant d is shown in figure 3. At low temperatures it rises linearly with temperature due to the dominant effects of anharmonicity, but near T_c it increases rapidly. This increase is attributed to a relatively low, ~10%, concentration of anion defects which are effectively static relative to the phonon frequencies, and cause a breakdown in the wavevector selection rules[27]. The dynamic effects of hopping anions are found to contribute much less to the line widths. This interpretation is supported by the coherent quasielastic data discussed below.

4. COHERENT QUASIELASTIC DIFFUSE SCATTERING

The most marked effect of the dynamic anion disorder on the neutron scattering spectrum is the appearance and build up of quasi-elastic diffuse scattering in the region of T_c. This was first observed in $SrCl_2$[28] and shown to be coherent in nature by chlorine isotopic substitution experiments. Confirmation of the coherent nature comes from its subsequent observation in CaF_2[29], PbF_2[30] and BaF_2[16,31], since fluorine only scatters coherently. This scattering is now established as characteristic of the fluorites in the fast ion phase. Its energy spectrum can be described to a good approximation by a Lorentzian shape function $S(\underline{Q}, \omega)$ centred on zero energy transfer $\omega = 0$, with energy width 2Γ (FWHM), as shown for CaF_2 in figure 4. Here $\underline{Q} = \underline{k}_i - \underline{k}_f$ and $\hbar\omega = E_i - E_f$ are the neutron wavevector and energy transfer respectively, where $\underline{k}_i$ (E_i) and $\underline{k}_f$ (E_f) are the initial and final neutron wavevector (energy) respectively. $S(\underline{Q}, \omega)$ is related by a space and time Fourier transform to $G(\underline{r}, t)$, the time-dependent total pair-correlation function, and it has been measured for single crystals of $SrCl_2$, CaF_2 and PbF_2 in some detail using 3-axis spectrometers at PLUTO, Harwell, and TAS1 and TAS7 at Risö. In general the energy-resolution of these instruments varied between 0.1 and 1 meV. The resolution function was convolved with $S(\underline{Q}, \omega)$ in fitting to the data, after allowance had been made for scattering from phonons and sample can, and for instrumental efficiencies.

The integrated intensity $S(\underline{Q}) = \int S(\underline{Q}, \omega) d\omega$ taken over the Lorentzian lineshape increases rapidly in the region of T_c as shown for CaF_2 in figure 5.

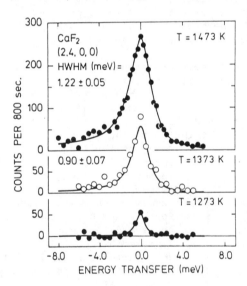

Fig.4. Typical scattering intensity as a function of energy transfer for CaF_2 at the point (2.4,0,0). The solid lines are best fits of a single Lorentzian shape-function plus background (ref.29).

Similar behaviour is found in $SrCl_2$[28] and PbF_2[30], in the former the intensity increases in parallel with the enthalpy increase due to anion disorder as deduced from specific heat data. $S(\underline{Q})$ has a very characteristic variation with $\underline{Q}$. It is most intense on an anisotropic shell of radius 2-3$\AA^{-1}$ and peaks in certain directions, as shown in figure 8 for CaF_2. $S(\underline{Q})$ for $SrCl_2$ is similar, with maxima in the same regions of reciprocal space, but the data are less extensive. $S(\underline{Q})$ for PbF_2 shows a somewhat wider shell than CaF_2, with less pronounced maxima.

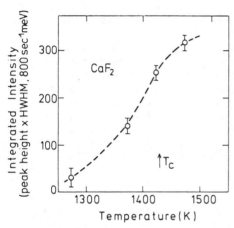

Fig.5. Integrated quasielastic intensity in CaF_2 at (2.4,0,0) as a function of temperature (ref.29).

$S(\underline{Q},\omega)$ has been examined in most detail for CaF_2 and PbF_2[30] along the [100] direction at points (2+q,0,0). The principal results may be summarised as follows: At the highest temperatures, above T_c, $S(\underline{Q})$ peaks in the region of (2.4,0,0). However, below T_c this peak occurs near (200) and moves to larger q as the temperature increases. The energy width 2Γ increases with temperature. In PbF_2, for which the width is almost independent of q in the range $0.15 < q < 0.35$, an activation energy ~ 0.5 eV may be deduced as shown in figure 6. The width shows a greater variation with q for CaF_2, and appears to tend to zero at the (200) point. No single activation energy can be deduced in this case.

It should be noted that although a single Lorentzian scattering function fits the coherent scattering data well, there is always a possibility of the existence of a further narrow component within the instrumental resolution. However in the case of $SrCl_2$ the scattering near (2.4,0,0) was investigated with the very good resolution (FWHM $\sim 1\mu eV$) of IN11 at I.L.L., and the only observable coherent scattering had a width $2\Gamma > 0.3$ meV[32].

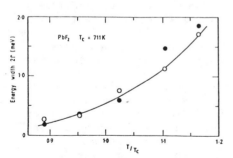

Fig.6. Variation of quasielastic scattering line-width with reduced temperature for PbF_2, at two values of wavevector. The solid line is calculated from an activation energy of 0.50 eV (ref.30).

5. INTERPRETATION OF THE COHERENT QUASIELASTIC SCATTERING

A number of general theoretical approaches to the calculations of $G(\underline{r},t)$ and $S(\underline{Q},\omega)$ for fast-ion conductors have now been given[33]. Before attempting to explain the finer details of the data a simple model is currently being developed which can account for their main features, in the first instance the distribution of $S(\underline{Q})$[30] as given by the instantaneous distribution of the disordered anions. The observed scattering can in principle contain contributions from both anion-anion and cation-anion correlations, but molecular dynamics calculations suggest the former dominate[34]. As mentioned in Section 2, we are led to consider various arrangements of the defective anions in clusters comprising true interstitials and relaxed neighbours. $S(\underline{Q})$ can be calculated relatively simply if the correlations between different clusters are neglected so that the scattering intensity from each can be summed. All possible configurations of the instantaneous cluster are averaged over. The simplest cluster model is that of Model B in Section 2, with one interstitial near the mid-anion position and two relaxed nearest neighbours. Better agreement with $S(\underline{Q})$ is obtained however from Model C, with two true interstitials near the mid-anion position and two relaxed neighbours. The position of the two associated vacancies is not critical. Such a cluster configuration is shown in figure 7, and is similar to the 2:2:2 clusters of anions previously proposed for Y-doped fluorites and hyperstoichometric UO_2[35]. Static energy calculations suggest such clusters could be stable in the fast ion phase of fluorites[36]. $S(\underline{Q})$ calculated from Model C is compared with the observed distribution for CaF_2 in figure 8. It is seen that the main features of an anisotropic shell is reproduced and the general agreement is very good. In the case of PbF_2 an estimate of the concentration of defects has been made by comparison of the experimental and theoretical ratio of the quasielastic and acoustic phonon scattering. This indicates at 678K about 20% of the anions are displaced from their regular sites, in very good agreement with the value from diffraction. Although we have no guarantee of the uniqueness of the cluster model, it seems very probable that in the complex dynamic disorder of the fluorites these 2:2:2 type clusters predominate in an instantaneous picture of the anions.

The dynamical nature of the fast-ion phase is reflected in the observed energy widths of $S(\underline{Q},\omega)$, the maximum values of which are summarised in Table 1. From the energy widths, the average cluster lifetime is $\sim 10^{-12}$ sec, that is larger than a typical phonon vibration period, as mentioned in Section 3. The decreases in the observed linewidth in CaF_2 near the (200) point suggest that the clusters appear and disappear in a manner that conserves the symmetry of the anion lattice - a salient feature of the cluster model centred on the mid-anion position. Calculations of $G(\underline{r},t)$ from first principles, for a system of clusters with a mean life-time for the cluster and a residence time of the interstitial ions which hop between the I sites, are currently being carried out[37].

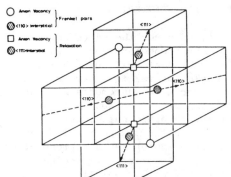

Fig. 7. Typical arrangement of two interstitials, vacancies and relaxed anions giving the best agreement with the experimental quasi-elastic data for CaF$_2$. The displacements of the interstitial ions is 0.18 lu, from the centre of the complex in <110> directions, and of the relaxed anions 0.11 lu from their normal sites in <111> directions, (ref. 30).

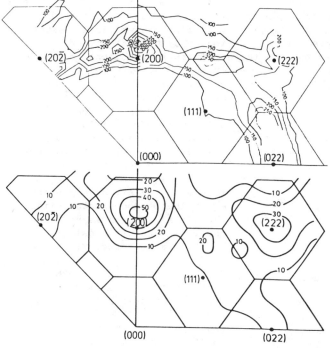

Fig. 8 Comparison of observed (upper) variation of S($\underline{Q}$) in CaF$_2$ with that calculated (lower) from the simple cluster model C (ref. 30).

Another approach to a detailed understanding of G($\underline{r}$,t) is by comparison with the results of molecular dynamics calculations[34]. Recently these have been analysed to give S($\underline{Q}$,ω), and do show a peak in S($\underline{Q}$) in the [100] direction close that observed.

6. INCOHERENT QUASIELASTIC SCATTERING

$SrCl_2$ offers a unique opportunity to study the anion self-diffusion directly by observation of the incoherent quasielastic scattering which relates to the time-dependent self correlation function $G_s(\underline{r},t)$. This is, in principle, easier to interpret than the total pair correlation function. Recent measurements have shown that the strontium incoherent scattering cross section is almost zero, 0.0 ± 0.2 barns, whereas that for chlorine nuclei is relatively large, ~ 7 barns[38]. Incoherent quasielastic scattering from a 3cc single crystal of $SrCl_2$ was first observed using the high-resolution spectrometer (1μV FWHM) IN10 at I.L.L., but was more conveniently studied using the 3-axis spectrometer IN12 with resolution of 27μeV FWHM[39]. At 987K and below no broadening of the incoherent peak was observable, but at 1053K its width had increased sufficiently to be measurable over a range of $\underline{Q}$ from 0 to 1.8 Å^{-1}. This is in a region in which the coherent-quasielastic diffuse scattering is weak. The lineshape of $S_i(\underline{Q},\omega)$ is well fitted by a single Lorentzian convolved with the resolution function, and the FWHM of the Lorentzian is plotted for the three principal directions in figure 9.

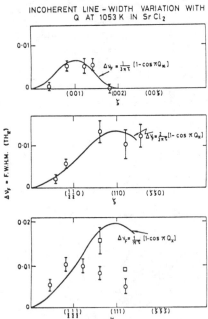

INCOHERENT LINE-WIDTH VARIATION WITH Q AT 1053 K IN Sr Cl₂

It should be stressed that these results are of a preliminary nature, and the data in the [111] direction are poor, however the data in [100] and [110] directions can be fitted quite well by a simple model[40] of instantaneous uncorrelated hops in < 100> directions between nearest-neighbour regular anion sites, or between equivalently coordinated sites. The hopping length is $a_o/2$ and the residence time estimated to be $\tau \sim 30 \times 10^{-12}$sec. This corresponds to a diffusion constant of 7×10^{-10} m^2sec^{-1} at 1053K, compared to the maximum value measured by tracer techniques of about $3 \times 10^{-9} m^2sec^{-1}$. Further experiments to measure this scattering in more detail are in progress. It is interesting to note that molecular dynamics calculations[6,7] on CaF_2 and on $SrCl_2$ suggest that diffusion in the fast-ion phase takes place via hopping along <100> between the regular sites.

Fig.9. Incoherent quasielastic scattering linewidth for $SrCl_2$ in 3 directions. The solid lines are calculated from the expressions shown with $\tau=30\times10^{-12}$ sec. (1THz = 4.14 meV) (ref.39).

7. SUMMARY AND CONCLUSIONS

The principal results of an investigation of coherent diffraction, inelastic scattering, and quasielastic diffuse scattering from
several fluorites in their fast-ion phase can be interpreted consistently in terms of a model of the dynamic anion disorder in which
simple fluctuating clusters of anions similar to the 2:2:2 type,
centred at the mid regular-anion position and having a lifetime of
$\sim 10^{-12}$sec, are the predominant configuration of defective anions.
In the case of $SrCl_2$ preliminary data are also available on the
incoherent scattering cross section, and these suggest that the main
diffusion process is a hopping between regular anion sites, or sites
with the same symmetry and spacing. The residence times deduced are
an order of magnitude larger than the lifetime of the clusters, and
one possible interpretation is that the incoherent scattering
reflects diffusion of vacancies whereas the coherent scattering is
dominated by the motion of interstitials and relaxed anions. The
combination of model calculations and molecular dynamics simulations,
and their comparison with the experimental data, should help to
clarify this conjecture, and lead to a full understanding of the
details of fast-ion diffusion in fluorites in the not too distant
future.

ACKNOWLEDGEMENTS

The collaboration of many colleagues in this work is gratefully
acknowledged. Part of this work was supported by a NATO Research
Grant.

REFERENCES

1. A.Hooper, Contemp. Phys. 19, 147 (1978).
2. W.Hayes, Contemp. Phys. 19, 469 (1980).
3. W.Schröter and J.Nölting, J.de Physique, Colloque C6, 41, C6-20
   (1980).
4. V.M.Carr, A.V.Chadwick and R.Saghafian, J.Phys.C11, L637 (1978)
   and references therein.
5. C.R.A.Catlow, Comments in Solid State Physics 9, 157 (1980).
6. M.J.Gillan and M.Dixon, J.Phys C 13 1901 and 1919 (1980).
7. G.Jacucci and A.Rahman, J.Chem Phys 69, 4117 (1978).
8. S.M.Shapiro and F.Reidinger, Physics of Superionic Conductors Ed.
   M.B.Salamon, Topics in Current Physics (Springer-Verlag, Berlin,
   (1979) p.45.
9. E.Perenthaler and H.Schulz, Solid State Ionics 2, 43 (1981).
10. A.B.Walker, M.Dixon and M.J.Gillan, Proc. Int. Conf. Fast Ion
    Transport in Solids, Gatlinburg 1981 (to be published),
11. H.Schulz, Proc. Int. Conf. Fast Ion Transport in Solids, Gatlinburg
    (to be published).
12. M.J.Gillan and D.D.Richardson, J.Phys.C12, L61 (1979).
13. M.H.Dickens, W.Hayes, M.T.Hutchings and C.Smith, J.Phys. C12, L97
    (1979).

14. K.Koto, H.Schulz and R.A.Huggins, Solid State Ionics $\underline{1}$, 355 (1980).

15. M.H.Dickens, W.Hayes, C.Smith and M.T.Hutchings, Fast Ion Transport in Solids, Eds.P.Vashista et al.(Elsevier North-Holland, 1979)p.225.

16. C.Smith, D.Phil Thesis Oxford (1980).

17. M.H.Dickens, W.Hayes, M.T.Hutchings and C.Smith, Harwell Report MPD/NBS/153, and to be published.

18. B.Dawson, A.C.Hurley and V.W.Maslen, Proc. Roy.Soc. A$\underline{298}$, 289 (1967).

19. W.C.Hamilton, Acta Cryst. $\underline{18}$, 502 (1965).

20. M.H.Dickens, W.Hayes and M.T.Hutchings, J. de Physique, Colloque C7, $\underline{37}$, C7-353 (1976).

21. M.H.Dickens and M.T.Hutchings, J. Phys. C$\underline{11}$ 461 (1978).

22. M.H.Dickens and M.T.Hutchings, Lattice Dynamics, Ed. M. Balkanski (Flammarion, Paris 1977) P.540.

23. M.H.Dickens and M.T.Hutchings, Neutron Inelastic Scattering 1977 (I.A.E.A. Vienna, 1978), p.285.

24. M.H.Dickens and M.T.Hutchings (private communication).

25. M.H.Dickens, M.T.Hutchings and J.B.Suck, Solid State Commun. $\underline{34}$, 559 (1980).

26. L.L.Boyer, Phys. Rev. Letters $\underline{45}$, 1858 (1980).

27. M.H.Dickens, W.Hayes, M.T.Hutchings and W.G. Kleppmann, J.Phys. C$\underline{12}$, 17 (1979).

28. M.H.Dickens, M.T.Hutchings, J.Kjems and R.E.Lechner, J.Phys. C$\underline{11}$; L583 (1978).

29. M.H.Dickens, W.Hayes, C.Smith, M.T.Hutchings and J.Kjems, Fast-Ion Transport in Solids, Eds. P.Vashishta et al.(Elsevier North Holland, 1979), p.229.

30. K.Clausen, W.Hayes, M.T.Hutchings, J.K.Kjems, P.Schnabel and C.Smith, Proc. Int. Conf. on Fast-Ion Transport in Solids, Gatlinburg 1981 (to be published).

31. M.H.Dickens, W.Hayes, C.Smith, M.T.Hutchings and R.E.Lechner, Phys. Letter $\underline{80A}$, 337 (1980). (Note the widths given here are half-widths at half maximum not FWHM as stated.)

32. M.H.Dickens, M.T.Hutchings and F.Mezei (private communication).

33. W.Dieterich, P.Fulde and I.Peschel, Adv. Phys. $\underline{29}$, 527 (1980).

34. M.J.Gillan and M.Dixon, J. Phys. C$\underline{13}$, L835 (1980).

35. See for example A.K.Cheetham, Chemical Applications of Neutron Scattering, Ed. B.T.M.Willis (Oxford University Press, 1973) p.225.

36. C.R.A.Catlow and W.Hayes (private communication).

37. K.Clausen (private communication).

38. F.Mezei (private communication).

39. M.H.Dickens, M.T.Hutchings, R.E.Lechner, B.Renker and P.Schnabel (private communication).

40. C.T.Chudley and R.J.Elliott, Proc. Phys. Soc. (London) $\underline{77}$, 353 (1961).

NEUTRON SCATTERING FROM ORIENTATIONALLY DISORDERED SOLIDS

B.M. Powell, V.F. Sears and G. Dolling

Atomic Energy of Canada Limited, Chalk River, Ontario, KOJ 1JO, Canada

ABSTRACT

In an orientationally disordered solid the equilibrium positions of the molecular centres of mass lie on an ordered three-dimensional lattice, but the molecular orientations are disordered. The effects of this disorder can be observed in the scattered neutron distributions in several ways. The intensity of the Bragg reflections, particularly those with high Miller indices, is greatly reduced. Intense, diffuse scattering appears which may be distributed throughout reciprocal space. Well-defined librational phonons are difficult to observe and the translational phonons are often very broad. A brief outline of the theory of neutron scattering from orientationally disordered solids is given and the information which may be derived from neutron scattering is discussed with reference to results on CBr_4, SF_6 and $\beta-N_2$.

INTRODUCTION

In an ordered crystalline molecular solid the molecules have both well-defined equilibrium positions and well-defined equilibrium orientations. The molecules lie on a three-dimensional lattice and execute small amplitude translational and librational displacements about their equilibrium positions and orientations respectively.

If the molecular displacements become large or the molecules are disordered, then both elastically and inelastically scattered neutrons are affected. The intensity of the elastic (Bragg) scattering at the reciprocal lattice points is greatly reduced and diffuse scattering appears which is distributed throughout reciprocal space. The peaks in the inelastic scattering (due to phonons) have a large spread in frequency and particular phonons may become unobservably broad.

In this paper we are concerned with a particular type of molecular disorder in which the equilibrium positions of the molecular centres of mass form a well-ordered three dimensional lattice, but the molecular orientations are disordered. This disorder may also be accompanied by large amplitude translational displacements of the centres of mass. Orientational disorder of this type usually occurs in solids formed by molecules of generally "globular" shape and often results in a crystal symmetry which is cubic.

THEORY

The theory of elastic neutron scattering from orientationally disordered crystals has been considered by several authors. Here we summarize the derivation due to Dolling et al.[1] The scattered neutron intensity in a diffraction experiment is proportional to the static structure factor $S(\vec{Q})$, and

$$S(\vec{Q}) = S_c(\vec{Q}) + S_d(\vec{Q}) \tag{1}$$

$\vec{Q}$ ($|\vec{Q}| = (4\pi \sin\phi/2)/\lambda)$ is the neutron wave-vector transfer, ϕ is the scattering angle and λ is the neutron wavelength. The coherent elastic (Bragg) scattering $S_c(\vec{Q})$ arises from the mean value of the total scattering amplitude of the crystal, while the diffuse scattering $S_d(\vec{Q})$ arises from the fluctuations in the scattering amplitude. For a crystal with cubic symmetry, the coherent scattering can be written as:

$$S_c(\vec{B}) = |\sum_m <\beta_m> \exp(-\frac{1}{6}B^2 <U^2_m>) \exp(i\ \vec{B}\cdot\vec{r}_m)|^2 \tag{2}$$

where $\vec{B}$ is the reciprocal lattice vector for the Bragg reflection. $<U^2_m>$ is the mean square translational displacement of the centre of mass of molecule m. $<\beta_m>$ is the rotational form factor for molecule m. For "normal" crystals $<\beta_m>$ may be evaluated as a cumulant expansion in the molecular librational angle. For orientationally disordered crystals an expansion of $<\beta_m>$ in terms of symmetry-adapted functions is more appropriate. Assuming no correlation between the translational and librational displacements and assuming the molecule has only one "shell" of atoms we can write:

$$< \beta_m^{(o)} > = 4\pi\ b_{ma} \sum_\ell i^\ell j_\ell(Br_a)C_\ell K_\ell(\Omega_B) \tag{3}$$

$j_\ell(Z)$ is the ℓth order spherical Bessel function
$K_\ell(\Omega_B)$ is the ℓth order Kubic Harmonic
Ω_B represents the polar coordinates of B
C_ℓ are expansion coefficients
b_{ma} is the coherent neutron scattering length of atom "a" in molecule "m".
r_a is the radius of the "shell" of atoms "a".

The zeroth order term ($\ell=o$) describes atom "a" as uniformly distributed over a spherical shell of radius r_a. This corresponds to a randomly oriented molecule. High order terms represent localization of atoms "a" on this spherical shell and the degree of localization due to the ℓth harmonic is specified by the magnitude of C_ℓ.

One of the approximations made in the above derivation is the neglect of correlations between the translational and librational displacements. For larger molecules (e.g. CBr_4) this approximation is probably not valid. Press and al.[2] have given the formalism necessary for the inclusion of such correlations. They show that $<\beta_m^{(o)}>$ is the leading term in an expansion. For the case of tetrahedral molecules at a cubic site (appropriate for CBr_4) the leading correction term is:

$$\beta_m^{(1)}> = < \frac{1}{3} U^2_m> Bj_3(Br_a)\ c_3^{(1)}K_4(\Omega_B)$$
$$+ < \frac{1}{3} U^2_m> Bj_7(Br_a)[c_{71}^{(1)}\overline{K}_1(\Omega_B) + c_{72}^{(1)}\overline{K}_2(\Omega_B)]+\ldots \tag{4}$$

where $c^{(1)}$ are additional expansion coefficients and $\bar{K}_1$, $\bar{K}_2$ are particular linear combinations of Kubic Harmonics.

The diffuse scattering term $S_d(\vec{Q})$ has two contributions: one is due to the centre of mass translational displacements and the other due to fluctuations in β_m. If we again neglect translational-rotational correlations we can write

$$S_d(\vec{Q}) = (\Delta\beta)^2 + |<\beta>|^2 (1-\exp(-2W)) \qquad (5)$$

where $\exp(-2W)$ is the translational Debye-Waller factor and $(\Delta\beta)^2$ is the fluctuation term.

With certain modifications, the above formalism can be applied to crystal symmetries other than cubic. In particular, for the case of hexagonal close packed crystals the primitive unit cell contains two molecules rather than the single molecule assumed for the cubic case. The mean square translational displacement $<U^2>$ is not isotropic but has components $<U^2_{||}>$, $<U^2_{\perp}>$ parallel and perpendicular respectively to the hexagonal axis. Finally, the symmetry adapted functions in which $<\beta_m>$ is expanded must have hexagonal rather than cubic symmetry.

EXPERIMENTAL RESULTS

(a) Carbon tetrabromide

Carbon tetrabromide (CBr_4) has an orientationally disordered phase from its melting point (365 K) down to 319 K. In this phase the centres of mass of the tetrahedral molecules lie on a face centred cubic lattice. Neutron diffraction measurements were made on a polycrystalline sample of CBr_4 in its disordered phase by Dolling et al.[1]. The discrete Debye-Scherrer peaks were observed to lie on an intense, oscillatory background of diffuse scattering. The data were analysed with equations 2,3 and 5, but the theory could not describe an intense diffuse peak observed at $|\vec{Q}| \sim 2.1\text{Å}$. An extensive set of Bragg intensities was obtained by More et al.[3] from a single crystal and were analysed by Press et al.[2] to examine the effects of rotational-translational correlation. Their results are given in Table I.

Table I Parameters of the fits to the experimental Bragg intensities from CBr_4 at 325 K (from ref. 2)

$R_W(\%)$	$<U^2>$ (Å^2)	$c_4^{(0)}$	$c_6^{(0)}$	$c_8^{(0)}$	$c_{31}^{(1)}$	$c_{71}^{(1)}$	$c_{72}^{(1)}$
7.7	0.540 (60)	0.06 (2)	-0.83 (7)	0.14 (14)			
5.1	0.531 (36)	-0.12 (4)	-0.75 (5)	0.33 (11)	0.04 (1)		
3.4	0.576 (30)	-0.37 (9)	-0.57 (8)	0.85 (17)	0.10 (2)	0.05 (1)	-0.13 (4)

If correlations are neglected ($R_W=7.7\%$) the parameters are in good agreement with those of Dolling et al.[1] and More et al.[3]. They show a very large translational displacement $<U^2>$ and the relative magnitudes of the coefficients of the Kubic Harmonics, C_ℓ, show the four Br atoms are distributed over the twelve <110> directions. If rotational-translational correlation is included the R-factor imroves and the existing parameters re-adjust in magnitude. The additional parameters introduced by this correlation are all small. It is clear that extremely high quality structural data are required if the rotational-translational correlation is to be specified accurately.

With the same crystal More et al.[3] measured the distribution in reciprocal space of the diffuse scattering peak observed at $|\vec{Q}| \sim 2.1\text{Å}$ in polycrystalline CBr_4 by Dolling et al.[1]. Some results are shown in fig. 1. The scattering is highly anisotropic, with equi-intensity contours elongated along the [111] directions lying in the plane perpendicular to [220], but not along the [111] directions which project out of this plane. The shape of the scattering is very similar to that observed by Hüller and Press[4] in CD_4, but, in contrast to the behaviour there, no temperature dependence was observed in CBr_4.

The molecule of CBr_4 is much larger, relative to the unit cell, than the molecule of CD_4. Consequently steric hindrance is expected to be important in determining correlations between neighbouring molecules. Coulon and Descamps[5] have developed a statistical model to take account of steric hindrance and have calculated the diffuse scattering in CBr_4. They show that, due to steric hindrance, only three relative configurations of a CBr_4 molecule and its nearest neighbour molecule can be distinguished. The expression for $S_d(\vec{Q})$ contains a sum over the normalised probability for a particular set of molecular configurations and Coulon and Descamps[5] evaluate this function by the method of "weak graphs". Their calculated diffuse scattering contours are

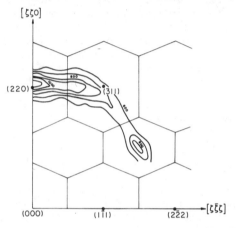

Fig. 1. Diffuse elastic scattering in the $(1\bar{1}2)$ plane of CBr_4 at 325K (from ref. 3.)

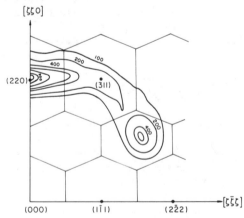

Fig. 2. Calculated diffuse scattering in the $(1\bar{1}2)$ plane of CBr_4 (from ref. 5.)

shown in fig. 2. The agreement with the observed distribution of fig. 1 is clearly quite good. Both the shape of the diffuse distribution and its intensity agree well with observation. Despite the simplifying assumptions of discrete molecular orientations and strict steric hindrance (i.e. hard sphere atoms) the agreement suggests that steric hindrance is important in the disordered phase of CBr_4.

(b) Sulphur hexafluoride

Another molecule which exhibits an orientationally disordered phase is sulphur hexafluoride (SF_6). In this phase, extending from the melting point (223 K) to 96 K, the molecular centres of mass lie on a body centred cubic lattice. Previous neutron diffraction measurements have been made on this phase of SF_6 with a polycrystalline sample close to the transition at 96 K. No diffuse scattering was observed and the intensities of the Debye-Scherrer lines were analysed with equation 3. More recent diffraction measurements have been made at 200 K and the results are shown in fig. 3. The experimental distribution shows the expected Debye-Scherrer lines, with the higher index lines being rather weak. But these discrete lines are superposed on an intense background of diffuse scattering which shows the now familiar distinct oscillations arising from orientational disorder.

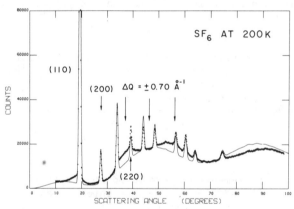

The intensities of the Debye-Scherrer peaks were fitted by the expressions given in equations 2 and 3. The resulting parameters are compared with those at 100 K in Table II.

Fig. 3. Neutron diffraction pattern from polycrystalline SF_6 at 200 K. The solid line is calculated as explained in the text.

Table II Comparison of the parameters at 100 K and 200 K for polycrystalline SF_6 in its orientationally disordered phase. The parameters at 100 K are from ref. 1.

TEMPERATURE	$\langle u^2 \rangle (\mathring{A}^2)$	C_4	C_6	C_8
100 K	0.084 (6)	1.36 (3)	0.36 (3)	0.73 (5)
200 K	0.195 (20)	1.26 (4)	0.25 (11)	0.52 (34)

The mean square translational displacement at 200 K is much larger than the value at 100 K, while the expansion coefficients C_ℓ are all smaller at 200 K. This implies a greater tendency for free molecular rotation at 200 K.

The parameters at 200 K have been used to calculate the diffuse structure factor $S_d(|\vec{Q}|)$ from equation 5. The total structure factor $S(|\vec{Q}|)$ was evaluated and convoluted with the experimental resolution. The result, after normalising to the integrated intensity of the experimental distribution, is shown as the solid line in fig. 3. The discrete lines are described very well and the calculation shows oscillations in the diffuse background. The calculated oscillations arise from the molecular form factor, but the experimental distribution shows two peaks displaced from the broad calculated peak at $\phi \sim 47°$. The presence of these observed peaks is evidence for orientational correlations, which are neglected in the calculations. The separation of the observed peaks from the calculated peak ($\Delta|\vec{Q}| \sim 0.7 Å^{-1}$) can be viewed as an indication of the range of the correlations. This correlation range is $\sim 9 Å$, equivalent to ~ 1.5 unit cells of SF_6.

(c) β-nitrogen

Solid nitrogen has an orientationally disordered phase (β-phase) from its melting point at 63 K down to 36 K (at zero pressure). The structure of $β-N_2$ was first investigated by X-ray diffraction and the centres of mass of the linear molecules were found to lie on an hexagonal close packed lattice[6]. However, the orientational disorder of the molecules could not be unambiguously determined from the X-ray data. Press and Hüller[7] suggested that an interpretation of the data in terms of symmetry adapted harmonics might give a clearer picture of the disorder.

Neutron diffraction measurements were made on polycrystalline samples at pressures of 0, 0.55 and 2.5 kbar[8]. The intensities of the Debye-Scherrer peaks were analysed by equation 3 suitably modified for application to hexagonal crystals.

Table III Coefficients of the hexagonal harmonics and mean square translational displacements ($Å^2$) for $β-N_2$ at 4 kbar and 55 K.

$\langle u^2_{||}\rangle (Å^2)$	$\langle u^2_{\perp}\rangle (Å^2)$	c_{20}	c_{40}	c_{60}	c_{66}
0.056 (6)	0.055 (5)				
0.056 (7)	0.055 (6)	-0.008 (32)			
0.056 (7)	0.055 (6)	-0.011 (34)	-0.044 (126)		
0.056 (7)	0.055 (6)	-0.012 (35)	-0.028 (140)	0.219 (875)	
0.055 (6)	0.056 (6)	-0.012 (35)	-0.033 (142)	0.178 (902)	-0.211 (1014)

The values determined for the expansion coefficients indicated that, despite its linear shape, the N_2 molecule has a spherically symmetric orientational distribution. In addition to the discrete lines intense diffuse scattering was also observed. The oscillations in this scattering suggest significant orientational correlations exist between a molecule and its nearest neighbours at all these pressures.

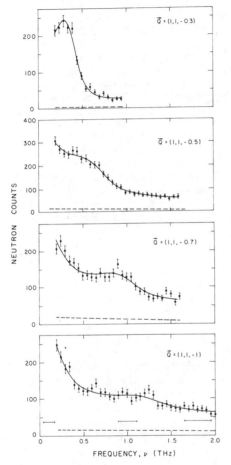

Fig. 4 Comparison of experimental and fitted scattered neutron distributions for TA phonons propagating along C in β-N_2 at 4kbar and 55K. The wavevector transfers $\vec{Q}$ are given in units of the reciprocal lattice cell. The dashed lines show the background from the Aℓ pressure cell and the horizontal bars show the experimental resolution.

In an attempt to determine the expansion coefficients more accurately a single crystal of β-N_2 was grown at 55 K and a pressure of 4 kbar. Integrated intensities of 34 Bragg reflections were measured in the (hoℓ) and (hhℓ) planes and were again analysed with equation 3. The quality of the fit is reasonable (R-factor = 9.6%) and the parameter values are shown in Table III. It is clear that none of the expansion coefficients can determined. The conclusion from the polycrystalline data remains unchanged - the orientational probability distribution for the N_2 molecule is essentially spherical, with an hexagonal distortion no greater than 1%.

Measurements were also made to locate regions of intense diffuse scattering in the planes (hoℓ) and (hhℓ). In particular, a search was made around the point in reciprocal space (1.75, 1.75, 0.5) at which a strong diffuse peak was observed as a function of temperature at zero pressure by Kjems and Dolling[9]. Only weak diffuse scattering was observed in this region of reciprocal space in either of the two accessible scattering planes.

Measurements of the inelastic neutron scattering were also made on this single crystal to determine the phonon dispersion curves along the major symmetry directions at 4 kbar. As in the case of β-N_2 at zero pressure[9] no well-defined librational phonons could be observed, and although translational acoustic phonons could be observed at small wave-vectors, near the zone boundary even these excitations become broad, weak and poorly defined. Some of the measured scattered neutron groups corresponding to transverse acoustic modes propogating along the hexagonal axis are shown in fig. 4.

A theory to analyse scattered neutron distributions such as these has been developed by Michel and Naudts[10]. The translational centre of mass displacements are treated conventionally, but the orientational potential is expanded in terms of symmetry-adapted functions and a rotation-translation interaction term is included. The orientational "displacements" are taken to be diffusive in nature and exponential decay of the rotational correlation functions is assumed. The neutron scattering cross-section derived from the theory was convoluted with the experimental resolution, and fitted to the data shown in fig. 4. All parameters in the cross-section could not be varied simultaneously, because the strong correlations between several parameters cause instabilities in the fitting procedure. The results of the fitting with the phonon-phonon damping fixed at zero are shown as the solid line in fig. 4. The agreement between the observed and fitted distributions is clearly quite good and the values of the best-fit parameters are shown in an extended zone scheme in fig. 5. We see that λ is always less than or equal to ω_o and so, at least for this particular phonon branch, the excitations are in the "slow-relaxation" regime i.e. the orientational correlations are long lived compared with the phonon frequency. The theory predicts the existence of a central-peak in this regime, but elastic scattering from the Aℓ pressure cell prevents direct observation of the peak.

The neutron scattering results thus suggest that in the β-phase at 4 kbar the N_2 molecules has a spherical orientational distribution but that pronounced and long-lived correlations exist between neighbouring molecules.

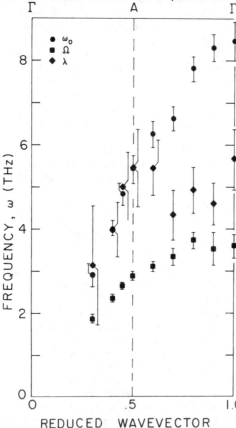

Fig. 5 The wave-vector dependence of the phonon frequencies ω_o, Ω and of the relaxation frequency λ for orientational correlations.

CONCLUSIONS

We have discussed the effects which orientational disorder in a crystal may have on its neutron scattering distribution using data from CBr_4, SF_6 and $\beta-N_2$ as examples. A formalism exists with which the experimental intensities of Bragg reflections (or Debye-Scherrer powder peaks) can be analysed to give information not only about the orientational distribution function of a molecule but also about its rotational-translational correlations. The diffuse scattering component also contains information about the orientational correlations among neighbouring molecules. But to derive quantitative information from detailed experimental intensity measurements of the diffuse scattering a specific model must be assumed for the form of the correlations. Inelastic neutron scattering provides further information about orientational correlations, particularly their time scale compared with that of the phonon frequencies. Theoretical work to derive the phonon lineshapes for alternative models of the rotation-translation interactions would be very desirable.

REFERENCES

1. G. Dolling, B.M. Powell and V.F. Sears, Mol Phys. 37, 1858 (1979).
2. W. Press, H. Grimm and A. Hüller, Acta Cryst. A35, 881 (1979).
3. M. More, J. Lefebvre, B. Hennion, B.M. Powell and C.M.E. Zeyen, J. Phys. C. (Solid State) 13, 2933 (1980).
4. A. Hüller and W. Press, Phys. Rev. Lett. 29, 266 (1972).
5. G. Coulon and M. Descamps, J. Phys. C. (Solid State) 13, 2847 (1980).
6. W.E. Streib, T.H. Jordan and W.N. Lipscombe, J. Chem. Phys. 37, 2962 (1962).
7. W. Press and A. Hüller, J. Chem. Phys. 68, 4465 (1978).
8. B.M. Powell, H.F. Nieman and G. Dolling, Chem. Phys. Lett. 75, 148 (1980).
9. J.K. Kjems and G. Dolling, unpublished.
    G. Dolling, Proc. Conf. on Neutron Scattering ed. R.M. Moon CONF-76-601-P1 (National Technical Information Service). P. 263 (1976).
10. K.H. Michel and J. Naudts, J. Chem. Phys. 68, 216 (1978).

Neutron Scattering Experiments on Partially Disordered Materials

C. Riekel

Max-Planck-Institut für Festkörperforschung

Heisenbergstrasse 1, D - 7000 Stuttgart 80

ABSTRACT

Neutron scattering methods to characterize intercalation compounds
of layered, conducting compounds are presented. It is shown that
complex structural and chemical processes exist, which must be
understood before an "in depth" study of selected systems can be
started.

1. Introduction

In this review, neutron scattering experiments on electronically
conducting materials, capable of taking up atoms or molecules
from gas- or liquid-phase are mentioned. Generally, the "host"-
lattice may have a one-, two- or three dimensional frame work.
Examples will, however, be restricted to layered host-lattices.
A more general review on the chemistry of such systems is given
in ref.1.

2. Materials

Graphite, the prototype material, consists out of regularly
stacked carbon planes. Donors (e.g. K) or acceptors (e.g. AsF_5)
may be inserted (intercalated) between these planes which implies
an electronic interaction[2]. A second class of compounds, capable
of intercalation, are transition metal disulphides[1]. In these
materials, 2D-slabs of two sulphur layers, sandwiching a metal
layer, are stacked in c-direction.
The structural processes occuring during intercalation are not
well understood as they are usually deduced via a post-mortem
analysis of stable or meta-stable phases. This statement holds for
most solid state reactions.

ISSN:0094-243X/82/890230-09$3.00 Copyright 1982 American Institute of Physics

231

3. Real-Time Diffractometry

Real-time neutron diffractometry has been developed to study the
dynamics of the structural evolution[3]. Due to the reduction of
measuring time per spectrum, few reflections are observed and
the main application is the establishment of phase-limits via
low angle reflections. In addition, kinetic information is
obtained from the temporal variation of selected reflections.
Model calculations, although limited, are still useful for
complicated phase-sequences or highly disordered phases.

4. Discussion of Specific Systems

4.1 Staging during Graphite Reactions

In fig.1, the development of the intensity along c^* ($\perp$ to the
planes), during the reaction of a 100 mm^3 graphite-crystal (HOPG)
with potassium-vapour, at 280°C is shown[5]. Spectra of 30 sec.
each were recorded by a 1D-multidetector while "rocking" the
crystal stepwise around an axis, normal to c^*.

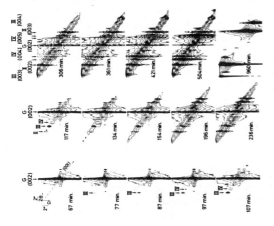

Fig.1 Temporal develop-
ment of intensity along
c^* during the reaction
$K + xC \to KC_x$.
Measuring time per Ω-
step = 30 sec. 17 Ω-steps,
i.e. 9 min./frame.
The intensity is given in
logarithmic scale between
$10^2 - 4 \cdot 10^5$ counts.

Satellites, due to different KC_x-phases, appear next to the
graphite-(002) reflection.
X-ray data, obtained under equilibrium conditions, suggest the
existence of a regular series of phases, called "stages", where
n, n-1, n-2 ... 1 graphite planes separate every two intercalate
layers (n_{max} = 12, ref.6). Under dynamic conditions, however,
a regular sequence of stages develops only for n $\leq$ 5. This may
be explained by intergrown domains of different stages at the
beginning of the reaction as observed for the graphite/FeCl$_3$
system by electron microscopy[7]. A hitherto unobserved feature
is, however, the crystallisation of stages n $\geq$ 3 at an early
time and in a reverse way than expected. A closer study of this
phenomenon might give more information on the domain interaction.
Graphite intercalation reactions are also studied for practical

reasons in order to find the optimum conditions for the fabrication of large d-spacing monochromators[8].

4.2 Staging during Reactions of Transition Metal Dichalcogenides

Real time neutron diffractometry has also been used to study inter-calation reactions of NbS_2- and TaS_2-powders. Staging, as found for graphite intercalation, has been observed for most systems[9,10,11]. The structural and chemical complexity of such reactions will be shown for the electrochemical formation of $K_x(D_2O)_yMeS_2$ acc. to:

$$xe^- + xK^+ + yD_2O + MeS_2 \rightarrow K_x^+(D_2O)_y[MeS_2]^{x-} \qquad (1)$$

This reaction can be easily studied under galvanostatic conditions, using compressed powder electrodes[11,12]. The homogenity of the charge transfer may be shown via the linear increase of the a-axis during a real time neutron diffraction experiment (fig.2, time-resolution: 15 min./spectrum)

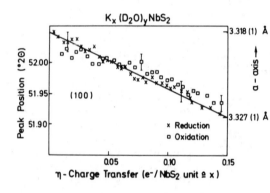

Fig.2 Change of a-axis during formation of $K_x(D_2O)_yNbS_2$. Time-resolution: 15 min./spectrum. The time-scale has been converted into charge-transfer NbS_2 unit.

This increase is due to an increase of the electron concentration in the nonbonding $Me-d_{z^2}$ orbitals which leads to a repulsion of neighbouring metal atoms[13,14]. A similar expansion has been observed for the reaction of stage-2 KC24 with deuterobenzene (C_6D_6)[15] and which may be interpreted as a charge-transfer from the benzene rings to the graphite planes.

Characteristic c-axis reflections, indicating consecutive phases, were found for the $K_x(D_2O)_yTaS_2$ formation (P1, P2 in fig.3).

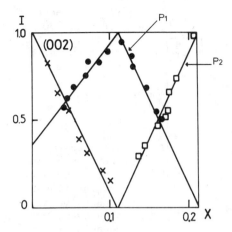

Fig.3 Observation of reflections P1 and P2, due to consecutive phases, during electrochemical $K_x(D_2O)_y TaS_2$ formation. Individual reflections normalized to 1.0 at maximum intensity. Time resolution: 15 min./spectrum[3,11].

Acc. to fig.3, the P1-phase is transformed into the P2-phase which is not the case for TaS_2 and the P1-phase. A closer analysis of position and width of the P1-reflection suggests that it "hides" two successive and closely neighbouring reflections. These are attributed to distinct phases as a constant interlayer separation is derived from each reflection[11]. A schematic model, shown in fig.4, suggests that the reaction starts with stage 3 and proceeds to stage 1.

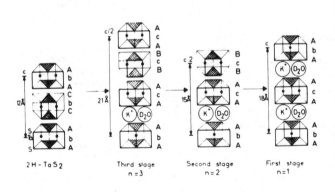

Fig.4 Model of the structural evolution during $K_x(D_2O)_y MeS_2$ formation[11]. It is based on the lattice spacings of (ool) reflections (P1, P2 in fig.3), and the known dimensions of intercalate- and host lattice-layers in c-direction.

Two further, nonstoichicmetric phases (I and II) exist in the range of the stage 1 phase up to x = 1/3 (fig.5)[3].

234

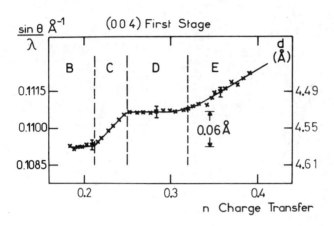

Fig.5 Change of peak-position of stage 1 (004) reflection (P2 in fig.3) upon $K_x(D_2O)_yTaS_2$ formation[11]. Two nonstoichiometric phases (I, II) are identified via a constant interlayer separation. Note that the interlayer separation decreases with increasing K^+-content.

Model calculations, based on the (002)/(004) intensity ratio, suggest that both phases observe a stochiometry $K_x(D_2O)_{2/3}TaS_2$ (fig.6).

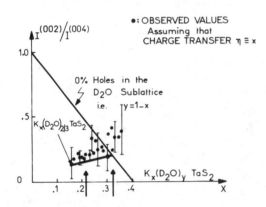

Fig.6 Observed and calculated (solid lines) ratios of (002)/(004) reflections. At $x = 1/3$, all lattice sites in the intercalate layers are assumed to be occupied, i.e. $K_{1/3}(D_2O)_{2/3}TaS_2$. For $x < 1/3$, two models are considered with (i) every lattice site occupied, i.e. $K_x(D_2O)_{1-x}TaS_2$ and (ii) vacancies, depending on x, i.e. $K_x(D_2O)_{2/3}TaS_2$[11]. Time resolution: 15 min./spectrum.

As the vacancy concentration depends on x, the intercalate layers must be stabilized by the $K^+...D_2O$ interaction. Geometric arguments suggest that condensed $K(D_2O)_6$ clusters exist for phase II at $K_{1/3}(D_2O)_{2/3}TaS_2$. Each D_2O molecule belongs to the solvation shell of 3 different cations. In contrast, isolated $K(D_2O)_6$ clusters may exist for $K_{0.15}(D_2O)_{2/3}TaS_2$ in the range of phase I. As the $K^+...$ D_2O distance in crystalline hydrates is somewhat smaller than the distance of two lattice sites on the sulphur planes, an incommensurate to commensurate transition with increasing x, might explain the stepwise decrease of the interlayer separation (fig.5).

4.3 Single Crystal Structural Studies

Intercalation of single crystals often produces a considerable misalignment of the host lattice planes, which makes 3D-data collection difficult. Although the orientation of molecular intercalates may be determined from (ool) reflections[16], this method is of limited use for studies of the intercalate/intercalate and intercalate/host-lattice interactions.

Electrochemical intercalation appears to be the best way to obtain crystals with sufficiently aligned planes. Data collection remains, however, tedious and only few results have been obtained until now. Hydrogen positions in the metal sublattice were found for $H_{0.76}NbS_2$[12] which may be obtained by electrochemical reduction in sulphuric acid[17]. The hydrogens are located in the centers of Nb_3-equilateral triangles.

A single crystal study has also been done for $K_x(D_2O)_yNbS_2$ with $x > 1/3$ at 145 K[18]. Real time neutron diffractometry suggests that for $x > 1/3$ a further phase with smaller interlayer separation is formed (fig.5). This agrees to electrochemical results[11]. The unit cell is shown in fig.7. Considerable disorder in the intercalate layers is evident from the high, apparant thermal mobility. Furthermore K^+ and O can't be separated. The deuterons are located in the middle plane between the NbS_2 slabs. This contrasts single crystal NMR experiments on $K_{1/3}(H_2O)_{2/3}MeS_2$, according to which the hydrogen atoms are located at the sulphur layers[19] and may be explained by the formation of OD^- acc. to:

$$K^+_{1/3}(D_2O)_{2/3}[NbS_2]^{1/3-} + 1/3K^+ + 1/3e^- \rightarrow K^+_{2/3}(OD)^-_{1/3}[NbS_2]^{1/3-} + 1/3\ D_2O + 1/6D_2 \quad (2)$$

Evidence for this reaction has recently been obtained from NMR experiments[20].

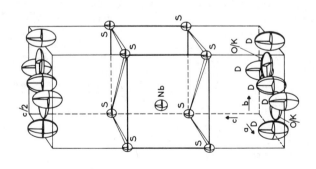

Fig.7 Unit cell of $K_x(OD)_yNbS_2$[12]. Atoms are symbolized by 50% probability ellipsoids. Due to disorder in the a/b planes, two equivalent lattice sites appear to be occupied statistically, allthough only one site per unit cell can be occupied for geometric reasons.

4.4 Diffusion of Molecular Intercalates

Mobile intercalates have been found by NMR and electrochemical methods[21]. No information on the microscopic diffusional steps, which should reflect the interactions intercalate/intercalate and intercalate/host lattice, are obtained from these experiments. High resolution quasielastic neutron spectroscopy (QNS) has been shown to give information on the jump vectors of hydrogen in metals[22] and consequently has been applied to hydrogen containing molecular intercalates.

Jump diffusion, with jump vectors, corresponding to jumps to equivalent lattice sites in adjacent unit cells was found for $TaS_2 \cdot NH_3$[23]. NH_3-jumps to neighbouring sites in the same unit cell could be excluded. This excludes likewise a cooperative diffusion via the movement of domain walls[21]. This distinction in the occupation of lattice sites is not possible by diffraction methods due to the disorder in the a/b plane (e.g. fig.7).

$TaS_2 \cdot NH_3$ is a nonstoichiometric material with $\simeq 10\%$ vacancies at room temperature and jumps via vacancies are therefore probable. A statistical distribution of the vacancies, in the temperature range of the QNS experiments, is also suggested by a real time neutron diffraction experiment on the $TaS_2 \cdot ND_3$ decomposition[3].

The QNS experiment is also supported by real time neutron diffractometry on the $TaS_2 \cdot NH_3/ND_3$ exchange at 215 and 223 K[24]. A temporal variation of the (ool) reflection is observed upon exchange. This experiment excludes a rapid proton exchange. At T = 215 K a value of $3 \cdot 10^{-9}$ cm^2 s^{-1} is determined for the self diffusion coefficient D^*.

$MeS_2 \cdot NH_3$ and $K_x(H_2O)_{2/3}MeS_2$ are similar materials as NH_4^+ cations, probably formed during intercalation[25] were found by neutron spectroscopy in the optic range[26]. The true composition is thus $(NH_4^+)_x(NH_3)_yMeS_2$ with x = 0.1 - 0.3 [25,27]. The $NH_4^+ \ldots NH_3$ interaction has been used to explain the position of the nitrogen atom on a plane, midway between the MeS_2-slabs[25]. Condensed clusters may exist as in the hydrate phase (see above).

2D-intercalate layers, with jump diffusion via statistically distributed vacancies, resemble hypercritical 2D-fluids (e.g. CH_4 on graphite[28]). According to this model, D^* should depend on the number of vacancies. This has not been studied by QNS until now. NMR results[29] on $TaS_2(NH_3)_x$ with x = 0.8, 0.9, 1.0 have not shown a dependance of the relaxation rate on the vacancy concentration. No diffraction data on the invetigated materials have, however, been reported. Thus the purity of the phases is not certain[3]. A better defined system, where the number of vacancies can be controlled via the electrochemical potential is $K_x(H_2O)_{2/3}MeS_2$. Preliminary QNS-experiment on $K_{1/3}(H_2O)_{2/3}NbS_2$ suggest that $D^* = 3 \cdot 10^{-7}$ cm^2 s^{-1} at T = 310 K[30]. Evidence for jump diffusion was found, too.

5. Conclusions

Low resolution, real time neutron diffraction studies of intercalation reactions have revealed a number of phases hitherto unknown."In deep"studies of the structures and the dynamic processes
are now possible.

References

1. R.Schöllhorn, Angew.Chem. I.E. 92, 983 (1980)
2. M.S.Dresselhaus and G.Dresselhaus, Advances in Phys.30, 2,
                        139 (1981)
3. C.Riekel, "Progress in Solid State Chemistry", Vol.13, N° 2,
        p. 89 (1980)
4. Neutron Scattering Facilities at the HFR (1981)
   Copies may be obtained from the scientific secretary,
   I.L.L., Grenoble, France
5. A.Hamwi, C.Riekel, P.Touzain, to be published
6. E.Matuyama, Nature, 4810, p. 61 (1962)
7. J.M.Thomas, G.R.Millward, R.F.Schlögl, H.P.Boehm,
   Mat.Res.Bull., 15, pp. 671-676 (1980)
8. A.Hamwi, P.Touzain, L.Bonnetain, A.Boeuf, A.Freund, C.Riekel,
   3 International Carbon Conference, Baden-Baden (1980)
9. C.Riekel and R.Schöllhorn, Mat.Res.Bull., Vol.11, pp. 369-
                        376 (1976)
10. C.Riekel and C.O.Fischer, J.Solid State Chem. 29, No.2, 181
                        (1979)
11. C.Riekel, H.G.Reznik, R.Schöllhorn, J.Solid State Chem., 34,
                        253 (1980)
12. C.Riekel, H.G.Reznik, R.Schöllhorn and C.J.Wright, J.Chem.
    Phys. 70, (11), 5203 (1979)
13. A.J.Bourdillon, R.F.Pettifer, E.A.Marseglia, Physica 99 B,
                        64 (1980)
14. N.J.Doran, B.Ricco, D.J.Titterington, G.Wexler, J.Phys.C.
    Solid State Phys. 11, 685 (1978)
15. A.Hamwi, P.Touzain, C.Riekel, Synthetic Metals, 2, 153 (1980)
16. C.Riekel, D.Hohlwein, R.Schöllhorn, J.Chem.Soc.Chem.Comm.,
                        863 (1976)
17. D.W.Murphy, F.J.Di Salvo, G.W.Hull, J.V.Wascak, S.F.Mayer,
    G.R.Stewart, S.Early, J.V.Acrivos, T.H.Geballe, J.Chem.Phys.,
                        62, 967 (1975)
18. C.Riekel, H.G.Reznik, R.Schöllhorn, to be published
19. U.Röder, W.Müller-Warmuth, R.Schöllhorn, J.Chem.Phys. 70,
                        2864 (1979)
20. U.Röder, W.Müller-Warmuth, R.Schöllhorn, J.Chem.Phys. 75 (1),
                        p. 412 (1981)
21. M.S.Whittingham, Progr.Solid State Chem., 12, 41 (1978)
22. T.Springer in "Springer Tracts in Modern Physics", Vol. 64,
                        Springer-Verlag, Berlin (1972)

23. C.Riekel, A.Heidemann, B.E.F.Fender and G.C.Stirling, J.Chem. Phys. <u>71</u> (1), 530 (1979)

24. C.Riekel, Solid State Comm., Vol. 28, pp. 385-387 (1978)

25. R.Schöllhorn and H.D.Zagefka, Angew.Chem. <u>3</u>, 193 (1977)

26. C.Riekel, R.Schöllhorn, J.Tomkinson, Z.Naturf. 35 a, 590 (1980)

27. T.Butz, A,Vasquez, H.Saitovitch, R.Mühlberger, A.Lerf, Physica, 99 B, 69 (1980)

28. J.P.Coulomb, M.Bienfait, P.Thorel, J.de Physique, Colloque N^O 4, C 4-31 (1977)

29. R.L. Kleinberg and B.G.Silbernagel, Solid State Comm., Vol.33, 867 (1980)

30. C.Riekel and R.Schöllhorn, to be published

MAGNETIC CORRELATIONS NEAR THE CRITICAL CONCENTRATION FOR
FERROMAGNETISM

B.D. Rainford
Physics Department, Southampton University, SO9 5NH, England

S.K. Burke
Institut Laue Langevin, Grenoble Cédex 38042, France

J.R. Davis
Physics Department, Caulfield Institute of Technology, Caulfield
VIC 3162, Australia

W. Howarth
Thorn EMI Central Research Labs., Hayes, UB3 1HH England

ABSTRACT

The onset of ferromagnetism in Fe Cr alloys and amorphous
Fe Mn alloys has been studied using neutron small angle scattering.
The technique allows the magnetic phase boundaries to be delineated
and reveals the evolution of the magnetic correlations as a function
of temperature and concentration.   The data for Fe Cr are consistent with the behaviour expected at a percolation multicritical
point.   The amorphous Fe Mn alloys undergo a transition from
ferromagnetism to a spin glass phase with a re-entrant phase
boundary.   In both systems the correlation lengths fail to diverge
at the ferromagnetic critical temperatures.

INTRODUCTION

The onset of long range order in disordered magnets has attra-
cted a great deal of attention recently due to advances in the
understanding of scaling behaviour in the vicinity of the critical
concentration [1,2,3,4].   Much of the experimental work [5,6] has been
concerned with mixed insulating antiferromagnets which form good
model systems for comparison with theoretical predictions, since the
exchange is usually short ranged and well characterised.   By
contrast, there has been relatively little work on disordered
ferromagnets [7] since there are few insulating ferromagnetic systems
amenable to study.   We have chosen instead to investigate disordered metallic transition metal ferromagnets.   There are obvious
advantages in this approach : firstly ferromagnets may be studied
by the powerful technique of neutron small angle scattering (SAS)
with its high data collection rate and excellent resolution.
Secondly since measurements are made around the forward direction
it is not necessary to use single crystals.   This simplification
of specimen preparation means that the systematics of the scattering as a function of composition may be studied in fine detail.
Also for ferromagnets and near ferromagnets the neutron scattering
is more directly related to bulk magnetic measurements.   The dis-
advantage of studying metallic systems is that the interactions

240

are not well characterised, so that detailed comparison with theoretical models is not necessarily possible. However it may be that information about the nature of the microscopic interactions can be derived from these studies.

In the present paper we shall compare and contrast the behaviour of the magnetic correlations in two alloy systems in the vicinity of the onset of ferromagnetism. One of these systems $Fe_x Cr_{1-x}$, we will argue, is well described by a percolation model, corresponding to the simple dilution of near neighbour exchange coupled iron atoms by non magnetic chromium atoms. In the second system, amorphous $Fe_{1-x} Mn_x$ alloys, the ferromagnetism is rapidly destroyed by the competition with the antiferromagnetic coupling of the Mn atoms, leading to a spin glass phase. In both alloys systems there are well defined local moments, insensitive to their local environment, so that the complications of polarisation clouds and giant moments do not appear.

$Fe_x Cr_{1-x}$ ALLOYS

Iron and chromium form a continuous range of bcc solid solutions. The magnetism envolves with composition from the ferromagnetism of pure iron to the itinerant spin density wave (SDW) ordering of pure chromium. The details of the magnetic phase

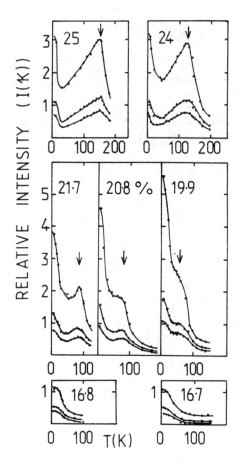

Fig.1. Temperature variation of small angle scattering for Fe Cr alloys measured on D17, ILL. Normalised intensities are shown for three values of κ: 0.019 $Å^{-1}$ (upper curves), 0.032 $Å^{-1}$ (middle curves) and 0.040 $Å^{-1}$ (lower curves). Curves are labelled by percentage of Fe.

diagram have only recently been established by us [8,9,10,11] using a combination of neutron diffraction, resistivity, low field magnetisation and neutron SAS. Commensurate SDW order persists up to a concentration of 16 ± 0.5% Fe, while the critical concentration for the onset of ferromagnetism x_f is 19 ± 0.5% Fe. In the narrow concentration range between the two regions of long range order, susceptibility measurements show a behaviour typical of spin glass order. The spin glass order appears to persist into the antiferromagnetic regime [12].

Some of the neutron SAS data used to establish the ferromagnetic phase boundary is shown in Fig.1, where the intensity at a number of scattering vectors is shown as a function of temperature. The data have been normalised to give the same relative scale. The ferromagnetic critical temperatures are given by the positions of the critical scattering peaks, shown by the arrows. These are in excellent agreement with those found from low field magnetisation measurements. It can be seen that the critical scattering weakens as x_f is approached, so that at 19.9% Fe it appears only as a shoulder. All the alloys show a remarkable increase in the scattering at low temperatures, the increase being most dramatic in the vicinity of x_f. This feature is due to a rapid growth in the

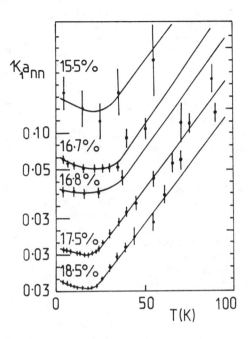

Fig.2. Inverse correlation range κ_1 times nearest neighbour separation a_{nn} for alloys below critical concentration. Solid lines are guides to the eye.

correlation range, as can be seen in Fig.2 and Fig.3, where the inverse correlation range κ_1 is plotted versus temperature for the alloys with $x < x_f$ and $x > x_f$ respectively.

Many aspects of the scattering are in qualitative agreement with a percolation model for the onset of ferromagnetism : we identify x_f with the percolation concentration x_p at which an infinite cluster of iron atoms coupled by nearest neighbour exchange interactions first forms. For $x < x_f$ only finite clusters of coupled spins can exist, but the average cluster size (and with it the susceptibility) diverges as x approaches x_f. For $x > x_f$ the infinite cluster co-exists with a distribution of finite clusters. As x increases beyond x_f the infinite cluster grows rapidly by incorporating the remaining finite clusters, until eventually when $x \lesssim 1$ there is a uniformly magnetised ferromagnet with individual defects. The magnetisation is proportional to the fraction of Fe atoms in the infinite cluster.

Returning to the SAS data, we can attribute the scattering at low temperatures to the finite clusters. Since these cannot support long range order, the correlations continue to grow towards T = 0 (i.e. κ_1 decreases as T is lowered). For $x < x_f$ the maximum correlation range cannot exceed the mean dimensions of the

242

clusters. It can be seen in Fig.2 that κ_1 decreases to a minimum value near 20 K. We take this minimum value of κ_1 to be the "geometrical" inverse correlation length $\kappa_G(x)$ and find that κ_G extrapolates to zero at x_f, a result consistent with the mean cluster size diverging at x_f. For the ferromagnetic alloys, the scattering from the finite clusters persists, but grows weaker as the alloys become more strongly ferromagnetic (Fig.1). At T = 0 the infinite cluster contributes only to the Bragg scattering, but gives rise to the critical scattering near T_c. The weakening of the critical scattering peak as x_f is approached (Fig.1) reflects the shrinking fraction of Fe atoms in the infinite cluster. Some justification for the applicability of a percolation model to Fe Cr alloys is provided by the constant magnitude of the iron moment and the vanishingly small chromium moment in this range of concentrations. It is reasonable then to regard Cr as a simple diluent. We note that the

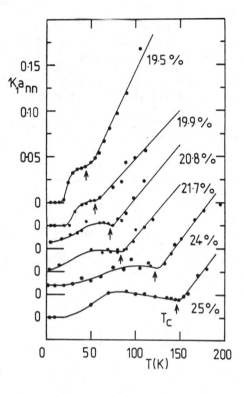

Fig.3. Inverse correlation range ($\kappa_1 a_{nn}$) for ferromagnetic Cr Fe alloys as a function of temperature. Solid lines are guides to the eye.

measured value of x_f (19%) is not very different from the theoretical value for nearest neighbour interactions in a bcc lattice (x_p = 24.3%).[14] The difference may be accounted for by a tendency towards clustering of the Fe atoms in Fe Cr alloys.

Recent theories [1,2,3,4] treat x_p as a multicritical point, since the susceptibility can diverge along the line of second order transitions $T_c(x)$ or along the line T = 0 as a function of (x_p - x). These theories are based on the assumption that the variables in the vicinity of the point (x = x_p, T = 0) may be expressed as scaling functions of the concentration difference (x_p - x) and a generalised temperature variable g, which is identified with the inverse correlation length of the pure system in one dimension. (It is argued that since clusters in the vicinity of x_p are highly ramified in form, correlations are propagated through essentially one dimensional paths). For x < x_p the essential results are as follows: if x_p is approached along the concentration axis at T = 0, the inverse correlation length κ_1 tends to zero according to (x_p - x)$^{\nu_p}$ while the forward scattering intensity I(κ = 0) diverges, as $(x_p - x)^{-\gamma_p}$. The percolation (geometrical) exponents are estimated by Stanley[15] to be $\nu_p \sim 0.82$ and $\gamma_p \sim 1.66$ for a three dimensional lattice. If x_p is approached along the temperature axis κ_1 tends to zero as g^ν , while $I(\kappa = 0)$ diverges as $g^{-\gamma}$. The thermal exponents ν ,γ are related to the geometrical exponents via

$$\phi \gamma = \gamma_p \qquad \text{and} \qquad \phi\nu = \nu_p \qquad (1)$$

where ϕ is termed the cross over exponent. Theoretical estimates for ϕ vary : Stinchcombe [4] suggests $\phi = 1$ for both Heisenberg and Ising systems. Birgeneau et al.[5] suggest an asymptotic form of the scattering intensity

$$I(\kappa) = \frac{B \kappa_1^\eta}{\kappa_1^2 + \kappa^2} \qquad (2)$$

with $\kappa_1 = \kappa_T + \kappa_G$, the sum of a thermal part ($\propto g^\nu$) and a geometrical part ($\propto (x_p - x)^{\nu_p}$).

If we take κ_G to correspond to the minimum values in Fig.2 we find the data to be adequately described by

$$\kappa_G a_{nn} = (2.5 \pm 0.5) \ (x_f - x) \qquad (3)$$

which implies an exponent $\nu \sim 1$. The linear temperature dependence of κ_1 above 20 K indicates Heisenberg like correlations. The tendency of κ_1 to saturate below 20 K and the weak minimum suggest a cross over to Ising behaviour at this temperature probably as a result of dipolar anisotropy. The proper analysis of the thermal part of κ_1 would require the exact form of the pure one dimensional inverse correlation range g(T) in a Heisenberg-Ising description. This is not possible in the present case since the appropriate exchange and anisotropy constants are unknown. As a first approximation g(T) is assumed to follow the asymptotic forms proposed by Stinchcombe.[4] For the data above 20 K, where the distinction between tranverse and longitudinal components of κ_1 is unimportant g(T) may be written

$$g(T) = a(b + T) \qquad T > 20 \ K$$

where a, b are undetermined parameters related to the exchange and anisotropy constants. The data in Fig.2 are well described by

$$\kappa_T a_{nn} = (1.75 \pm 0.1).10^{-3} (20 + T) = g(T)^\nu$$

which suggest an exponent $\nu \sim 1$.

By comparing the variation of $I(\kappa = 0)$ with the measured value of κ_1 (cf eq.2) we find an excellent fit to the data above 20 K with $\eta = -0.40 \pm 0.1$ (17.5% Fe) and $\eta = -0.41 \pm 0.05$ (18.5% Fe). This value for η is much larger in magnitude than that estimated

theoretically (η (theory) ~ 0.0). Percolation models have
$\eta = \eta_p$, and Stanley suggests that the effective dimensionality
of clusters near x_p is given by $d_p = 2 - \eta_p$. This relation
would give $d_p = 2.40 \pm 0.01$ in the present case.

The exponent γ may be estimated using the scaling relation
$\gamma = \nu (2 - \eta)$. With $\eta = -0.4$ and $\nu = 1$ this gives $\gamma \sim 2.4$.
The cross over exponent ϕ can be calculated from the pure percolation exponents $\gamma_p = 1.66$ and $\nu_p = 0.82$ to give $\phi = \nu_p/\nu = 0.82$
or $\phi = \gamma_p/\gamma = 0.69$. These estimates of exponents should be
treated with caution since they were deduced without an exact
knowledge of $g(T)$. However it is clear that the SAS data for
$x < x_f$ are in semiquantitative agreement with models of the
percolation multicritical point, and that the correlations cross
over from Heisenberg like behaviour at high temperatures to Ising
like behaviour at low temperatures, probably as a result of anisotropy effects.

The concentration and temperature dependence of the magnetic
correlations in the ferromagnetic alloys is more complex than that
observed for $x < x_f$. It can be seen from Fig.3 that κ_1 does not
go to zero at the ferromagnetic critical temperature i.e. the
correlation range does not diverge, as expected. While this may
be due to chemical inhomogeneities smearing the ferromagnetic
transition, we note that the same phenomenon is found in mixed
antiferromagnets, and appears to be a feature of isotropic
three dimensional magnets near the percolation concentration.
There may be a connection here to the conclusions of Imry and Ma[16]
that for systems with continuous symmetry the ordered state is
unstable against arbitrarily weak random fields (eg random anisotropy fields due to dipolar forces) in less than four dimensions.
Aharony and Pytte[17] predict, in this situation, that the magnetisation is zero in zero field at all temperatures, while the susceptibility is infinite at T_c and everywhere below T_c. They also
predict that the correlation functions follow a $1/\kappa^2$ behaviour
(ie $\kappa_1 = 0$). This is consistent with what we observe at lower
temperatures, though deviations from $1/\kappa^2$ behaviour are observed at
larger κ values.

Recent inelastic neutron scattering measurements on Fe Cr[18]
alloys near x_f show well defined spin wave excitations at temperatures above $0.4\ T_c$, but at lower temperatures the spin wave stiffness decreases and the excitations become over-damped. This
suggests that the degree of order in the infinite cluster decreases
at low temperatures. It seems then that the rapid rise in the
SAS signal at low temperatures is due not only to finite clusters,
but to the order within the infinite cluster breaking up. We have
suggested a possible mechanism by which this might occur[19]: since
the clusters (including the infinite cluster) near x_f are highly
ramified in character, the magnetostatic energy of the clusters will
be minimised if the spins are aligned along the local "chain" axis.
The ground state spin configuration is then determined by the
competition between this shape anisotropy within the cluster and the
exchange interaction, and the magnetic order within the cluster
might break up into "micro - domains".

The temperature dependence of κ_1 above T_c is well described by an expression of the form

$$\kappa_1 \, a_{nn} = (4.7 \pm 0.3) \; (x - x_f) \left\{ \frac{T - T_c}{T_c} \right\}^{\nu} + \kappa_R$$

with $\nu = 1.2 \pm 0.1$ and $\kappa_R = 0.026 \pm 0.01$. Except for the constant κ_R this form is consistent with the scaling laws proposed by Stinchcombe[4]. The occurrence of the prefactor $(x - x_f)$ demonstrates that the ferromagnetic alloys are 'aware' of the singularity at $x = x_f$. The value of κ_R suggests a correlation range near T_c of the order of 100 Å . The large value of ν is consistent with the anomalously large value of γ found in the bulk magnetisation measurements on Fe Cr alloys in this concentration range by Aldred and Kouvel.[20] Using their value for γ and the scaling relation $\gamma = \nu(2 - \eta)$ with η assumed zero gives $\nu = 1.06 \pm 0.01$ in reasonable agreement with the value above.

In summary we conclude that the magnetic correlations for Fe Cr alloys in the vicinity of x_f are well described by a percolation model, suggesting that nearest neighbour ferromagnetic exchange couplings are dominant. Bulk magnetisation measurements are consistent with this conclusion, as we report elsewhere.[11] However the nature of the magnetic order for the ferromagnetic alloys is complex and not fully understood at present.

AMORPHOUS $(Fe_{1-x} \, Mn_x)$ ALLOYS

A different type of critical concentration for ferromagnetism occurs in alloy systems with competing exchange interactions, where the second component of the alloy has anti-ferromagnetic exchange couplings. In this situation, as the concentration of the second component is increased, the ferromagnetic order becomes increasingly frustrated and there is a transition to either an antiferromagnetic state, or to a spin glass state. There have been a number of recent studies[21] on amorphous transition metal alloys where there is necessarily a transition to the spin glass state, since an ordered antiferromagnetic ground state is not possible in an amorphous material.

We have recently made neutron SAS measurements on a series of amorphous $(Fe_{1-x} \, Mn_x)_{80} \, P_{16} \, C_4$ alloys using the D17 instrument at ILL Grenoble. The alloys were prepared by the melt spinning technique. The intensities for three values of the scattering vector κ are shown as a function of temperature in Fig.4. The inset at the right hand side of the figure shows the data for 34% Mn on an expanded scale. Superficially the data look similar to the data for Fe Cr alloys in Fig.1, in that the critical scattering peaks at the Curie temperatures are clearly seen, T_c is seen to decrease progressively as the Mn concentration is increased, and there is additional scattering intensity at low temperatures. However there are important qualitative differences. Firstly the

246

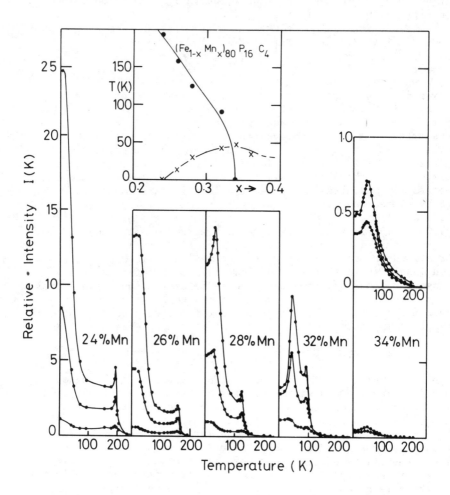

Fig.4. Temperature variation of small angle scattering from $(Fe_{1-x} Mn_x)_{80} P_{16} C_4$ amorphous alloys. Normalised intensities are shown for three values of κ : $.012$ $\mathring{A}^{-1}$ (upper curves), $.017$ $\mathring{A}^{-1}$ (middle curves) and $.037$ $\mathring{A}^{-1}$ (lower curves).

scattering below T_c for alloys with more than 26% Mn, a well defined maximum as a function of temperature. The temperature of this maximum increases steadily with Mn concentration, crossing the line of $T_c(x)$ near 33% Mn (upper inset, Fig.4). Secondly the intensity of the 'sub-critical' scattering decreases uniformly towards the concentration at which ferromagnetism disappears.

The occurrence of a well defined peak in the SAS intensity below T_c suggests a transition to another phase. Other authors have shown phase diagrams with re-entrant ferromagnetic - spin glass (F - SG) phase boundaries in these materials but these have always been based on broad features in magnetisation curves, or shoulders in the low field susceptibility. We believe the present data are the first to indicate a well defined transition to another phase.

The uniform decrease of the scattering intensity with increased concentration of Mn is what would be expected for a transition from ferromagnetic to spin glass order. The addition of
Mn corresponds to a shift in the average value of the distribution
of exchange energies $\bar{J}_o$ from a negative value (ferromagnetic
coupling) towards zero (equal proportions of ferro and antiferro-

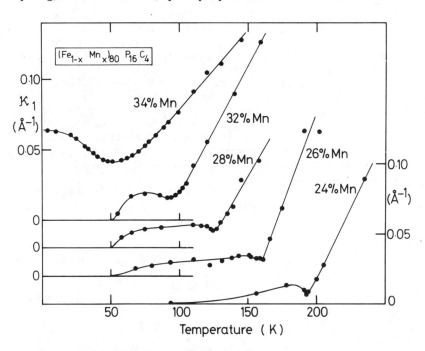

Fig.5. Inverse correlation range κ_1 for amorphous $(Fe_{1-x}Mn_x)_{80}$
$P_{16}C_4$ alloys as a function of temperature.

magnetic couplings). The magnetic response will simultaneously
shift from being peaked around $\kappa = 0$, to a flat response, independent of κ, characteristic of the spin glass phase. Indeed alloys
containing in excess of 36% Mn show very little scattering around
the forward direction. This behaviour is completely different
from that observed in Fe Cr where the sub-critical scattering was
greatest in the vicinity of x_f.

On the other hand the behaviour of the inverse correlation
ranges for these alloys is very similar to that observed in the
Fe Cr alloys. These are shown in Fig.5. The alloy which shows
only the spin glass transition (34% Mn) has a pronounced minimum
in κ_1, similar to the cross over effect found in Fe Cr for $x < x_f$.
The ferromagnetic alloys again have finite values of κ_1 at T_c,
though this value appears to decrease for the more strongly ferromagnetic alloys. κ_1 finally goes to zero some way below T_c, as in
Fe Cr, though for these alloys the κ dependence of the scattering

is much closer to $1/\kappa^2$, at least for the alloys closest to the critical concentration.

Salomon et al.[21] have compared their phase diagram of amorphous $Fe_{1-x}Mn_x$ alloys with those derived from the Sherrington-Kirkpatrick (SK) model[22]. They have also used scaling arguments, derived from the SK model using a Landau-Ginzburg approach, to analyse their magnetisation and susceptibility data. Their scaling laws predict that the susceptibility diverges both at T_c and at the lower (F-SG) transition. The double peak seen in the SAS data suggests that this might be the case, though the applicability of the SK model (an infinite range Ising model) to these alloys is rather dubious.

The recent theoretical predictions of Gabay and Toulouse[23] are probably much more pertinent to the present case. In an infinite range Heisenberg model they find two mixed phases close to the F-SG phase boundary, in which ferromagnetism co-exists with spin glass order due to the freezing of the transverse components of spins. Comparison with the present data will be possible when predictions are available for the behaviour of the susceptibility and correlation lengths in these mixed phases.

REFERENCES

1. D.Stauffer, Z.Phys. B22, 161 (1975)
2. T.C.Lubensky, Phys.Rev. B15, 311 (1976)
3. H.E.Stanley, R.J.Birgeneau, P.J.Reynolds and J.F.Nicholl, J.Phys. C 9, L553 (1976)
4. R.B.Stinchcombe, J.Phys.C 13, 3723 (1980)
5. R.J.Birgeneau, R.A.Cowley, G.Shirane and H.J.Guggenheim, Phys. Rev. B21, 317 (1980)
6. R.A.Cowley, G.Shirane, R.J.Birgeneau and E.C.Svensson, Phys.Rev. Lett. 39, 894 (1977)
7. H.Malletta and W.Felsch, Z.Phys. B 37, 55 (1980)
8. S.K.Burke, R.Cywinski and B.D.Rainford, J.Appl.Cryst. 11, 644 (1978)
9. S.K.Burke and B.D.Rainford, J.Phys.F 8, L239 (1978)
10. S.K.Burke, Ph.D. thesis, University of London (1980)(unpublished)
11. S.K.Burke, R.Cywinski, J.R.Davis and B.D.Rainford (to be published)
12. J.O.Strom-Olsen, D.F.Wilford,S.K.Burke and B.D.Rainford, J.Phys. F 9, L95 (1979)
13. A.T.Aldred,B.D.Rainford,J.S.Kouvel and T.J.Hicks,Phys.Rev. B14, 228 (1976)
14. J.W.Essam, Phase Transitions and Critical Phenomena Vol.2 eds. C.Domb and M.S.Green 197 (1972) (Academic Press : London)
15. H.E.Stanley, J.Phys.C 11, L211 (1977)
16. Y.Imry and S.Ma, Phys.Rev.Lett. 35, 1399 (1975)
17. A.Aharony and E.Pytte, Phys.Rev.Lett. 45, 1483 (1980)
18. C.R.Fincher, S.M.Shapiro, A.H.Palumbo and R.D.Parks, Phys.Rev. Lett. 45, 474 (1980)
19. S.K.Burke, B.D.Rainford and M.Warner,J.Mag.Mag.Mat. 15-18, 259 (1980)
20. A.T.Aldred and J.S.Kouvel, Physica 86-88B, 329 (1977)
21. M.B.Salomon,K.V.Rao and Y.Yeshurun, J.Appl.Phys. 52, 1687 (1981)
22. D.Sherrington and S.Kirkpatrick,Phys.Rev.Lett. 35, 1792 (1975)
23. M.Gabay and G.Toulouse, Phys.Rev.Lett. 47, 201 (1981)

NEUTRON INELASTIC SCATTERING BY AMINO ACIDS

C.L. Thaper, S.K. Sinha and B.A. Dasannacharya
Bhabha Atomic Research Centre
Trombay, Bombay 400 085.

ABSTRACT

Inelastic neutron scattering experiments on normal, N-deuterated and fully deuterated glycine, normal and N-deuterated alanine, L-valine, L-tyrosine and, L-phenyl alanine at 100 K, are reported. Coupling of the external modes to different hydrogens is discussed.

INTRODUCTION

Amino acids have been extensively studied by infrared, Raman and NMR spectroscopy, much less so by neutron scattering. In this paper we report neutron inelastic scattering measurements from several amino acids; in particular we use the change of intensity of lines on selective deuteration in α-glycine and L-alanine to elucidate the nature of some of their external vibrations.

EXPERIMENTS AND RESULTS

The experiments were performed on powder samples of normal, N-deuterated and fully deuterated glycine, normal and N-deuterated alanine, L-valine, L-tyrosine and L-phenyl alanine, at 100 K on a filter-detector spectrometer with a BeO filter and a Cu(111) monochromator. The spectra, corrected for room background and normalised to the same number of molecules is shown in Fig.1. Neutron time-of-flight data have been reported earlier on normal α-glycine[1] and L-alanine[2]. The present spectra are better resolved and show some additional features.

Our measurements confirm most of the earlier frequencies on α-glycine[1,3] and L-alanine[2,4]. The assignment of the frequencies, arrived at after comparison with normal coordinate analysis is given in Fig.1.

On the basis of observed intensity in normal and deuterated samples, it is seen that COO^-, $CC_\alpha N$ and $NC_\alpha CH_3$ modes are coupled to hydrogen atoms attached to N as well as C_α atoms. However, the $CC_\alpha N$ mode in glycine is predominantly coupled to hydrogens of N atom but in alanine it is to those of C_α. On the other hand, the COO^- wag is coupled to hydrogens of C_α in glycine but to N atoms in alanine. The COO^- bend in glycine is coupled to hydrogens of N but it is not seen at the

ISSN:0094-243X/82/890249-03$3.00 Copyright 1982 American Institute of Physics

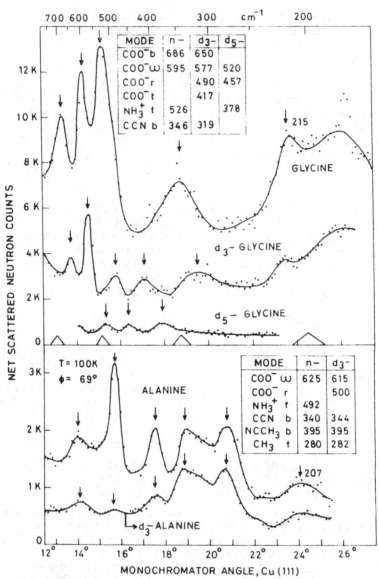

Fig.1 Neutron inelastic spectra from α-glycine and L-alanine. Inset gives the frequencies in cm⁻¹ (indicated by arrows) with their assignments; b, w, r and t denote bend, wag, rock and torsion modes respectively; n-, d₃- and d₅- stand for normal, N-deuteration and complete deuteration.

expected positions in alanine samples. The $NC_\alpha CH_3$ mode in alanine is mostly coupled to hydrogens of N.

Spectra of L-valine, L-tyrosine and L-phenyl alanine are quite complex. They show NH_3^+ torsions at 490, 485 and 550 cm^{-1} respectively. CH_3 torsion in L-valine is observed at 260 cm^{-1}.

Potential barriers calculated from the observed frequencies assuming 3-fold harmonic rotational potential are compared with those derived from diffraction and NMR[5] data (Table I). The large disagreement for the NH_3 barriers is believed to be due to large distortion of the top of the potential due to hydrogen bonding. Direct measurement of the reorientation rate of the NH_3 using quasi-elastic scattering will be of interest in this connection.

Table I   Comparison of barrier heights

Rotating group	Substance	Barrier heights (Kcal/mole)		
		neutron	NMR	diff.
$-NH_3$	α -glycine	14.7	6.8	
	L-alanine	13.1	9.2	20.0
	L-valine	12.8	8.9	
	L-tyrosine	12.7	8.8	8.3
	L-phenylalanine	16.2	12.2	
$-CH_3$	L-alanine	4.6	5.3	5.6
	L-valine	3.9	2.7	1,8, 2.0

1. V.D. Gupta and R.D. Singh, Chem. Phys. Letts. 5, 218 (1970).
2. V.D. Gupta and M.V. Krishnan, J. Phys. B3, 572 (1970).
3. S. Suzuki and T. Shimanouchi, Spectrochim Acta, 19, 1195 (1963). Also A.M. Dwivedi and V.D. Gupta, Ind. J. Biochem and Biophys., 10, 77 (1973).
4. R.B. Srivastava and V.D. Gupta, Ind. J. Pure and Appl. Phys. 10, 596 (1972). Also C.H. Wang and R.D. Storms, J. Chem. Phys. 55, 3291 (1971).
5. E.R. Andrew, W.S. Hinshaw, M.G. Hutchins and R.O.I. Sjoblom, Mol. Phys. 31, 1479 (1976); ibid 32, 795 (1976); ibid 34, 1695 (1977).

252

EXPERIMENTAL MODEL FOR **NEUTRON** SCATTERING IN DISORDERED SYSTEMS: STATIC STRUCTURE FACTOR DETERMINATION OF MODE-SOFTENING

Edward Siegel

Queen Mary College, University of London,London,U.K.
and
C.N.Pq., São Jose dos Campos, Brazil

ABSTRACT

The generalized-disorder collective-boson mode-softening univ-ersality-principle (GDCBMSUP) for collective-boson mode dispersion in disordered systems (liquids, quantum liquids, glasses, powders,disord-ered magnets, plasmas...),a unified qualitative and semi-quantitative descriptive prescription for treating the properties of very differe-ntly disordered systems, is directly dependent upon a measurement (or calculation) of the static structure factor S(k) determined from a frequency average of the dynamic structure factor S(k,w), a multiple of the inelastic differential neutron scattering cross section $d^2\sigma/dwd\Omega$. The prescription for this principle is given and, because of its uni-versal applicability to disordered systems of any type with any type and/or degree of disorder, the neutron scattering determination of S(k) takes on renewed importance.

INTRODUCTION

The generalized-disorder collective-boson mode-softening univ-ersality-principle is a universal response of physical systems,indep-endent of type and/or degree of disorder,to the existence of disorder via mode-softening with negative-dispersion of the collective-bosons that the system can support. As such the collective-boson dispersion relation is directly determined by the degree and type of disorder, measured via inelastic differential external radiation scattering (neutrons, X-rays, electrons...) cross section determined static str-ucture factor S(k)

$$S(k) = \int_0^\infty S(k,w)\ dw = \int_0^\infty (d^2\sigma/dw\ d\Omega)dw \qquad (1)$$

The relation is via the Feynman relation between phonon dispersion re-lation w(k) and static structure factor S(k)

$$w(k) = N\ \hbar\ k^2\ /\ 2\ m\ S(k) \qquad (2)$$

originally derived for liquid helium, but equally valid for classical liquids and glasses in trivially modified form by Egelstaff[2] and Hu-bbard and Beeby,[3] seems to be universally true independent of inter-particle force law and Hamiltonian, type and/or degree of disorder, type of collective-boson the system can support or bonds, bands,scr-eening etc. As such the static structure factor S(k), measured by in-elastic differential neutron scattering most aptly, is the physical measurement to be used as input to the collective-boson dispersion relation (2) from which a plethora of physical properties can be cal-culated.

CALC ULATIONAL PROGRAM

The static structure input to the collective-boson dispersion relation,(1) & (2) , yields direct results to many diverse physical properties of any disordered system. The concept of the static structure factor modulating the collective-boson dispersion relation is inherent in Brillouin's classical work on the formation of stopping and pass bands in geometric (crystal lattice, electric filter network, acoustic impedance network) lattices due to the breaking of the symmetry of the continuum (infinitely homogeneous, infinitely isotropic) by formation of the geometric lattice (heterogeneous, anisotropic). This is a mathematical property, independent of the particular system or how its symmetry is broken ie. how it is discrdered! Whatever the cause of the generalized-disorder in whatever system, the result is that the relevant collective-boson (phonon, magnon, plasmas...) dispersion becomes mode-softened with negative-dispersion at large enough wavevector. The dispersion relation develops a horizontal tangent-zero group velocity at the disorder parameter k_c, related to the inflection point in the static structure factor $S(\hat{k})$ before its first peak. This disorder parameter k_c is 1:1 mapped into a frequency disorder parameter $w_c=w(k_c)$. This is in turn mapped into an effective Debye-like temperature $\theta_c=(\hbar / k_B) w_c = (\hbar/k_B) \hbar k^2/2m\, S(k)$ which can be used to characterize the now disordered system. The wavevector disorder parameter is the threshold of the stopping band in wavevector due to the disorder introduced into the system,1:1 mapped into the threshold of the stopping band in frequency due to this same disorder via the dispersion relation (2) or its variants.

PROPERTIES

Consider a liquid, or glass, or powder, or heterogeneous mixture (slush). How is one to calculate its physical properties knowing its disorder relative to the crystalline ordered phase of the same material? A technique so successful in crystalline solids since Debye and Einstein is to characterize various diverse crystalline solids with complicated phonon spectra by one parameter, an effective (Debye or Einstein) temperature. Given that this technique has proven eminantly successful for all other crystalline solids aside from the early applications of Debye and Einstein strictly only to the specific heat of ionically bonded alkali halides, independent of bonds, bands, screening, metallic/ionic/semiconducting nature and force law/Hamiltonian, one would like to develop a similar prescription for disordered systems that is universal in character and scope as well.

But how is one to determine the numerical value of this parameter in a disordered system. In the crystal lattice one chooses some cutoff in the phonon spectrum frequency, related to the effective temperature via $\hbar\omega_c=k_B\theta_c$, but only if one can integrate the phonon density of states (Debye case) over the phonon wavevector spectrum. However this procedure is invalid in disordered systems; the collective-boson wavevector is ill defined at higher wavevector values. How then can one choose an appropriate effective temperature for a particular disordered systenm at a particular temperature, pressure, composition and grain size? The answer comes from the static structure factor $S(k)$, either theoretically calculated with inordinate effort or determined via the ideal analog computer, the inelastic differential scattering of external radiation off the system!

The GDCBMSUP gives a universal unique prescription for converting an external radiation (neutron,X-ray,electron...) scattering cross section into a dispersion relation which, because of the universal mode-softening property, uniquely defines an effective collective-boson temperature at any one external parameter group of temperature, pressure, composition and grain size. The existence of a stopping band threshold 1:1 mappable from the wavevector axis to the frequency (or equivalent effective temperature) axis uniquely defines the effective temperature. In the static structure factor $S(k)$ this threshold corresponds to the inflection point of the first peak in $S(k)$, the point where $\partial^2 S(k)/\partial k$ changes from negative to positive. This can be determined as accurately as experiment allows fron neutron scattering inelastic differential scattering cross section either averaged over frequency or measured without frequency detection. Siegel[4] has utilized this prescription in liquid metals and alloys with great success for such a simple method (hard core structure factors). Detailed carefully resolved small awhile scattering static structure factors are sorely required for liquids, glasses and powders of all types to further qualify this method and bring out its great potential for a universal experimental model technique of calculating various physical properties requiring carefully resolved, accurate neutron or X-ray static structure factors be measured with renewed purpose and application using

$$\theta_c(T,P,X_j,R...)=(\hbar/k_B)w_c(T,P,X_j,R...)=(\hbar/k_B)\,w(k_c;T,P,X_j,R...)$$
$$= (\hbar/k_B)\,\hbar k_c^2/2m\,S(k_c) \tag{3}$$

where k_c corresponds to the inflection point of the static structure factor in rising to its first peak. This essentially measures the effective deviation of $S(k)$ from the sum of Dirac delta functions characteristic of a crystal lattice.

REFERENCES

1. E.Siegel, J.Noncryst.Solids 40,453(1980); Int'l.Conf.Lattice Dynamics Paris (1977)
2. P. Egelstaff, An Introduction to the Liquid State,Academic Press(1967)
3. J.Hubbard and J, Beeby, J.Phys. C, 2,556(1969)
4. E.Siegel, J.Phys.Chem.Liquids 4(1975);ibid.9 (1976)

NEUTRON INELASTIC SCATTERING MEASUREMENTS ON THE METALLIC GLASS $Cu_{0.6}Zr_{0.4}$

T.M. Holden

Atomic Energy of Canada Limited, Chalk River, Ontario, K0J 1J0, Canada

J.S. Dugdale, G.C. Hallam and D. Pavuna

Physics Department, University of Leeds, Leeds, England

ABSTRACT

Neutron inelastic scattering measurements have been made on the metallic glass $Cu_{0.6}Zr_{0.4}$ at 106 K by triple-axis crystal spectrometry. From the high wavevector constant-Q scans the weighted density of vibrational states has been derived. The constant-Q scan at the lowest wavevector studied $(Q = 2\text{Å}^{-1})$ shows a definite inelastic neutron group. The results are compared in detail with a computer calculation for $Cu_{0.57}Zr_{0.43}$.

INTRODUCTION

Measurements of the dynamical structure factor $S(Q,\omega)$ of the metallic glass $Cu_{0.6}Zr_{0.4}$ at low temperatures by neutron inelastic scattering techniques are presented. Calculations[1] of the vibrational density of states of an alloy of composition $Cu_{0.57}Zr_{0.43}$ by the recursion method have also recently been carried out and it is interesting to make a direct comparison of theory and experiment. The calculation also indicates the presence of phonon-like longitudinal modes up to wavevectors approaching the maximum in the structure factor. The experiment gives evidence for a very broad distribution of inelastic scattering centred at 4 THz which probably corresponds to these phonon-like modes. A full account of the experiment is being published elsewhere[2]. Recently, there has been a measurement[3] of $S(Q,\omega)$ for the metallic glass $Cu_{0.46}Zr_{0.54}$ by neutron time-of-flight spectrometry and there is close agreement between the density of states obtained for the two compositions.

EXPERIMENTS AND RESULTS

The measurements were made with a triple-axis crystal spectrometer operating in the constant-Q mode with fixed scattered frequency (8 THz). The (111) planes of a Si crystal were used as analyser and the (113) planes of a Ge crystal were used as monochromator. The specimen temperature was held at 106 K to minimize multiphonon scattering. The sample was in the form of a vertical pillar made from ribbon material, prepared at Leeds University[4], stacked together and had a mass of 8.0 g. Measurements of the inelastic scattering made at 2.0, 2.8, and 6.5 Å^{-1} are shown in Fig. 1. Cross-sections were established by measurements on a cylindrical vanadium sample at the same position in the neutron beam. The inelastic scattering at high wavevectors has the form of a plateau in the frequency range

ISSN:0094-243X/82/890255-03$3.00 Copyright 1982 American Institute of Physics

256

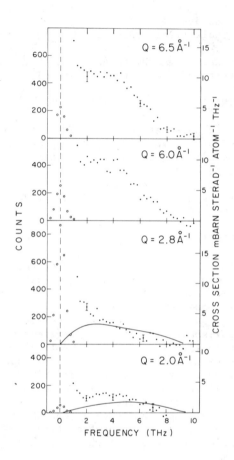

Fig. 1. Frequency distribution of
scattered neutrons from the metallic
glass $Cu_{0.6}Zr_{0.4}$ at 106 K obtained
from constant-Q scans at 6.5, 6.0,
2.8 and 2.0Å^{-1}. The open circles
represent the elastic scattering
scaled down by a factor 10^2. The
lines represent a calculation for
$Cu_{0.57}Zr_{0.43}$.

1–4.5 THz followed by a
monotonic decrease to zero
at about 9 THz. The plateau
region is absent at 2.8Å^{-1}
(close to the maximum in
S(Q)) and only an inelastic
tail is observed. At Q =
2Å^{-1} there is a weak inelastic peak centred on 4 THz
whose breadth is mainly due
to natural lifetime effects
since the experimental
resolution at 4 THz is only
1 THz. The density of
states derived from measurements at Q = 6.5Å^{-1} and
uncorrected for Debye–Waller
factor is shown in Fig.2.

The calculated density
of states for $Cu_{0.57}Zr_{0.43}$
is also shown in Fig. 2 and
is very similar in general
features (single peaked
character and correct frequency range) to experiment.
There are, however, significant discrepancies such
as (I) the low density of
states at the centre of the
distribution and (II) the
high density of states at
low frequencies. The cal-
culated curve is skewed to
the low frequency side and
is less symmetrical than
experiment. These effects
may stem from inaccuracies
in the parameters describing
the pair potentials between
Cu and Zr atoms.

The sum of the
calculated[1] transverse and
longitudinal scattering at
2.0 and 2.8Å^{-1} is shown as
solid lines in Fig. 1. The
curves are normalized to the
experimental results so that
their sum reproduces the
cross-section at 4 THz at
Q = 2.8Å^{-1}. The curve for

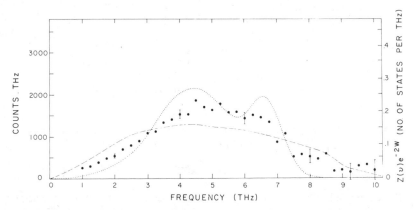

Fig. 2. Density of vibrational states derived from inelastic scattering measurements at 6.5Å^{-1}. The dotted curve represents the density of states of copper folded with the experimental resolution function. The dashed curve represents a calculation for $Cu_{0.57}Zr_{0.43}$.

2.0Å^{-1} is scaled by the Q^2 factor appearing in the neutron-scattering cross-section, but no account is taken of possible changes in the relative strength of longitudinal and transverse modes. With this, somewhat arbitrary, normalization scheme a peak would be expected in the inelastic scattering around 5 THz. However, any other normalization scheme would not significantly shift the peak. The experiment indicates that the peak position is about 4 THz.

It is concluded that the present methods of calculating $S(Q,\omega)$ for metallic glasses give a satisfactory overall description of the behaviour, but detailed agreement between theory and experiments is still lacking.

ACKNOWLEDGEMENTS

We would like to acknowledge the assistance of H.F. Nieman, D.C. Tennant, M.M. Potter, J.C. Evans and A.H. Hewitt at Chalk River, N. Hance and R. Hill at UKAEA, Harwell and D. Hainsworth and M.J. Walker at Leeds University.

REFERENCES

1. S. Kobayashi and T. Takeuchi, J. Phys. C: Solid St. Phys. 13, L969 (1980)
2. T.M. Holden, J.S. Dugdale, G.C. Hallam and D. Pavuna, J. Phys. F: Metal Phys. (In press)
3. J-B Suck, H. Ruden, H-J Guntherodt, H. Beck, J. Daubert and W. Glaser, J. Phys. C: Solid St. Phys. 13, L167 (1980)
4. D. Pavuna, J. Non-Cryst. Solids 37, 133 (1980).

258

TOPOLOGICAL AND CHEMICAL SHORT-RANGE ORDER IN AMORPHOUS Ni-Ti ALLOYS

C.N.J. Wagner
Materials Science and Engineering Department
University of California
Los Angeles, California 90024 USA

Henner Ruppersberg
Fachbereich Angewandte Physik, Universität des Saarlandes
D-6600 Saarbrücken, West-Germany

ABSTRACT

Neutron and x-ray scattering measurements were made on amorphous $Ni_{35}Ti_{65}$ and $Ni_{40}Ti_{60}$ alloys prepared by the melt-spinning process. The x-ray interference functions (structure factors) $S^x(K)$ are dominated by the topological short-range order $(TSRO) S_{NN}(K)$, but exhibited a small prepeak. The neutron structure factors $S^n(K)$ are dominated by the CSRO $S_{CC}(K)/(c_1c_2)$, describing the concentration fluctuations in the alloys. Assuming that the size effect term $S_{NC}(K)$ which describes the correlation between number density and concentration can be approximated by the Percus-Yevick hard sphere model, the TSRO $S_{NN}(K)$ and CSRO $S_{CC}(K)/(c_1c_2)$ were evaluated. From their Fourier transforms it became possible to evaluate the chemical short-range order parameter α which is of the order of -0.1 to -0.15 indicating a preference for unlike nearest neighbors in the amorphous Ni-Ti alloys.

INTRODUCTION

The total coherent scattering per atom, $I_a(K)$, from amorphous phases can be expressed as[1]

$$I_a(K)=<f>^2 S_{NN}(K)+2<f>(f_1-f_2)S_{NC}(K)+[<f^2>-<f>^2]S_{CC}(K)/(c_1c_2) \qquad (1)$$

$S_{NN}(K)$ is the topological short-range order (TSRO) term and describes the spatial correlation between particle density fluctuations. $S_{CC}(K)$ is the chemical short-range order (CSRO) term and represents the concentration fluctuations. The size effect term $S_{NC}(K)$ is a measure of the cross-correlation between particle density and concentration. $<f^2>$ is the mean square scattering amplitude, i.e., $<f^2> = c_1f_1^2 + c_2f_2^2$ and $<f>$ is the mean scattering amplitude, i.e., $<f> = c_1f_1 + c_2f_2$, where c_i and f_i are the atomic concentration and scattering amplitude, respectively, of element i.

In most alloy systems, it is very difficult to observe $S_{CC}(K)$ because the Laue monotonic scattering $<f^2>-<f>^2$ is usually small compared to $<f>^2$ which is the scattering factor for $S_{NN}(K)$ in Eq. (1). The term $<f^2>-<f>^2$ is large only if f_1 and f_2 are very different from each other. The optimum case results when $<f> = 0$ which yields an equation for the composition of the so-called "zero alloy", i.e., $c_1^o =-f_2/(f_1-f_2)$. A zero alloy can be prepared if f_1 and f_2 have different signs. This is **only possible** in neutron diffraction experiments.

ISSN:0094-243X/82/890258-03$3.00 Copyright 1982 American Institute of Physics

Whereas $f \equiv b$ is positive for most nuclei, it is negative for ^7Li, nat. Mn, nat. Ti, and ^{62}Ni. If one of these isotopes or elements is mixed with an element, for which f is positive, in the composition c_i^0, the interference function $S(K)$, also called structure factor, defined as:

$$S(K) = I_a(K)/<f^2> \qquad (2)$$

becomes directly equal to $S_{CC}(K)/(c_1c_2)$. The zero alloy composition in the Ni-Ti system containing natural elements is found at 24 at. %, Ni, whereas in the Cu-Ti system the zero alloy composition is 30 at. % Cu.

Chemical short-range order has been observed in an amorphous Cu-Ti alloy[2] and a $Ni_{40}Ti_{60}$ alloy[3]. We will report on the topological and chemical short-range order in an amorphous $Ni_{35}Ti_{65}$ alloy and compare it to $Ni_{40}Ti_{60}$.

EXPERIMENTAL PROCEDURES

Amorphous $Ni_{35}Ti_{65}$ and $Ni_{40}Ti_{60}$ alloys have been prepared by the melt-spinning process. The variable 2θ transmission method was applied using Ag-Kα and Mo-Kα radiation in conjunction with a Li-drifted Si solid state detector.[4] The neutron scattering experiments were carried out on the D4 instrument of the Institute Laue-Langevin in Grenoble using a wavelength of 0.695 Å.

RESULT AND DISCUSSION

The x-ray and neutron interference functions $S(K)$, defined in Eq. (2), are shown in Figs. 1 and 2 for the amorphous $Ni_{35}Ti_{65}$ and $Ni_{40}Ti_{60}$ alloys, respectively. The x-ray functions $S^x(K)$ are characterized by a small prepeak at $K \sim 1.8$ Å^{-1} which becomes the main peak in the neutron curve $S^n(K)$ at $K \sim 1.9$ Å^{-1}. In the x-ray case, $a = <f>^2/<f^2>$ is 0.986, and consequently $S^x(K)$ is dominated by $S_{NN}(K)$, whereas in the neutron case a is 0.046 and 0.088 for the $Ni_{35}Ti_{65}$ and $Ni_{40}Ti_{60}$ samples, respectively, and $S^n(K)$ is dominated by $S_{CC}(K)/(c_1c_2)$ assuming that $S_{NC}(K)$ is small.

Since only two experiments could be carried out on the same amorphous sample yielding the interference functions $S^x(K)$ and $S^n(K)$, a third function $S(K)$ must be known in order to evaluate the number-concentration partial structure factors $S_{N-C}(K)$ [Eq. (1)]. In the absence of a third experiment, we make the assumption that the term $S_{NC}(K)$ can be approximated by the Percus-Yevick hard-sphere model[5]. Using an atomic size ratio of 0.87 and a packing fraction 0.55, the hard-sphere $[S_{NC}(K)]_{HS}$ was calculated and used to evaluate $S_{NN}(K)$ and $S_{CC}(K)/(c_1c_2)$ which are shown in Figs. 1 and 2.

The Fourier transforms of $K[S_{NN}(K)-1]$, $K\{[S_{CC}(K)/(c_1c_2)]-1\}$, and $K S_{NC}(K)$ yielded the radial correlation functions $4\pi r^2\rho_{NN}(r)$, $4\pi r^2\rho_{CC}(r)$ and $4\pi r^2\rho_{NC}(r)$, respectively, which were used to evaluate the integrals Z_{N-C} over the range of r between zero and the first minimum in $4\pi r^2\rho_{NN}(r)$. The following values were obtained: $Z_{NN}=12.8$ and $Z_{NC}=0.39$ for both Ni-Ti alloys, but $Z_{CC} = -1.40$ for $Ni_{35}Ti_{65}$ and $Z_{CC} = -2.2$ for $Ni_{40}Ti_{60}$.

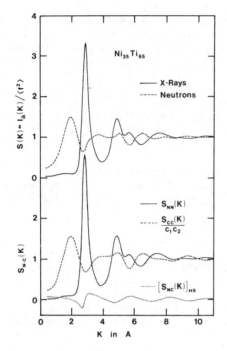

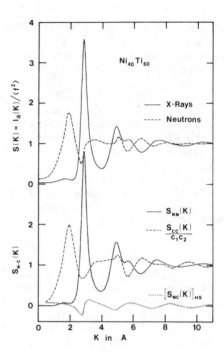

Fig. 1 Interference functions (structure factors) of $Ni_{35}Ti_{65}$

Fig. 2 Interference functions (structure factors) of $Ni_{40}Ti_{60}$

Using the definition of the Warren chemical short-range order parameter[6] α, i.e.,

$$\alpha = Z_{CC}/[Z_{NN} + Z_{NC}(c_2-c_1)/(c_1c_2)] = 1-Z_{12}/[c_2(c_2Z_1 + c_2Z_2)] \qquad (7)$$

the values of α were found to be $\alpha = -0.10$ for $Ni_{35}Ti_{65}$ and $\alpha = -0.15$ for $Ni_{40}Ti_{60}$ indicating a preference for unlike nearest neighbors in amorphous Ni-Ti alloys.

REFERENCES

1. A.B. Bhatia and D.E. Thornton, Phys. Rev. B2, 3004 (1970).
2. M. Sakata, N. Cowlam, and H.A. Davies, J. Physique 41, C8-190 (1980).
3. H. Ruppersberg, D. Lee, and C.N.J. Wagner, J. Phys. F 10, 1645 (1980).
4. C.N.J. Wagner, J. Non-Cryst. Solids 31, 1(1978).
5. N.W. Ashcroft and D.C. Lengreth, Phys. Rev. 156, 685 (1967).
6. C.N.J. Wagner and H. Ruppersberg, Atomic Energy Review (in press).

The research leading to this paper was supported by the grant DMR80-07939 from the National Science Foundation and by the Institute Laue-Langevin and the Deutsche Forschungsgemeinschaft.

THE OBSERVATION OF CLUSTER FORMATION AND CHEMICAL SHORT-RANGE ORDER IN LIQUID Bi-I ALLOYS

Kazuhiko Ichikawa and Takaaki Matsumoto*

Dept. of Chemistry and Dept. of Nuclear Engineering*,

Hokkaido University, Sapporo - 060, Japan

ABSTRACT

The differential cross-sections for neutron scattering from the I-rich melts of liquid Bi-I alloys have been measured by time-of-flight neutron diffractometry on the electron linac. The compositional dependences of the pair correlation functions G(r) have been reinvestigated for the melts up to 40 at% Bi. The flattening saddles and the second peaks are located at the short-distance side of about 4 Å or the I-I distance and at about 6.5 Å, respectively, for the compositions of 29.1 and 37.5 at% Bi. A tendency towards preference of like nearest neighbours Bi-Bi or formation of clusters consisting of Bi species may be achieved even at the I-rich side of the composition at 40 at% Bi. The metal-nonmetal transition quite well documented in the region of 40 and 50 at% may originate from the clusters.

INTRODUCTION

Current interest in the molten metal-salt solutions has arisen within the context of nuclear magnetic resonance (NMR)[1-3] and of neutron small angle scattering,[4] which are of considerable value in providing microscopic details. The observation of chemical short-range order and cluster formation can be expected in these solutions.

This paper reports a pulsed neutron scattering study of the binary solutions Bi_xI_{1-x} at the I-rich region up to 40 at% Bi, while the solutions at the compositions having liquid-liquid immiscibility and the Bi-rich solutions have been also investigated in our recent work[5] on the neutron scattering.

ISSN:0094-243X/82/890261-03$3.00 Copyright 1982 American Institute of Physics

EXPERIMENTAL

Neutron diffraction measurements were conducted on the time-of-flight (TOF) neutron diffractometer with the aid of the 45 MeV Hokkaido University Electron Linac, while experimental details are shown in the previous paper.[5]

RESULTS AND DISCUSSION

The measurements of the structure factors $S(Q)$ of the Bi_xI_{1-x} melts at 506°C were carried out. The distribution functions $G(r)$ in real space calculated from the Fourier transform of $S(Q)$ are plotted for the different concentrations versus the interatomic distance r in fig. 1. Figure 2 shows the compositional dependences of the nearest neighbour distance r_1 (o) and a flattening saddle ([) or

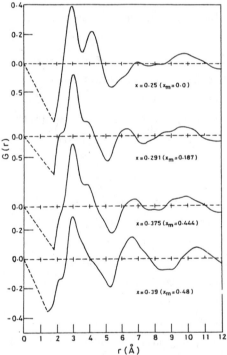

Fig. 1. The pair correlation functions $G(r)$ of Bi_xI_{1-x} melts at 506 ±5°C. The dashed lines were calculated from $-4\pi\rho r$ (ρ:number density)

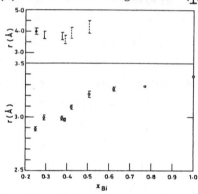

Fig. 2. The nearest neighbour Bi-I or Bi-Bi distance (o) and the second nearest neighbour I-I distance (•) versus atomic fraction of Bi. The signals of and are explained in this paper.

a sloping shoulder (⌐) located at the long-distance side of the first peak, together with the previous data.[5] The flattening saddle observed for the two solutions at 29.1 and 37.5 at% may be due to a supposed peak located between the first nearest neighbour Bi-I distance and the I-I distance. The probable peak ranges from 3.1 Å to 3.4 Å as well as a rather strong peak located at about 6.5 Å requires the existence of a number of clusters consisting of Bi species, which may be considered as probable configurations, such as $(Bi_3)^{3+}$, $(Bi_5)^{3+}$, $(Bi_8)^{2+}$ and $(Bi_9)^{5+}$. These polybismuth cations with oxidation states less than that of the familiar bismuth (III) were characterized in crystalline phase[6]: bonding distances range from about 3.0 to about 3.2 Å and according to MO bonding arguments for such homonuclear species it may be recognized that some valence-electrons are accommodated in the bonding MO's with the gap between the highest bonding and the lowest antibonding levels. Figure 2 elucidates an abrupt increase of r_1 between 40 and 50 at% Bi. It may be proved in relation to the strong concentration dependences of the nearest neighbour Bi-Bi distance in such clusters as well as their size and number density. At the I-rich side of about 50 at% Bi the transport coefficients[7,8] and NMR[1] show an important variation with concentration. At about 50 at% Bi it seems to be quite all right to consider the structural and bonding pattern in the clusters to be characteristic of a delocalized state of valence electrons: the clusters may give rise to microscopic inhomogeneity and the metal-nonmetal transion.

REFERENCES

1. R. Dupree and W. W. Warren, Proc. Conf. on Liquid Metals (The Inst. Phys., London, 1977) p.454.
2. K. Ichikawa and W. W. Warren, Phys. Rev. B20, 900 (1979).
3. S. Sotier and W. W. Warren, J. Physiq. Suppl. 8, 32 (1980).
4. J. F. Jal, P. Chieux and J. Dupuy, J. Appl. Cryst. 11 (1978) 610.
5. K. Ichikawa and T. Matsumoto, Phys. Lett. 83A, 35 (1981).
6. J. D. Corbett, Inorg. Chem. 7, 198 (1968).
7. L. F. Grantham and J. Yoshim, J. Chem. Phys. 38, 1671 (1963).
8. K. Ichikawa and M. Shimoji, Philos. Mag. 19, 33 (1969); Ber. Bunsenges, Phys. Chem. 73, 302 (1969).

264

QNS MEASUREMENTS ON WATER IN BIOLOGICAL AND MODEL SYSTEMS*

E. C. Trantham and H. E. Rorschach
Rice University, Houston, TX 77001

J. C. Clegg
University of Miami, Coral Gables, FL 33124

C. F. Hazlewood
Baylor College of Medicine, Houston, TX 77030

R. M. Nicklow
Oak Ridge National Laboratory+, Oak Ridge, TN 37830

ABSTRACT

Results are presented on the quasi-elastic spectra of 0.95 THz neutrons scattered from pure water, a 20% agarose gel and cysts of the brine shrimp (Artemia) of hydration 1.2 gms H_2O per gm of dry solids. The lines are interpreted with a two-component model in which the hydration water scatters elastically and the "free" water is described by a jump-diffusion correlation function. The results for the line widths $\Gamma(Q^2)$ are in good agreement with previous measurements for the water sample but show deviations from pure water at large Q for agarose and the Artemia cysts that suggest an increased value of the residence time in the jump-diffusion model.

INTRODUCTION

The diffusive motion of protons in water solutions has often been studied by NMR spin-echo techniques. Present pulse technology requires a measuring time of a few milliseconds, which for water gives a diffusion distance $\sqrt{2Dt} \stackrel{\sim}{>} 1$ micron. This introduces difficulties in the interpretation of the results for a heterogeneous system such as the biological cell where the scale size is also of the order of a micron. The measured diffusion coefficient is then an average over the cellular environment.

The method of quasi-elastic neutron scattering (QNS) offers the possibility both to reduce the "measuring distance" to a few angstroms and to obtain details about the diffusion dynamics not available from NMR data. The results of measuring QNS spectra are usually interpreted within the framework of the Van Hove theory,[1] which relates the scattering law $S(Q,\omega)$ to the space-time (self) correlation function $G_S(\vec{r},t)$ of the scattering nuclei. $G_S(\vec{r},t)$ can be calculated for various models chosen to represent the motion of the diffusing particles, and the parameters of the models can be determined by a fit to the experimental spectra.

*Supported in part by ONR Contracts N00014-79-C-0492 and N00014-760 C-0100, and by ORAU Participation Agreement S-2016.
+Operated by Union Carbide Corp. under U.S. Dept. of Energy Contract W-7405-eng-26.

We have measured the QNS spectra of a 20% gel of the poly-saccharide <u>agarose</u> and the cyst of the brine shrimp (<u>Artemia</u>) and compared them with the spectrum of pure water.

EXPERIMENTAL

Agarose gel is a polymer $(C_{12}H_{17}O_9)_n$ of known structure.[2] It contains 4 OH groups and 13 covalently bonded portons in each repeat unit. It was selected for a pilot study to determine if the spectro-meter resolution was sufficient to separate the quasi-elastic line due to the water protons from the elastic line of the polymer protons. The brine shrimp cysts have a diameter of $\sim$ 0.2 mm and consist of an inner mass of about 4000 cells surrounded by a complex non-cellular shell. Their biochemical and physical properties have been studied,[3] and the diffusion coefficient of water for various degrees of hydra-tion has been measured by NMR methods.[4] The degree of hydration can be varied over a wide range (0.02-1.5 gm H_2O/gm dry cyst) while still retaining viability for the extended periods needed for the neutron spectrometer scans. The neutron spectra were determined on the triple-axis spectrometer HB-2 at ORNL. The samples were contained in alumi-num sample holders sealed with an indium gasket. The sample thickness was 1 mm for the water and agarose gel and 2 mm for the brine shrimp cysts. Spectrometer scans of neutron energy change were made at five Q values between 0.7 and 1.9 A^{-1}. The lines were analyzed by a least-squares fit with a 5-parameter function which included a linear back-ground and the convolution of the spectrometer resolution function with an elastic line and a Lorentzian quasi-elastic line. The width Γ of the Lorentzian line and the intensities of the two components were determined at each value of Q from the fitting parameters. The contribution from the agarose protons was determined by scattering from a D_2O gel. For the brine shrimp, the contribution from the protons of the "dry" cyst was determined from an extrapolation of line intensity vs. hydration to zero water content.

RESULTS

Figure 1 shows the line width Γ as a function of Q^2 for pure water (solid curve) and for the 20% agarose and the 1.2 g H_2O/g dry solids cysts. The data points are not shown for water. Their error bars are similar to those for the agarose and brine shrimp results.

DISCUSSION AND CONCLUSIONS

The line width $\Gamma(Q^2)$ for water and the agarose gel could be fitted by a jump diffusion model[5] with Γ given by

$$\Gamma = \frac{Q^2 D}{1 + Q^2 D \tau_0} .$$

For water, a fit of $\Gamma(Q^2)$ gives D = 2.4×10^{-5} cm^2/sec and $\tau_0 \approx 10^{-12}$ sec. For agarose, the initial slope of $\Gamma(Q^2)$ is nearly the same as for pure water, although data at smaller Q are needed to be certain

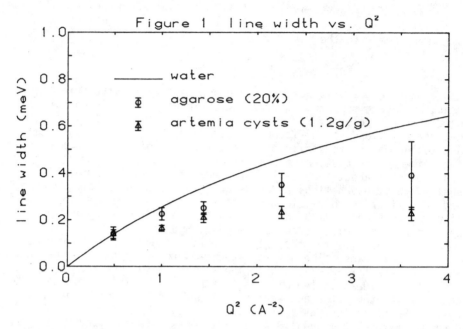

Figure 1   line width vs. Q²

of this. There are deviations from the water curve at large Q, indicating an increased value of τ_0.

The brine shrimp data show substantial deviations from pure
water. The line width is relatively independent of Q and is much
narrower than that of pure water. This behavior is consistent with
the NMR data on translational diffusion[4] which shows that the diffusion coefficient is greatly reduced from the pure-water value.
The constancy of the line width with Q is also consistent with a rotational diffusion model,[6] but we have not yet developed a model
which is consistent with all the features of the spectra.

REFERENCES

1. L. Van Hove, Phys. Rev. 95, 259 (1954).
2. C. R. Noller, "Chemistry of Organic Compounds" (Philadelphia,
    1965), p. 432.
3. J. S. Clegg in "Cell-Associated Water," ed. by W. Drost-Hansen
    and J. S. Clegg, Academic Press (N.Y. 1979) pp. 165-259.
4. P. K. Seitz, D. C. Chang, C. F. Hazlewood, H. E. Rorschach and
    J. S. Clegg, "The Self-Diffusion in Water in Artemia Cysts,"
    Arch. Biochem. and Biophys., in press.
5. T. Springer, "Quasi-Elastic Neutron Scattering for the Investi-
    gation of Diffusive Motions in Solids and Liquids (Springer-
    Verlag, NY 1972) pp. 41-44.
6. Ibid., pp. 64-69.

THE ROTATIONAL POTENTIAL OF THE
AMMONIUM MOLECULE IN $(NH_4)_2SnCl_6$

K. Vogt, W. Prandl and D. Hohlwein

Institut für Kristallographie der Universität Tübingen
Charlottenstr. 33, D-7400 Tübingen, West-Germany

ABSTRACT

Neutron Bragg scattering data taken at 140 and 293 K are used to compute the rotational potential of the NH_4 molecule in $(NH_4)_2SnCl_6$. In an expansion of the potential into rotator functions which are invariant under the direct product group of the site and the molecular point group only terms with $l \leq 6$ are found to be necessary. The potential does not depend on the temperature.

INTRODUCTION

$(NH_4)_2SnCl_6$ has the K_2PtCl_6 structure with the space group Fm3m. NH_4 molecules occupy the c position where the site group $\bar{4}3m$ is identical to the molecular point group. The orientation of the NH_4 molecules is determined by the rotational potential $V(\omega)$ where ω represents a set of Eulerian angles (α,β,γ) between a coordinate system Σ' fixed in the molecule and a system Σ fixed in the crystal.

Activation energies for reorientation of the NH_4 molecules, energies of the first and the second excited librational state and the splitting of the tunnel states have been extracted from NMR [1], heat capacity [2] and inelastic neutron scattering measurements [3]. The results of the neutron experiment [3] have been used to calculate the coefficients of a rotational potential expanded in cubic rotator functions [4,5].

In this paper we describe a method to determine the rotational potential from neutron Bragg scattering experiments.

METHOD [6,7,8]

We take the origins of Σ and Σ' to coincide with the center of mass of the NH_4 molecule. The NH_4 contribution to the structure factor $F(\underline{Q})$ is the Fourier transform of the averaged scattering length density $a(\underline{r})$ which may be expanded in symmetry adapted functions [9,10] (SAF) $P_{1\gamma}(\vartheta,\varphi)$ of the point group $P = \bar{4}3m$ of the NH_4 position:

$$a(\underline{r}) = \Sigma_s b_s \delta(r-r_s)/r^2 \Sigma_{1\gamma} c_{1\gamma} P_{1\gamma}(\vartheta,\varphi) \tag{1}$$

r_s is the radius of a shell of equivalent atoms with a scattering length b_s, l is the order of the SAF and greek indices label unit representations. In an analysis of the Bragg scattering $c_{1\gamma}$ may be determined directly.

In the primed system the scattering length density $b(\underline{r}')$ may be expanded in SAF's $\Pi_{1\tau}(\vartheta',\varphi')$ of the molecular point group $\Pi=\bar{4}3m$ and transformed to the unprimed system with the aid of rotator functions

$\mathcal{M}_{\gamma\tau}^{1}(P,\Pi;\omega)$. With the molecular constants $\bar{\bar{\Pi}}_{1\tau}^{s}= \sum_{i}\Pi_{1\tau}(\boldsymbol{\vartheta}_{si}',\boldsymbol{\varphi}_{si}')$

$$b(\underline{r},\omega) = \sum_{s} b_{s}\delta(r-r_{s})/r^{2} \sum_{1\gamma\tau}\bar{\bar{\Pi}}_{1\tau}^{s}P_{1\tau}(\boldsymbol{\vartheta},\boldsymbol{\varphi})\mathcal{M}_{\gamma\tau}^{1}(\omega) \qquad (2)$$

The averaged scattering length density is obtained by

$$a(\underline{r}) = \int f(\omega)b(\underline{r},\omega)d\omega \qquad (3)$$

where $d\omega = \sin\beta d\beta d\alpha d\gamma$ and $f(\omega)$ is the probability density function for the relative orientations of Σ' and Σ expressed in Eulerian angles. In the region of classical statistics

$$f(\omega) = Z^{-1} \exp\{-\beta V(\omega)\} \qquad (4)$$

with $\beta = 1/k_{B}T$ and $Z = \int\exp\{-\beta V(\omega)\}d\omega$. The rotational potential $V(\omega)$ is expanded in rotator functions [8]. For $P =\Pi= \bar{4}3m$ and, tentatively, $1<8$ only the rotator functions with $l=0,3,4,6,7,8$ are symmetry allowed. The $l=0$ function is constant and may be omitted. So we use:

$$V(\omega) = V_{3}\mathcal{M}_{11}^{3}(\omega) + V_{4}\mathcal{M}_{11}^{4}(\omega) + V_{6}\mathcal{M}_{11}^{6}(\omega) \qquad (5)$$

A comparison of (3) and (1) yields

$$c_{11}^{s}=\bar{\Pi}_{11}^{s}Z^{-1}\int\mathcal{M}_{11}^{1}(\omega)\exp\{-\beta V(\omega)\}d\omega \qquad (6)$$

Now instead of c_{11}^{s} the potential coefficients V_{1} may be determined from the experiment.

RESULTS

Data were taken with the instruments P32 and P110 at the FR2 reactor at Karlsruhe. Two data sets at T = 140 K and 293 K with $h^{2}+k^{2}+l^{2} \leq 283$, and ≤ 387, respectively were corrected for absorption taking into account the crystal shape. In the structural model parameters for isotropic extinction, isotropic temperature factors for Sn and NH_{4}, anisotropic temperature factors for Cl, a third order temperature factor for NH_{4} and the bond lengths Sn-Cl and N-H were refined together with the coefficients V_{1} of the rotational potential up to $l=8$. V_{7} and V_{8} turned out to be strongly correlated with the other V_{1}. Their introduction did not improve the R-values nor did they change significantly the values of $e_{11}^{H}=c_{11}^{H}/\bar{\Pi}_{11}^{H}$ and with this the orientational distribution of the H-atoms calculated from (1). For three selected rotational axes the potential with the computed errors is shown in fig. 1. The barriers to rotation range from 800 K to 900 K which is in agreement with observed reorientational activation energies between 600 K and 750 K [1,2,3]. The R-value for the data set at T=140 K is the same whether the potential parameters are refined freely or kept fixed on their 293 K-values. In any case the coefficients e_{11}^{H} are nearly identical. That the potential at 140 K is not equally well defined as at 293 K may be due to quantum effects since the first excited librational state corresponds to 156 K [3].

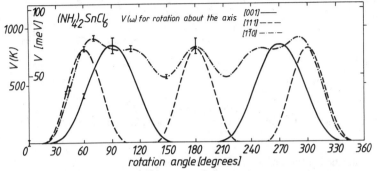

Fig. 1.

The rotational potential

Table I Coefficients of the rotational
potential and of the density distribution

	Smith [4]		Hüller, Müller [5]		this work	
1	V_1/meV	e_{11}^H *	V_1/meV	e_{11}^H *	V_1/meV	e_{11}^H *
3	-95.9	0.427	-69.0	0.379	-35.7±1.9	0.258
4	0	0.236	-20.1	0.223	-38.1±3.1	0.195
6	30.8	0.027	33.8	-0.008	32.1±2.1	-0.047
7	--	-0.003	--	-0.005	---	-0.001
8	--	-0.002	--	0.013	---	0.025

*calculated from the potential $V(\omega)$ with eq. (6)

We are grateful to Professor Hüller for making available results
prior to publication. We acknowledge financial support by the Bundesministerium für Forschung und Technologie under project no. KNF
03-41E03P; 04-45E03I.

REFERENCES

1. A. Watton, A. R. Sharp, H. E. Petch and M. M. Pintar, Phys. Rev.
   B5, 4281 (1972).
2. R. G. S. Morfee, L. A. K. Staveley, S. T. Walters and D. L. Wig-
   ley, J. Phys. Chem. Solids 13, 132 (1960).
3. M. Prager, W. Press, B. Alefeld and A. Hüller, J. Chem. Phys. 67,
   5126 (1977).
4. D. Smith, Chem. Phys. Letters 66, 84 (1979).
5. A. Hüller, W. Müller, private communication.
6. W. Press and A. Hüller, Acta Cryst. A29, 252 (1973).
7. W. Press, Acta Cryst. A29, 257 (1973).
8. A. Hüller and W. Press, Acta Cryst. A35, 876 (1979).
9. C. J. Bradley and A. P. Cracknell, The mathematical theory of
   symmetry in solids (Clarendon Press, Oxford, 1972), pp. 51-76.
10. W. Prandl, Acta Cryst. A37 (1981) in the press.

REORIENTATIONAL MOTIONS OF NH_2 GROUPS IN POTASSIUM AMIDE

L. Tielemans, W. Wegener
Materials Science Department, S.C.K./C.E.N., B-2400 Mol, Belgium

A. Dianoux
Institut Laue-Langevin, Grenoble, France

P. Vorderwisch
Hahn-Meitner-Institut für Kernforschung Berlin GmbH, Germany

L. Van Gerven
Katholieke Universiteit Leuven, Belgium

ABSTRACT

Rotational motions of NH_2 groups have been observed in the three known solid phases of KNH_2 by means of quasielastic neutron scattering. The observed intensities of the elastic and quasielastic components can be accounted for by various reorientational jump models between equivalent equilibrium orientations of the NH_2 ion: 180° jumps about the molecular axis (monoclinic phase); 90° jumps about the fourfold crystal axis (tetragonal phase); most probably 120° jumps about the threefold axes (cubic phase). In some cases preliminary values of residence time have been obtained from line-shape fits.

INTRODUCTION

The structure of KNH_2 and KND_2 has been determined by X-ray and neutron diffraction[1,2] to be cubic for $T > 348$ K with the space group Fm3m; to be tetragonal between 348 K and 326 K with the space group P4/nmm; to be monoclinic for $T < 326$ K with the space group $P2_1/m$. In all phases the NH_2 ion is surrounded by an octahedron of K ions. In that work it was concluded that the NH_2 groups are freely rotating in the cubic phase; that these motions are restricted to strong precessions in the tetragonal phase and to large rocking vibrations in one plane in the monoclinic phase.

MEASUREMENTS

The experiment was performed on a flat polycrystalline KNH_2 sample, using the IN5-spectronometer at the Institut Laue-Langevin in Grenoble with an incident wavelength of 5.14 Å at 341 K, 320 K, 353 K, 373 K, 393 K and of 3.47 Å at 373 K. The energy resolution (FWHM) was 0.12 meV and 0.4 meV for those two incident wavelengths. The scattering angles ranged from 13° to 124°. Furthermore empty can and vanadium measurements were performed in order to correct for container scattering, background and detector efficiency.

ISSN:0094-243X/82/890270-03$3.00 Copyright 1982 American Institute of Phys

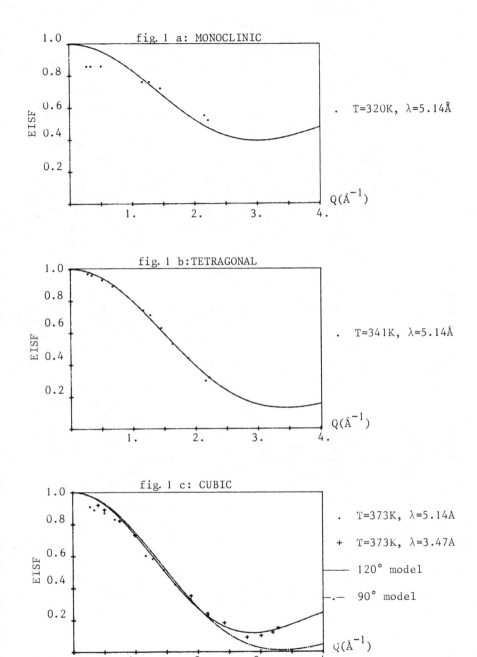

Fig. 1. Elastic incoherent structure factor versus momentum transfer Q for the three solid phases of KNH_2.

272

ANALYSIS

In opposition with the conclusions of the structural study, we have chosen to describe the rotational motions of the NH_2 groups by reorientational jump models. The structure of the monoclinic phase allows only one equivalent equilibrium orientation and therefore only 180° jumps about the molecular axis can occur. In the tetragonal and cubic phase no equivalent equilibrium orientations have been found by neutron diffraction work[1]. In the tetragonal phase we can assume an alignment of the molecular axis with the fourfold crystal axis, which leads to two equivalent equilibrium orientations and suggests 90° jumps about the fourfold axis. In the cubic phase, based on steric considerations and calculations with Born-Meyer potentials, we have selected twelve equivalent equilibrium orientations: the molecular axis aligned with a fourfold axis [001] and the molecular plane lying in the (110) symmetry plane. Two random models have been considered: 1) 90° jumps about the fourfold axes; 2) 120° jumps about the threefold axes.

RESULTS

In a preliminary analysis we have extracted the intensities of the elastic and quasielastic components, from which the elastic incoherent structure factor EISF can be obtained. For the three phases the experimental values of the EISF are shown as a function of momentum transfer Q in fig. 1. In the monoclinic phase we have fitted the theoretical EISF to the experimental points in fig. 1a and we got a H-H distance of 1.5±0.2 Å. From lineshape fits we obtained a residence time of about 25 ps. In the tetragonal phase, from analogous fits, we obtained a H-H distance of 1.65±0.05 Å and a residence time of about 8ps. In the cubic phase we only fitted the model EISF to the experimental points on fig. 1c. There we see that the 120° jump model gives the best description of the results. From this fit we obtained a N-H distance of 1.015±0.05 Å and a H-N-H angle of 109±18°, values which are in good agreement with the results of the neutron diffraction work[1].

In continuing the analysis of these data we will make multiple scattering corrections and fit with models wich have different residence times for jumps about different crystal axis in the cubic phase.

REFERENCES

1. M. Nagib, E. von Osten, H. Jacobs, Atomkernenergie 29 p. 41 (1977)
2. E. von Osten, Doctoral thesis, Aachen (1978)

ACKNOXLEDGEMENTS

We are grateful to Dr. S. Hautecler and Dr. E. Legrand for suggesting this subject and for useful discussions at various stages of this study. One of the authors (L.T.) is indebted to I.I.K.W. for granting financial support.

APPLICATION OF AB INITIO CALCULATIONS TO DEFECT INDUCED DISTORTION FIELDS

Klaus Werner*
Solid State Division, Oak Ridge National Laboratory[†]
Oak Ridge, Tennessee 37830

George Solt
Université de Lausanne
CH-1015 Lausanne
Switzerland

ABSTRACT

Using the diffuse-elastic neutron scattering (DENS) technique, the defect-induced strain fields of several Al-based dilute binary alloys were experimentally determined. It turned out that the measured fine structure could not be satisfactorily interpreted using force-constant methods (phenomenological Kanzaki model). In contrast, ab initio calculations based on the framework of the perturbed electron liquid-pseudopotential formalism could reproduce both the experimentally observed q-dependence of the Fourier transform of the distortion field and the known value of the volume change.

INTRODUCTION

Even very small defect concentrations in pure metals may change the properties of the system drastically. Thus, to understand the properties of some alloy systems, it is essential to understand first the properties of point defects. Point defects in a nonmagnetic matrix give rise to impurity induced distortions, which can be accurately determined by employing the technique of DENS.[1] The systems under investigation were dilute Al-based alloys, e.g., AlZn, AlMg, AlLi, or AlGa. Aluminum was used as host material because of its small absorption for thermal neutrons, and its small incoherent cross section; moreover, it is possible to grow comparably large alloy single crystals (20 mm diameter, 100 mm length) of relatively good mosaic spread (10' - 30' FWHM). The impurities were chosen to enable the application of ab initio pseudopotential calculations, which require simple s-p bond metals.

EXPERIMENT

The experiments were carried out using the spectrometer DNS1 at the Jülich FRJ2 reactor. Applying TOF analysis, reference measurements with pure Al crystals, and V-calibration, it was feasible to

*Permanent address: IFF, Kernforschungsanlage Jülich, D-5170 Jülich, West Germany.
[†]Operated by Union Carbide Corporation under contract W-7405-eng-26 with the U. S. Department of Energy.

obtain DENS cross sections in absolute units. Due to the relation

$$\frac{d\sigma}{d\Omega} = c \left| b_i - b_h + b_h \, i\underline{Q} \, \tilde{\underline{u}}(\underline{Q}) \right|^2 \tag{1}$$

(b_i and b_h: nuclear scattering amplitudes of the impurity and the host, c: defect concentration in at.% and Q: momentum transfer of the neutrons) it is possible to determine the Fourier Transform (FT) of the distortion field $\tilde{\underline{u}}(\underline{Q})$ directly.

Results for $d\sigma/d\Omega$ are plotted in Fig. 1 (for $\underline{Al}Mg$) and Fig. 2 ($\underline{Al}Li$).

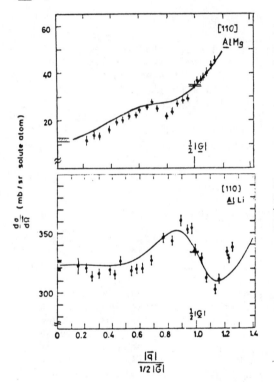

Fig. 1. DENS cross-section for $\underline{Al}Mg$ in [110] direction. Solid line: Pseudopotential calculation.

Fig. 2. DENS cross-section and pseudopotential calculation for $\underline{Al}Li$. q parallel [110].

THEORY

Minimizing the total energy E_{tot} of the lattice with respect to the individual displacement one obtains the equation (basis equation of lattice statics)[2]

$$\tilde{\underline{u}}(\underline{q})_\alpha = i \sum_\beta g_{\alpha\beta}(\underline{q}) \sum_\ell f(\underline{\ell})_\beta \, \sin(\underline{q}\cdot\underline{\ell}), \tag{2}$$

(g: FT of the lattice Green's function) which combines the FT of the distortion field $\tilde{\underline{u}}(\underline{q})$ with the defect induced forces $\underline{f}(\underline{\ell})$. As these forces are not known, several models were employed for their determination. While in the past the phenomenological Kanzaki model was widely used,[3] it turned out that this approach is not able to reproduce the fine structure in the measured DENS cross section[4] (unless many fit parameters are adjusted to the data). A different approach to determine the defect induced forces was developed recently[5]: the total energy of the system under investigation is expanded in perturbation series in terms of the ion-electron potential, for which one uses a Heine-Abarenkov type pseudopotential. Introducing the Toigo-Woodruff screening function and expanding the series up to the fourth order in the potentials, one gets expressions for the interatomic forces and the coupling constants. As the two parameters in the HA potential (range and well depth) are determined by using only the lattice constant and the c_{44} value of the pure elements involved, we speak about an ab initio calculation. Comparison with (1) and (2) gives the ab initio calculated results as shown in Figs. 1 and 2 by solid lines.

DISCUSSION

It is seen from the two figures that it is possible with DENS experiments to determine the defect-induced distortion field fairly accurately. Further, one notices that ab initio calculations are able to reproduce the absolute value (in particular the Q→0 value) and the structure of the experimentally evaluated data satisfactorily, a result which to our knowledge, has not been found before.

REFERENCES

1. G. S. Bauer, in Treatise on Materials Science and Technology, edited by H. Herman (Academic Press, New York, 1979), Vol. 15.
2. H. Kanzaki, Phys. Chem. Sol. 29, 24 (1957).
3. G. S. Bauer, E. Seitz and W. Just, J. Appl. Crystallogr. 8, 162, (1975).
4. K. Werner, W. Schmatz, G. S. Bauer, E. Seitz, H. J. Fenzl, and A. Bavatoff, J. Phys. F8, L207 (1978).
5. G. Solt and K. Werner, Phys. Rev., in press October 1980.

SHORT RANGE ORDER DETERMINATION IN A TERNARY ALLOY

$$Fe_{0.56}Ni_{0.23}Cr_{0.21}$$

F. Bley, P. Cénédèse
ILL, 156 X Centre de Tri, 38042 Grenoble Cedex, France

S. Lefebvre
C.E.C.M., 15, rue Georges Urbain, 94400 Vitry-sur-Seine, France

ABSTRACT

The local atomic arrangement in a ternary alloy has been, for the first time, investigated by neutron diffuse scattering from three single crystals of three different isotopic compositions.

The static displacements terms are shown to be small. The local order is very weak . The three sets of Cowley's parameters are measured.

INTRODUCTION

- The chosen alloy is a stainless steel of composition $Fe_{0.56}Ni_{0.23}Cr_{0.21}$. Owing to the closeness of the values of the X-Ray scattering factors for the three elements, only a neutron scattering experiment can enable the determination of pair correlation parameters. However we need three independant measurements on three alloys with different isotopic composition.

- The chosen composition lies in a domain of the ternary diagram where the austenite phase (γ fcc) is stable on cooling from the melt to room temperature. This particular composition has been used in cryogenic engineering and thus a significant amount of structural data is available (e.g. melting points (1), martensitic transition temperature and magnetic properties (2,3)).

- Local order, the nature of which is unknown, has been demonstrated by some experiments : a) dislocations pairs have been observed by electron microscopy studies in alloys with composition extending from $Fe_{0.56}Ni_{0.23}Cr_{0.21}$ to $Fe_{0.40}Ni_{0.40}Cr_{0.20}$ (4)
b) resistivity of these alloys increases with increasing annealing time at 500°C after quenching from 700°C.
c) magnetic properties may be accounted for by inhomogeneities (2-3) in local order.

ISSN:0094-243X/82/890276-03$3.00 Copyright 1982 American Institute of Phy

EXPERIMENT

- Three single crystals are grown together in the same furnace by a Bridgman method. The alloys are heated under vacuum to the melting temperature and then cooled very slowly under hydrogined argon. This avoids variations of Cr composition due evaporation of Cr close to the melting point. The crystals are annealed at 1000°C for one hour then annealed at 500°C for ten hours and quenched.

- The different isotopic compositions are given in table I.

Table I : Laue values and isotopic composition

Alloy	Composition	average value	Laue of each pairs		
			Ni-Cr	Fe-Cr	Fe-Ni
1	Fe $Ni^{62}Cr^{52}$	0.51	0.080	0.026	0.403
2	$Fe^{54}Ni^{62}Cr^{52}$	0.285	0.080	0.0005	0.195
3	Fe $Ni^{58}Cr^{52}$	0.100	0.043	0.026	0.029

- The diffuse intensities are measured at room temperature on the four circle D_{10} diffractometer at ILL (Grenoble) as described elsewhere (5). The diffractometer is equiped with a graphite analysor allowing a very good energy analysis. The wave length is 1.26 Å . The experimental intensity is corrected for background intensity and normalised using the diffuse intensity from a vanadium sample of the same shape as the crystal.

RESULTS

The diffuse intensity of each alloy shows maxima at 100 and 110 positions and seems to increase near the Bragg peaks.

Neglecting displacement terms, the diffuse intensity of a ternary alloy whose component have scattering length b_j and atomic composition c_j is the Fourier transform

$$I \, SRO \, (\vec{k}) = FT \, \left| \Sigma_{ij} \atop j>i \right. \, c_j c_j (b_i - b_j)^2 \, \alpha^{ij} \, (\vec{r})$$

$c_i c_j (b_i - b_j)^2$ is the Laue monotonic scattering for the paires of atoms i and j. The different values are given in table I. The resolution of the three linear equations give the values of α'_s parameters of each pairs of atoms, Table II.

Table II : α' parameters for each pairs of atoms
1

	Fe-Ni	Fe-Cr	Ni-Cr
000	1.052	1.409	0.645
110	0.021	0.021	− 0.138
200	0.019	0.148	0.078
211	− 0.005	0.035	0.019
220	− 0.0017	− 0.007	0.012
310	− 0.005	0.001	0.009
222	0.009	− 0.025	− 0.003
321	0.003	− 0.0015	− 0.005
400	0.013	0.101	− 0.03
411	0.005	0.017	− 0.01
330	− 0.002	0.004	0.011

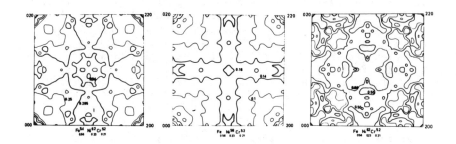

REFERENCES

1. E. Schurmann, J. Brauckmann, Arch. Eisenhüttenwes. 48,1977,p.3-7
2. L.A.A. Warnes, H.W. King, Cryogenics, 16, 1976, p. 473-481
   L.A.A. Warnes, H.W. King, Cryogenics, 16, 1976, p. 659-667
3. A.Z. Men'Shihov, A.Y.E. Teplykh, Fiz Metal Metalloved, 44, n° 6,
   1977, p. 1215-1221
4. W. Bell, W.R. Roser, G. Thomas, Acta Met.,12, 1964,p.1347-1253
5. S. Lefebvre, F. Bley, M. Bessiere, M. Fayard, M. Roth, Acta
   Cryst. A 36, 1980, p. 1-7
   S. Lefebvre, F. Bley, M. Fayard, M. Roth, Acta Met. 29, 1981,
   749-761

THE UNCONVENTIONAL EXCITATIONS AND CRITICAL BEHAVIOUR
IN ACTINIDE COMPOUNDS

T.M. Holden, W.J.L. Buyers, E.C. Svensson, A.F. Murray and J.A. Jackman
Atomic Energy of Canada Limited, Chalk River, Ontario, KOJ 1J0, Canada

P. de V. DuPlessis
Rand Africaans University, Johannesberg, South Africa

G.H. Lander
Argonne National Laboratory, Argonne, Illinois, USA 60439

O. Vogt
ETH, Zürich, Switzerland

ABSTRACT

The low lying electronic states of the uranium ion, at present imperfectly understood, have been probed by neutron inelastic scattering methods. In the rock-salt structure pnictides and chalcogenides the character of the magnetic excitations and the magnitudes of the ordered moments are highly sensitive to the U-U distance. For example, in the ferromagnetic chalcogenides, US with the smallest U-U distance exhibits only a broad distribution of scattering at low temperatures, whereas UTe with the largest U-U spacing exhibits several distinct branches of excitations. Similarly, for the antiferromagnetic pnictides those with small U-U distance (UN, UAs) exhibit only a broad magnetic response with no sign of the sharp collective features seen in USb. The pnictides exhibit very anisotropic longitudinal spin correlations above T_N. For UN a dynamical scaling model has been obtained which incorporates the anisotropy and describes the linewidths of the critical scattering accurately.

INTRODUCTION

The characteristic feature of the electronic structure of the uranium rock-salt structure pnictide and chalcogenide compounds is the presence of hybridized 5f- and 6d- electrons very close to the Fermi energy. It seems likely that these groups of electrons dominate the magnetic behaviour, such as the magnitude of the ordered magnetic moments and the nature of the excitation spectrum, and also affect the elastic properties. The static properties of these compounds are listed in Table I. The low temperature magnetic structure of UN has recently been shown[1] to be a single-k structure (type I antiferromagnet), while the magnetic structure of UAs has been shown to be double-k, and that of USb to be triple-k with net staggered magnetization in [001] directions. The chalcogenide ferromagnets have a [111] easy direction.

Neutron inelastic scattering methods have been applied in recent years to study the magnetic inelastic response and the lattice vibrations of several of the antiferromagnetic pnictide compounds

ISSN:0094-243X/82/890279-09$3.00 Copyright 1982 American Institute of Physics

Table I

	a Å	T_{crit} K	Magnetic Structure	$<\mu>_n^{\dagger}$ (μ_B)	$<\mu>_m^{\dagger}$ (μ_B)
UTe	6.155	103	F	2.25±0.05	1.91±0.05
US	5.489	172	F	1.70±0.03	1.55±0.02
USb	6.176-6.209	217-246	AF I	2.82±0.05	
UAs	5.779	126	AF I,T>0.5T_N	1.93	
UN	4.890	49.5±0.5	AF I	0.75±0.10	

† Subscripts n and m denote ordered moments derived from neutron scattering and magnetization measurements respectively.

(UN[2,3], UAs[4,5], USb[6,7]) and two of the ferromagnetic chalcogenides[3,8] (US, UTe). Above the ordering temperature all the compounds exhibit paramagnetic scattering centred on zero frequency. As the temperature is lowered towards the ordering temperature the pnictides exhibit highly anisotropic "critical scattering" whose form demonstrates the presence of long-range correlations within sheets of spin parallel to cube faces, but much weaker correlations between sheets i.e. quasi two-dimensional behaviour. Furthermore, the critical scattering is entirely longitudinal. Below the ordering temperature the two compounds with the largest lattice parameters (USb, UTe) do exhibit some sharp collective features which can be identified as spin-waves, but the other compounds only show a broad frequency distribution of very weak scattering.

CHARACTERISTIC RESPONSE

The scattering observed at the ordering wavevector for UTe, US and UN at 300 and 4.2K is shown in Fig. 1. At 300 K the distributions at low frequency have the form typical of paramagnetic scattering: that is to say, a quasi-elastic peak whose integrated intensity gives the magnitude of the effective, randomly oriented, local moments and whose width is determined primarily by interatomic coupling. For UTe the paramagnetic scattering is replaced at 4.2 K by a sharp excitation at 3.45±0.10 THz which is identified as a spin-wave by its structure factor and by its quadratic dispersion law above the large gap frequency. The optic phonon at 4.75±0.20 THz is coupled to the magnetic excitation since it has increased in intensity at low temperatures. UTe thus exhibits well-defined collective spin excitations characteristic of conventional localized or itinerant ferromagnets.

On the other hand, the paramagnetic scattering in US far exceeds the low temperature response. At 4.2 K there is no evidence up to 9 THz for sharp excitations like those in UTe. Instead the scattering, obtained by subtracting the instrumental background, shown

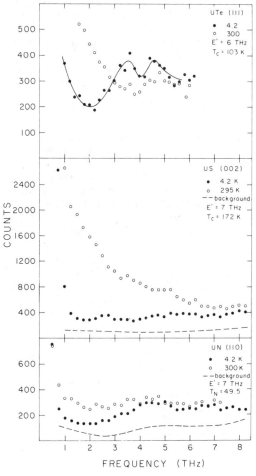

as a broken line in Fig. 1, has the form of a continuum extending to at least 9 THz. Since two-phonon processes can also give a continuum of scattering, the magnetic character has to be established by several measurements at equivalent wavevectors.

For UN the paramagnetic scattering is narrower in frequency than for UTe and US but it is still larger than the low frequency, low temperature response. Beyond 4.5 THz, however, the scattering is nearly temperature independent. Because the scattering is proportional to the temperature factor, $1+n_\nu$, and to the imaginary part of the susceptibility, $\chi''(Q,\nu)$, the temperature dependence of χ'' may be derived from the results. For antiferromagnetic UN, χ'' decreases with temperature indicating a significant change in the population of states from which the magnetic transitions are made. Curiously, the paramagnetic scattering below 2 THz in UN is stronger at the nuclear reciprocal lattice point (002) than at the ordering wavevector.

Fig. 1 Frequency distribution of scattered neutrons at 300 and 4.2 K for UTe, US and UN.

CRITICAL SCATTERING

The scattering in the paramagnetic phase, which precedes the development of the ordered structure, has been studied in detail for UN[3,9,10], USb[7] and UAs[4]. All three antiferromagnetic pnictides are characterized by highly anisotropic critical magnetic scattering even though the lattice is cubic. The scattering is much wider along the [00ζ] direction of the ordering wavevector than along the perpendicular [$\zeta\zeta$0] direction. Results for UN[10] at $T/T_N = 1.065$ and for UAs[4] at $T/T_N = 1.01$ are shown in Figs. 2 and 3. UAs is the most

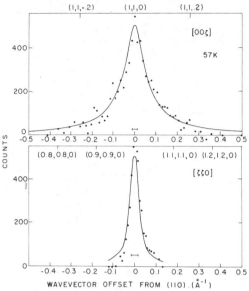

Fig. 2 Anisotropic critical scattering in UN near (110)

extreme case since the precursor scattering peaks near $(1,1,\eta)$ with $\eta = 0.3$. However, the formation of an incommensurate spin density wave is pre-empted by a first-order phase transition to a type I antiferromagnetic[1] structure with $\eta = 0$. If similar behaviour occurred for UN, the incommensurate wavevector would be close to $\eta = 0$. There is no evidence, for any of the pnictides, for any critical scattering near (001), a wavevector which is equivalent to (110). Hence the susceptibility which is diverging, assuming that it is the part at $\eta = 0$ which diverges, is entirely longitudinal and the transverse component is suppressed. The anisotropy of the critical scattering is assumed to originate in the anisotropy of the coupling between the magnetic moments. Mean field analyses indicate that, for spins lying in an (001) correlated sheet, the exchange between spin components perpendicular to the sheet is stronger than the exchange between spin components in the sheet by factors of 49,37 and 13 for USb, UAs and UN respectively.

Only for UN have the dynamic aspects of the critical scattering been studied. The intrinsic width of the Lorentzian lineshape was found[10] to have the form

$$\Gamma(q,\kappa_1) = A^2 (\kappa_1^2 + q_\perp^2 + Cq_{\parallel}^2)$$

Here κ_1 is the inverse correlation length and $q_\perp$ and $q_{\parallel}$ are the wavevector displacements perpendicular and parallel to the ordering wavevector. C accounts for the anisotropy of the distribution. The prefactor A measures the stiffness of the antiferromagnet and has a value that is three times less than the itinerant antiferromagnet Cr^{11}. The variation of κ_1 with temperature is shown in Fig. 4. The value of κ_1 appears to tend to zero at a higher temperature, T_N^*, = 54.5 K than that at which the order parameter vanishes, (T_N = 49.5 ± 0.5 K). With this reference temperature, T_N^*, the inverse correlation length in UN has an exponent $\nu = (0.84 ± 0.06)$ which lies between the value 1 for the 2d Ising model and 0.7 for the 3d Heisenberg model. The inverse correlation lengths are, however, between 5 and 7 times less than those for a 3d Heisenberg model at the same reduced temperature. Because κ_1 is small, the inelasticity at low wavevector offsets from the ordering wavevector

283

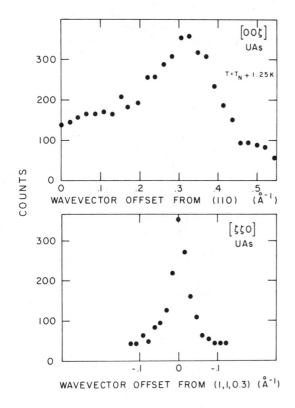

Fig. 3 Anisotropic
critical scattering in
UAs near (110)

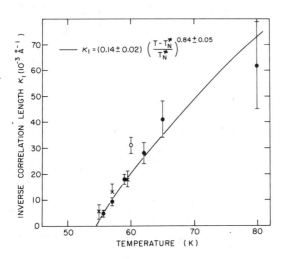

Fig. 4 Temperature
dependence of inverse
correlation length, κ ,
in UN. κ tends to zero
at $T_N^* = 54.5$ K.

284

is also small, so that the scattering looks, at first sight, completely elastic; in the wings of the peaks the inelasticity is important, however. In the temperature region between T_N and T_N^* the true critical scattering, corresponding to the development of the three dimensional ordered magnetic structure presumably develops.

LOW TEMPERATURE BEHAVIOUR

In the ordered phase a sharp distinction emerges between the compounds with the largest distance between uranium atoms, UTe and USb, and the other compounds. Fig. 5 shows that UTe exhibits at least two branches of sharp magnetic excitations in addition to lattice vibrations. Although the data for UTe has not been interpreted, an approach based on crystal fields and exchange can probably give a satisfactory description. In the hatched region extending from $4\frac{1}{2}$ THz at the zone centre to 8 THz at the zone boundary the peaks are considerably broader than the experimental resolution. There may be several branches of excitations superposed in this region or, alternatively, there may be a continuum of scattering as observed in the other compounds (Fig. 1). A possible schematic explanation for the situation in US is that d- and f-electrons are coupled together at low temperatures to produce a ground state with a low moment and low matrix elements for excitations. Above the ordering temperature these groups of electrons are uncoupled and can scatter independently.

A contour plot of the scattering near (110) for UN, corrected for the instrumental background, is shown in Fig. 6. From structure factor information the inelastic distribution in Fig. 6 has been found, like the critical scattering, to be longitudinal

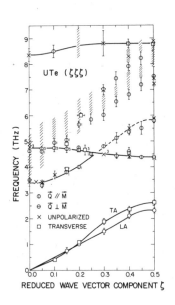

Fig. 5 Dispersion relation for UTe in the [ζζζ] direction at 4.2 K.

i.e. to correspond to fluctuations of the magnitude of the
magnetic moment. The most difficult feature of UN to understand is
the fate of the transverse spin-wave excitations which characterise
conventional magnets.

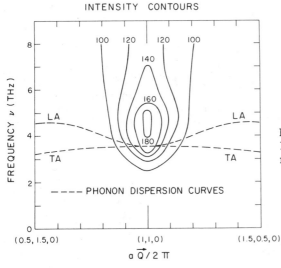

Fig. 6 Contour map of the
low temperature magnetic
response in UN.

LATTICE DYNAMICS

The results of a preliminary force-constant analysis of the
lattice vibrations of uranium compounds (UX) with a rigid ion model
have already been published[12]. The U-X force was found to be
largest when U-U and X-X forces were both small. Where the magnetic
moment is small (UN) or completely suppressed (UC) the U-U force is
comparable with the UX force. The bulk modulus and Poisson's
Ratio have been calculated from the force constants, as shown in
Fig. 7. With increasing cell parameters both pnictides and chalcogenides are more compressible, while for the same lattice parameter
the chalcogenides are more compressible than the pnictides. UAs,
US and UTe exhibit a negative value of Poisson's Ratio (stemming
from the negative values of the C_{12} elastic constant). A similar
effect[13] has been observed in the intermediate valent compound
$Sm_{.75}Y_{.25}S$. Thus, under pressure the 5-f electron may readily be
promoted into the 6-d band leading to a contraction of the atomic
volume. Other evidence for intermediate valency of UX compounds
comes from the systematics of the lattice constants[8]. However, the
effects are likely to be smaller than for $Sm_{.75}Y_{.25}S$ for which the
bulk modulus is smaller and Poisson's Ratio is more negative than in
the uranium compounds.

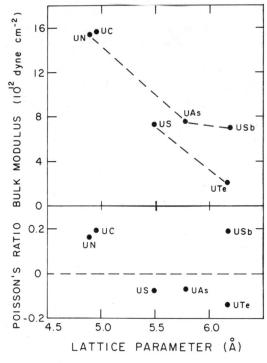

Fig. 7 Dependence of the bulk modulus and Poisson's Ratio on lattice parameter for UX compounds

CONCLUSION

Uranium compounds with the largest cell parameters (UTe, USb) have magnetic properties at low temperatures like conventional magnetic materials. The other compounds show unconventional behaviour such as the absence of transverse magnetic scattering in UN and the dramatic suppression of the magnetic scattering as the temperature tends to zero leaving only a weak continuum as in US. Near the Néel temperature the spin correlations are anisotropic and long ranged. Finally, several of the compounds exhibit a negative value of Poisson's Ratio suggesting incipient instability of the ground-state electronic structure.

ACKNOWLEDGEMENTS

We are grateful to J.C. Evans, H.F. Nieman, M.M Potter and D.C. Tennant for expert technical assistance.

REFERENCES

1. J. Rossat-Mignod, P. Burlet, S. Quezel, and O. Vogt, Physica 102B, 237 (1980).

2. G. Dolling, T.M. Holden, E.C. Svensson, W.J.L. Buyers, and
    G.H. Lander, Proc. Int. Conf. on Lattice Dynamics, Paris,
    5-7 Sept. 1977 (Flammarion, 1978) p. 81.
3. W.J.L. Buyers, A.F. Murray, T.M. Holden, E.C. Svensson,
    P. de V. DuPlessis, G.H. Lander, and O. Vogt, Physica
    102B, 291 (1980).
4. S.K. Sinha, G.H. Lander, S.M. Shapiro, and O. Vogt, Phys. Rev.
    B23, 4556 (1981).
5. W.G. Stirling, G.H. Lander,and O. Vogt, Physica 102B, 249
    (1980).
6. G.H. Lander and W.G. Stirling, Phys. Rev. B21, 436 (1980).
7. G.H. Lander, S.K. Sinha, D.M. Sparlin, and O. Vogt, Phys.
    Rev. Lett. 40, 523 (1978).
8. W.J.L. Buyers, T.M. Holden, A.F. Murray, J.A. Jackman,
    P.R. Norton, P. de V. DuPlessis, and O. Vogt, Proc. Int. Conf.
    on Valence Fluctuations in Solids. Santa Barbara, Jan 27-30 1981.
    (North Holland in press); and to be published.
9. W.J.L. Buyers, T.M. Holden, E.C. Svensson,and G.H. Lander
    in "Neutron Inelastic Scattering 1977" Vol. II, (International
    Atomic Energy Agency: Vienna) p. 239 (1978).
10. T.M. Holden, W.J.L. Buyers, E.C. Svensson, and G.H. Lander, to
    be published.
11. J. Als-Neilson, J.D. Axe, and G. Shirane, J. Appl. Phys. 42,
    1666, (1971).
12. W.J.L. Buyers, A.F. Murray, J.A. Jackman, T.M. Holden, P.
    de V. DuPlessis,and O. Vogt, J. Appl. Phys. 52, 2222 (1981).
13. H.A. Mook and R.M. Nicklow, Phys. Rev. B20, 1656 (1979).

OBSERVING SOLITONS IN ONE DIMENSIONAL MAGNETIC SYSTEMS

George Reiter

Brookhaven National Laboratory, Upton, New York 11973

ABSTRACT

Classical models of one dimensional magnetic systems show that in addition to the linear spin wave excitations, there should exist localized, large amplitude excitations, that can move along the chains while retaining their integrity. It is expected that these excitations, "solitons", exist in real materials. We will discuss the progress that has been made to date in observing solitons in one dimensional magnets by means of neutron scattering, and the difficulties that still remain in unambiguously identifying the soliton contributions to $S(q,\omega)$.

INTRODUCTION

That solitons, mobile, localized non-linear excitations that retain their identity upon collisions, exist in classical models of one dimensional magnetic systems is clear mathematically[1,2,3,4]. Considerable effort has gone into observing these excitations in real materials by means of neutron scattering, particularly TMMC and CsNiF$_3$[5,6], and it is my purpose to provide an introduction to the physics of solitons in magnetic systems and review the extent to which the efforts to observe them have been successful.

The theoretical intepretation of the results in these systems is based upon the sine Gordon equation. It is clear[2] that solitons exist in one dimensional magnetic systems that cannot be described by the sine Gordon equation, for instance, CsNiF$_3$ in the absence of a field.

THEORY AND EXPERIMENT FOR SINE GORDON-LIKE SYSTEMS

The sine Gordon equation

$$\frac{\partial^2 \phi}{\partial t^2} - \frac{1}{c^2}\frac{\partial^2 \phi}{\partial x^2} = \frac{m^2}{\beta}\sin\beta\phi \qquad (1)$$

is an example of a completely integrable system . By means of non linear transformation, an arbitrary solution may be regarded as being composed of three distinct types of excitations. These are unbounded (in space) oscillatory solutions closely related to the phonons of the linearized theory, and going over to these phonons as their amplitude decreases, bounded oscillatory solutions of finite amplitude that have envelopes that can move along the chain, called breathers, and fixed amplitude kink solutions, corresponding to a 2π change in ϕ, that are also free to move along the chain.

These latter are usually called solitons, and will be so denoted in the rest of the paper. The large amplitude breathers may be thought of as bound states of a kink and anti-kink, while the small amplitude ones can be thought of as bound states of a pair of phonons. The solitons of the theory are sometimes called topological solitons. The limit of ϕ as $x \to \pm\infty$ is different for such solitons. They are distinguished from pulse solitons, such as the breathers, for which the system returns to the same state as one passed through the soliton. It is characteristic of the system that what one regards as the elementary excitation and what composite depends upon the perspective one wishes to adopt. This is even more apparent in the quantum version of the model.

Classically, the parameter β is irrelevant. By defining $\phi' = \phi\beta$, it could be eliminated from the equation of motion. The Hamiltonian for the system is

$$\mathscr{H} = \frac{\hbar}{c} \int dx \left[\frac{1}{2c^2}\phi_x^2 + \frac{1}{2}\phi_t^2 - \frac{m^2c^2}{\beta^2}\cos\beta\phi \right] \tag{2}$$

and in calculating the thermodynamics, this scaling would simply change the effective temperature. In the quantum theory, however, $\hbar$ sets an intrinsic scale for the variations of ϕ, and the magnitude of β is an essential parameter of the theory, determining the nature of the spectrum by providing a measure of the anharmonicity[8,9]. For $\beta^2 \ll 4\pi$, the spectrum consists of a ground state, whose energy is mc and that corresponds to a phonon, a series of excited states corresponding to the breathers, and finally a state corresponding to an unbound kink anti-kink solution at an energy of $16\ mc/\beta^2$. As β increases, the number of breather states decreases, until at $\beta^2 = 4\pi$, there are none. For TMMC and CsNiF$_3$, it turns out that $\beta^2 \simeq .7$. The smaller the value of β, the greater the correspondence between the classical solutions and the quantum theory, and it is appropriate to use the classical theory as a first approximation in interpreting the experiments on TMMC and CsNiF$_3$.

The calculation of the contribution to correlation functions from the solitons in the classical model was done first by Kawasaki[10] in the limit of a dilute gas of solitons, and this has been extended by Mikeska[11] and Allroth to include the effect of the presence of the phonons on the soliton contribution at higher temperatures. Both of these calculations contain heuristic elements, and it is not clear whether they are in fact exact, even asymptotically as the temperature approaches zero. We shall be interested in the extent to which they are confirmed by experiment.

Neither CsNiF$_3$ nor TMMC have Hamiltonians that are obviously of the sine Gordon form, which is to be regarded as an effective Hamiltonian, valid at low temperatures. The correspondence, due first to Mikeska[3,4] is most easily seen for CsNiF$_3$, which is describable by the Hamiltonian

$$\mathcal{H} = -J\sum_i \vec{S}_i \cdot \vec{S}_{i+1} + DS_i^{z^2} - g\mu_B H\sum_i S_i^x \tag{3}$$

with S=1.

At low temperatures, i.e. $KT < \sqrt{DJ} S^2$, the spins will lie preferentially in the x,y plane. If we identify ϕ as the angle of departure of the spin direction (for the classical system) from the x axis in the x-y plane, and θ as the angle out of the plane, and observe further that ϕ and S_Z are canonical variable, so that $\dot{\phi}_i = (\delta H)/(\delta S_Z) = (2S_i^Z)/(D)$, we find in the continuum limit, and to lowest order in θ,

$$\mathcal{H} = \int \frac{dx}{a} \left[JS^2 a^2 \frac{1}{2}(\phi_x)^2 + \frac{1}{4D}\dot{\phi}^2 - g\mu_B HS\cos\phi \right] \tag{4}$$

which is the sine Gordon Hamiltonian with

$$c^2 = \frac{2DJS^2a^2}{h^2} \quad , \quad m^2 = \frac{g\mu_B H}{JSa^2} \quad , \quad \beta^2 = \sqrt{\frac{2D}{JS^2}}$$

The antiferromagnet requires a slightly more complicated derivation and differs in a fundamental way. The spins are nearly perpendicular to the field, most of the time, and a soliton corresponds to a rotation by π, not 2π. As a consequence, the soliton connects regions of the lattice that are in the two inequivalent ground state configurations. For the ferromagnet, the two configurations on either side of the soliton are identical. For a fixed density of solitons in a chain, the disordered region in the ferromagnet extends only over the distance occupied by the solitons, and vanishes with the density. For an antiferromagnet, the entire chain is always disordered, and it is the inverse coherence length that vanishes with the density. Thus while the solitons show up in the ferromagnet as a decrease in the intensity of the Bragg plane and a contribution to the inelastic scattering, in the antiferromagnet, they can be seen in the modification of the Bragg plane into a lorentzian whose width in q is basically the inverse separation of the solitons and whose width in frequency is determined by the mean time it takes the solitons to move that spacing. It is much easier, as a result, to observe the effect of the solitons in the antiferromagnet, although in a sense, what one is observing is primarily the space between the solitons.

In the sine Gordon approximation, $S^x = S\cos\phi$, $S^y = S\sin\phi$ and the prediction of the theory of Kawasaki and Mikeska for

$$S_q^{\alpha\alpha}(\omega) \equiv \frac{1}{2\pi} \int_{-\infty}^{\infty} \langle S_q^{\alpha}(t) S_{-q}^{\alpha} \rangle \, e^{i\omega t} dt \quad \text{is}$$

$$S_q^{\left\{\begin{matrix} xx \\ yy \end{matrix}\right\}}(\omega) \equiv \frac{64}{\pi} \frac{1}{KT} \frac{1}{cq} e^{\frac{-8mJa}{KT}} \left(1 + 1/2 \omega^2 / (cq)^2\right) \left\{ \begin{matrix} \left(\frac{\pi q}{2m} / \sinh\frac{\pi q}{2m}\right)^2 \\ \left(\frac{\pi q}{2m} / \cosh\frac{\pi q}{2m}\right)^2 \end{matrix} \right\} \quad (5)$$

The factors in curly brackets are the form factors for the projections of the spin component of a single soliton on the x and y axes.

The integrated intensity is proportional to these factors and so provides a direct measure, within this theory, of the shape of the soliton. The density of solitons in this theory is proportional to $e^{-E_s/KT}$, where $E_s = g\mu_B HS$. Boucher et al. have shown quite convincingly that the width in $q(\Gamma_s)$ of the Bragg plane, and its width in $\omega(\Gamma_D)$ are proportional to such an exponential in the region of low soliton density, where the theory is expected to apply. The value for E_s that they obtain differs significantly from its classical value, which may be due to quantum effects[12]. They have also obtained another measure of Γ_D from NMR measurements of T_1^{-1}, and it agrees with the neutron scattering results.

Of course there is no genuine Bragg plane in a one dimensional system, and in the absence of a field, the widths Γ_D and Γ_s (the lineshape is not actually lorentzian) can be calculated from a theory that includes only the linear excitations, the spin waves[13]. It remains an unsolved problem to interpolate between the case that H=0 and large enough H that the soliton theory can be expected to apply, both conceptually and with a practical calculation.

The problem of distinguishing between the contribution to the correlation function that arise from the linear excitations, the spin waves, and the non-linear excitations, the breathers and solitons, is more serious in the ferromagnet. In fact, it seems unlikely that a clear theoretical distinction betwen the spin wave contribution and the breather contribution can be made at all, as both contributions can presumably be obtained in perturbation theory in the parameter β. The soliton contribution, proportional to $e^{\frac{-8m\varsigma}{\beta KT}}$ is non-analytic in β, and is theoretically distinguishable, but the problem is to separate the soliton contribution experimentally from the anharmonic spin waves. The original experiments interpreted the inelastic scattering seen near $\omega=0$ at small wavevectors as being due to scattering off a gas of solitons, the spread in frequencies being essentially the doppler shifts from the moving solitons. The variation of the intensity of this central peak with q is predicted by the theory of Kawaski and Mikeska to correspond to the structure factor for a single soliton, so that these observations appeared to provide very direct observation of the solitons in the chain. It soon became clear that the interpretation was not so simple.

Scattering from pairs of spin waves, or in other words, fluctuations of the length of the spin component along the field, provides an additional mechanism for producing a central peak. (In fact, it produces also scattering at twice the spin wave frequency, which has now been observed[14].) The variation of the intensity with field temperature and wavevector, are all well described by the two spin wave scattering[15]. It has also been argued from simulations , that the sine Gordon approximation is not valid at the temperatures and fields that the experiments were done[16]. Allroth and Mikeska have argued that the data is best fit by a sum of the two contributions, the two being roughly equal at the highest temperatures of observation, but the uncertainty in the absolute value of the observed intensity, which could presumably be eliminated by comparing with the spin wave intensities prevents any unambiguous statement from being made. They do indicate that at least for the spin wave intensities and central peak structure factor, that the sine Gordon theory is a satisfactory approximation at the temperatures and fields of interest.

To resolve these difficulties in interpretation, Steiner[15] et al. have done an experiment in which the scattering vector is aligned along the field, so that the two spin wave scattering in $S_q^{xx}(\omega)$ does not contribute. They observe additional intensity q around $\omega=0$ in this case as well, which they attribute to the presence of solitons. There is, however, an anharmonic contribution to this central peak intensity that comes from the spin waves.

The spin operator has a boson equivalent that, in leading order that contains the anharmonic effect, is

$$S_i^y = i\frac{\sqrt{S}}{2}(a - a^+) - i\frac{1}{2\sqrt{2S}}(a_i^+ a_i a_i - a_i^+ a_i^+ a_i) \qquad (6)$$

The second term contributes to the scattering a term which is, neglecting the anharmonicity in the Hamiltonian,

$$(7)$$

$$\frac{1}{8S}N^{-2}\sum_{q_1 q_2 q_3} \Delta(q - q_1 - q_2 - q_3)\left|\Gamma_{q_1 q_2 q_3}\right|^2 \{n_{q_1}(n_{q_2}+1)(n_{q_3}+1)\delta(\omega - \omega_{q_1} + \omega_{q_2} + \omega_{q_3})$$

$$+ n_{q_2} n_{q_3}(n_{q_1}+1)\delta(\omega + \omega_{q_1} - \omega_{q_2} - \omega_{q_3})\}$$

where $\Gamma_{q_1 q_2 q_3} = \alpha_{q_1}\alpha_{q_2}\alpha_{q_3} + \beta_{q_1}\beta_{q_2}\beta_{q_3} + 2\alpha_{q_1}\beta_{q_2}\beta_{q_3} + 2\beta_{q_1}\alpha_{q_2}\alpha_{q_3}$

The α_q, β_q are the coefficients of the Bogolubov transformation, the n_q occupation numbers and the ω_q the spin wave frequencies, all defined in ref. 15.

This is not the only anharmonic contribution, as there are terms coming as well from the anharmonic terms in the Hamiltonian that could also be considered. We expect them to make comparable contributions, and a rigorous calculation has not yet been done. For the parameters of the experiment, the solitons do give a significant contribution to the scattering near $\omega=0$, and the three spin wave scattering is considerably smaller. The integrated intensity of the solitons is proportional to the structure factor and has a peak at $q \simeq .06$ R.L.O. Much of the three spin wave scattering would be included in the spin wave cross section, with the procedure used by Steiner et al. for fitting the data, and the intensity of the remainder would appear as a nearly q independent background. We conclude that the q dependence of the intensity observed cannot be explained by the lowest order spin wave theory. The possibility remains that a more accurate calculation could explain it, and we think this unlikely. We would conclude that the soliton interpretation of scattering near $\omega=0$ is tenable.

Nevertheless, the situation in $CsNiF_3$ is far from satisfactory. The observed intensity in the central peak, in $S_q^{yy}(\omega)$ is about a factor of three[18] above what one expects from the lowest order soliton theory and the finite temperature corrections serve to reduce the theoretically predicted intensity significantly. If we take the experiment at face value, then we must conclude that the prefactors in the theoretical expression (5) are significantly in error.

A further test of the soliton theory in $CsNiF_3$ has been suggested. It has been observed by both Kumar[19] and Magyari and Thomas[20] that the solitons become unstable if the field exceeds 18 Kg. This is because the spins that are antialigned with the field can lower their energy by tipping out of the plane if the field is large enough, and 18 Kg is the boundary for this to occur. Magyari and Thomas argue that this should lead to characteristic singularities in the soliton contribution to the correlation function, and it would be very interesting to determine if these were observable.

We have been discussing so far x-y like systems, in which an external field breaks the symmetry. In Ising like systems, for which the sign of D in (3) is negative, Mikeska has shown that the classical continuum limit for the antiferromagnet leads again to the sine Gordon equation[4] the angle now being the deviation of the spin direction from the Ising axis. The solitons connect two distinct ground states, and so disorder the system in a manner similar to the x-y antiferromagnet. The external field can be zero. If it is not, a double sine Gordon equation results, with terms involving both $\cos\phi$ and $\cos2\phi$.

A quantum system, $C_sC_oCL_3$ exists, with spin 1/2, that has the symmetry, and furthermore, the exchange terms are sufficiently small as to be treatable by perturbation theory. The basic excitation is indeed a moving domain wall, as first pointed out by Villain[21]. Extensive experimental work, in the absence of a field has been done by Yoshikawa, Hirakawa, Satija and Shirane[22] and both numerical calculations using finite chains and analytic calculations using perturbation theory have been done by Ishimura and Shiba[23]. Together with the earlier analytic calculations of Villain that are valid for wavelengths short compared to a coherence length, these provide a clear picture and a quantitative understanding of many aspects of the system.

The transverse excitation spectrum can be interpreted as being due to exciting pairs of domain walls. The degree of freedom corresponding to the relative motion of the pair leads to a band of excitations rather than a definite relationship between ω and q. The spectral density in the band is more asymmetric than predicted, perhaps due to interchain coupling. The longitudinal scattering is primarily a band around $\omega=0$ that can be thought of again as the doppler shifting of the neutrons due to scattering off of individual domain walls. The theory is in satisfactory agreement with experiment if $q>K$, but the long wavelength phonons are sensitive to multiple domain walls, and there is no analytic theory available.

We want to emphasize that the theory for the system is very much a quantum theory,, and although it seems natural to associate the excitations observed in $C_sC_oCL_3$ with the classical solitons, this is a purely heuristic identification at this point. $\beta^2=8/S$ for the mapping onto the sine Gordon theory, and for spin 1/2, the quantum sine Gordon theory is still definable,, although there are no spin waves or breathers, since $\beta^2>4\pi$. This is in accord with the theory based directly on the spin Hamiltonian. Presumably, a system with $S=1$ would show both spin waves and solitons.

REFERENCES

1. M. Lakshmanam, Phys. Letts A 61, 53 (1977).
    L. A. Takatajan, Phys. Letts A 64, 235 (1977).

2. E. K. Sklyanin (Preprint).

3. H. J. Mikeska, J. Phys. C 11, 2913 (1980).

4. H. J. Mikeska, J. Phys. C 13, 2913 (1980).

5. J. K. Kjems and M. Steiner, Phys. Rev. Letts. 41, 1137 (1978).

6. J. P. Boucher and J. P. Renard, Phys. Rev. Letts. 45, 486 (1980).
    J. P. Boucher, L. P. Regnault, J. Rossat-Mignod, J. P. Renard, J. Bouillot and W. G. Stirling, Solid State Comm. 33, 171 (1980).

7. M. J. Ablowitz, D. J. Kaup, A. C. Newell and H. Segur, Phys. Rev. Lett. 30, 1261 (1973).
    E. Stoll, T. Schneider and A. R. Bishop, Phys. Rev. Letts 42, 937 (1979).

8. R. F. Dashen, B. Hasslacher and A. Neven, Phys. Rev. D, 10, 4114 (1974); Phys. Rev. D 11, 3424 (1975).

9. S. Coleman, Phys. Rev. D 11, 2088 (1975).

10. K. Kawasak, Prog. Theor. Phys. 55, 2029 (1976).

11. E. Allroth and H. J. Mikeska (Preprint).

12. K. Maki and H. Takayama, Phys. Rev. B 20, 5002 (1979).

13. G. Reiter and A. Sjolander, Phys. Rev. C 13, 3027 (1980).

14. M. Steiner, K. Hirakawa, G. Reiter and G. Shirane (Preprint).

15. G. Reiter, Phys. Rev. Letts. 46, 202 (1981).
    G. Reiter, J. Appl. Phys. 52, 1961 (1981).

16. J. M. Loveluck, T. Schneider, E. Stoll and H. R. Jauslin, Phys. Rev. Letts 45, 1505 (1981).

17. M. Steiner, R. Pynn, W. Knop, K. Kakurai and J. Kjems (Preprint).

18. M. Steiner (Private Communication).

19. P. Kumar (Preprint).

20. E. Magyari and H. Thomas (Preprint).

21. J. Villain, Physica B 79, 1 (1975).

22. H. Yoshizawa, K. Hirakawa, S. K. Satija and G. Shirane, Phys. Rev. B 23, 2298 (1981).

23. N. Ishimura and H. Shiba, Prog. Theor. Phys. 63, 743 (1980).

EXCITATIONS OF THE SPIN DENSITY WAVE IN PURE CHROMIUM

S. A. Werner
Physics Department
University of Missouri-Columbia, Columbia, Mo. 65211

G. Shirane, C. R. Fincher, B. H. Grier
Physics Department
Brookhaven National Laboratory, Upton, N.Y. 11973

ABSTRACT

This paper summarizes our recent investigations of the magnetic excitations of the spin density wave (SDW) in pure Cr in both the low temperature longitudinally polarized phase (T < 122K) and in the higher temperature transversely polarized phase (122K < T < 312K). In both phases we observe spin wave modes of very high velocity originating from the incommensurate Bragg points. In the transversely polarized SDW phase new additional excitations are observed, centered in reciprocal space at the (1,0,0) commensurate point. These excitations are not affected by a magnetic field. Inelastic scattering in the paramagnetic phase above the Neel point (312K) is observed in a reasonably well localized region of reciprocal space near (1,0,0) indicating that there are spin-spin correlations extending over many bcc unit cells and persisting to temperatures at least as high as 1.7 T_N.

INTRODUCTION

The itinerant magnetism in chromium metal is due to an Overhauser spin density wave (SDW).[1] As first pointed out by Lomer[2] this unique antiferromagnetic state occurs in chromium because of the nearly perfect nesting of portions of the jack-shaped electron Fermi surface surrounding the Γ-point (0,0,0) in reciprocal space with the flat portions of the octahedral-shaped hole surface surrounding the H-point (1,0,0). The discovery of the incommensurate nature of the antiferromagnetism in Cr was made 22 years ago by Corliss, Hastings and Weiss,[3] when they identified by neutron diffraction the magnetic satellite structure in which Bragg peaks are found at neutron scattering vectors

$$\vec{Q}_N = \vec{G} \pm \vec{Q}. \tag{1}$$

Here $\vec{Q}$ ($\approx .95 \frac{2\pi}{a}$) is the wave vector of the SDW and $\vec{G}$ is any reciprocal lattice vector of the body centered cubic lattice of Cr. The resulting (100) reciprocal lattice plane is shown in Fig. 1. There is now a very large body of literature on investigations of the magnetic properties of this extremely interesting metal.

There are three magnetic phase of Cr. At temperatures below the spin-flip temperature (T_{sf} = 122K) the SDW is longitudinally polarized. Above T_{sf} the SDW is transversely polarized up to the first order transition at the Neel point (T_N = 312K)[4]. As a result of the cubic symmetry of paramagnetic chromium above T_N, three types of

ISSN:0094-243X/82/890296-13$3.00 Copyright 1982 American Institute of Phys

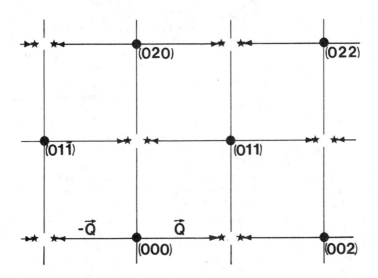

Fig. 1.   The (100) reciprocal lattice plane of antiferromagnetic Cr metal.   The solid circles are nuclear Bragg reflections.   The stars are magnetic Bragg points.

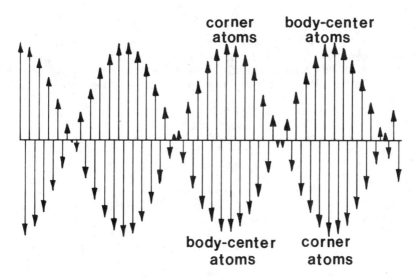

Fig.2.   Schematic diagram of the incommensurate magnetization wave in Cr.

domains develop below T_N with the Q vector of the SDW along any one of the three 100 -type directions in the crystal giving rise to three equivalent sets of satellite reflections. It is possible to produce a single-Q state of the crystal by cooling through the Neel temperature in a large applied magnetic field directed along a 100 axis. This procedure results in a crystal consisting of one type of domain with the modulation direction along the applied field. The reciprocal lattice shown in Fig. 1 is for a "field-cooled" crystal with the single-Q direction along 001 . A schematic diagram of the magnetization wave resulting from this SDW state in Cr is shown in Fig. 2.

The magnetic excitations of this static SDW are naturally of great interest and of fundamental importance in magnetism. Theories of the long wavelength spin waves have been developed, based on various simplifying two-band models by Fedders and Martin,[5] Liu[6] and by Sato and Maki.[7] These theories predict the occurence of very high velocity spin waves (comparable to the Fermi velocity). The predictions do not depend strongly on the extent of incommensurability of the SDW. Past measurements on commensurate Cr(Mn) alloys[8] give a spin wave velocity of about 1.5×10^7 cm/sec in rough agreement with theoretical predictions. Such high velocities present special problems for typical triple-axis neutron spectroscopy. This paper summarizes the results of our current experimental program aimed at understanding the magnetic excitations occuring in both the incommensurate longitudinal and transverse phases, and in the paramagnetic phase of pure chromium.[10]

EXPERIMENTAL RESULTS

The experiments were carried out on triple-axis spectrometers at the Brookhaven High Flux Beam Reactor. Two single crystal samples in the single-Q state were used in the experiments. The most extensive measurements have been made on a highly perfect, vapor-grown single crystal of approximately 0.5 cm^3. We have also taken data on a second larger (2 cm^3), less perfect, crystal grown by the strain and anneal method.

In Fig. 3 we show the results of a constant E (constant energy transfer) scan along 001 across the satellites positions near the (0,0,1) point in reciprocal space. As expected, strong scattering is observed near the satellite Bragg points, from which the high-velocity, probably conical, spin-wave dispersion surfaces originate. Single peaks are observed at .95 $\frac{2\pi}{a}$ and 1.05 $\frac{2\pi}{a}$ because the dimensions of this dispersion surface in momentum space are much smaller than the resolution ellipsoid. In addition to the spin-wave scattering, there is clearly an additional feature centered at the commensurate point (0,0,1). As the temperature is raised the intensity of this diffuse commensurate scattering increases very rapidly as seen in Fig. 3.

In Fig. 4, results of constant-Q scans at the commensurate (0,0,1) point are shown for a series of different temperatures. There is a well defined excitation at about 4 meV at the lower temperatures which shifts to about 3.5 meV at 300K. There is also a dramatic

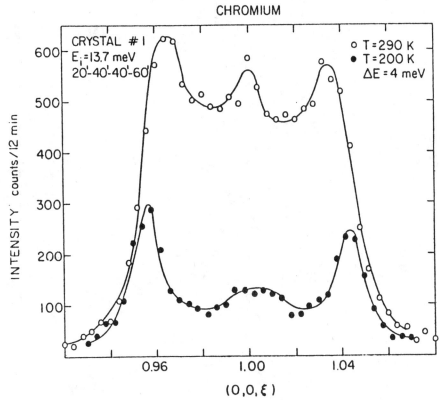

Fig. 3. Constant E scans in the region of the incommensurate magnetic Bragg reflections near (0,0,1). (C.R. Fincher, G. Shirane, S.A. Werner, ref. 10).

increase in the large sloping "background" scattering as the temperature is raised from 130K to 300K. We have carefully repeated this constant-Q scan at 290K and there appears to be an additional peak centered at about 7.5 meV (Fig. 5).

The temperature dependence of this "commensurate-diffuse" scattering is shown for selected energy transfers in Fig. 6. This scattering appears to be diverging at the Néel point. It is, if fact, increasing exponentially as shown by the plot on a log scale shown in Fig. 7. We have made no attempt yet to measure the detailed temperature dependence very close to T_N. This commensurate-diffuse scattering has recently been observed by Mikke and Jankowska[9] in a chromium alloy Cr (.18% Re). They also observe the rapid temperature dependence of this scattering, similar to the data of Fig. 6.

Earlier we had concluded that the fluctuations leading to both the spin wave scattering and this commensurate-diffuse mode were highly anisotropic as a result of a comparison of measurements made near (0,0,1) and (0,1,0).[10] Recently we have found that this

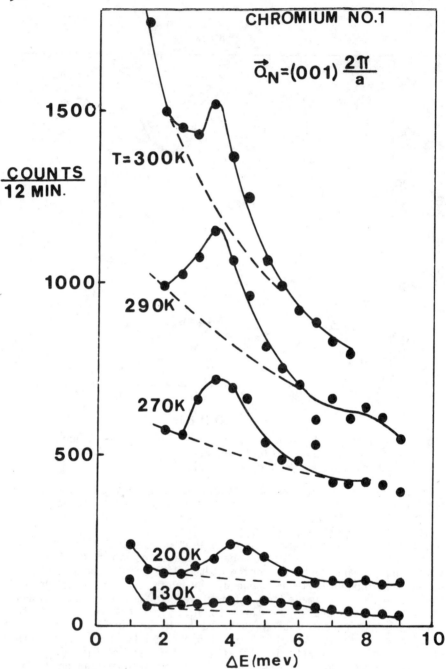

Fig. 4. Constant-Q scans at (0,0,1) at various temperatures showing the energy dependence of the commensurate-diffuse mode. (C.R. Fincher, G. Shirane, S.A. Werner, ref. 10).

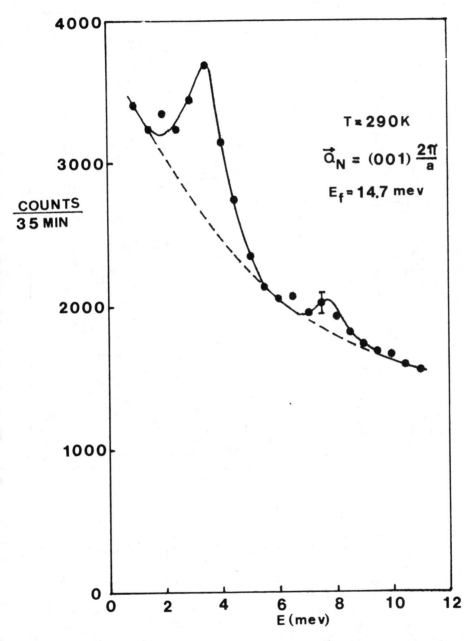

Fig. 5. Constant-Q scan at (0,0,1) at 290K showing an indication of an additional discrete excitation at E ≈ 7.5 mev. (B.H. Grier, G. Shirane, S.A. Werner, ref. 11).

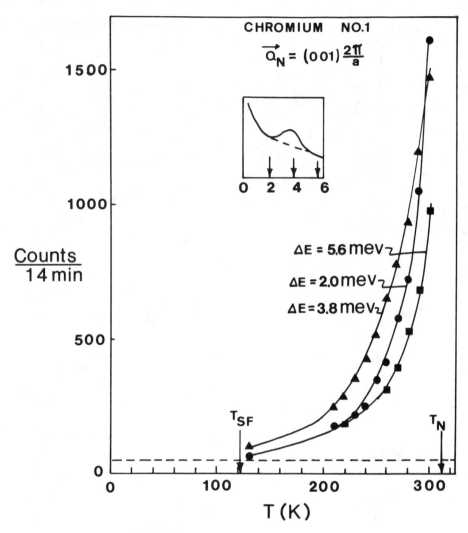

Fig. 6. Temperature dependence of commensurate-diffuse mode at selected energy transfers. (B.H. Grier, G. Shirane, S.A.
Werner, ref. 11).

conclusion is incorrect and that the earlier measurements were biased
by anisotropic sample capsule transmission effects. Representative
scans to resolve this question are shown in the upper portion of Fig. 8
taken at 125K (just above the spin flip temperature). Because the
extent of the resolution ellipsoid is large in comparison to the width
of the dispersion surface (at these low energy transfers), it is the
peak intensities of these scans which should be compared. From a large
amount of data of this type we now conclude that the fluctuations

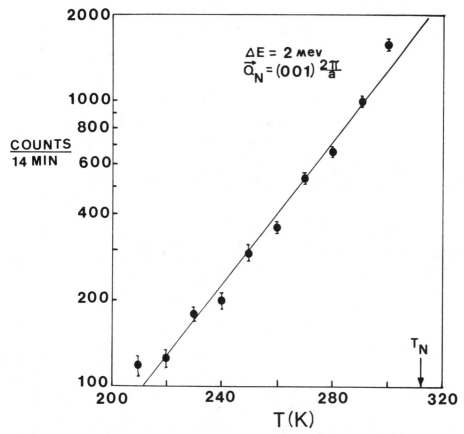

Fig. 7. Temperature dependence of the commensurate excitations at an energy transfer E = 2 mev on a log plot, showing that this intensity increases exponentially with temperature. (B.H. Grier, G. Shirane, S.A. Werner, ref. 11).

leading to the spin wave modes and also to the commensurate-diffuse mode are nearly isotropic in polarization directions. We have not yet resolved the question of whether the spin wave velocity is isotropic.

The data shown in the lower portion of Fig. 8 (taken at 119K, just below the spin flip temperature) is very interesting. The intensity of the scan along $[0,0,\xi]$ is much lower than along $[0,1,\xi]$. If only transverse fluctuations of the atomic spin were contributing to this scattering one would have expected the reverse intensity ratio. It therefore appears that the longitudianl susceptibility $\chi_{zz}(\vec{q},\omega)$, at least at these values of $\vec{q}$ and ω must be larger than the transverse susceptibility $\chi_{+-}(\vec{q},\omega)$ in the low-temperature longitudinally polarized SDW state. This is a surprising observation.

304

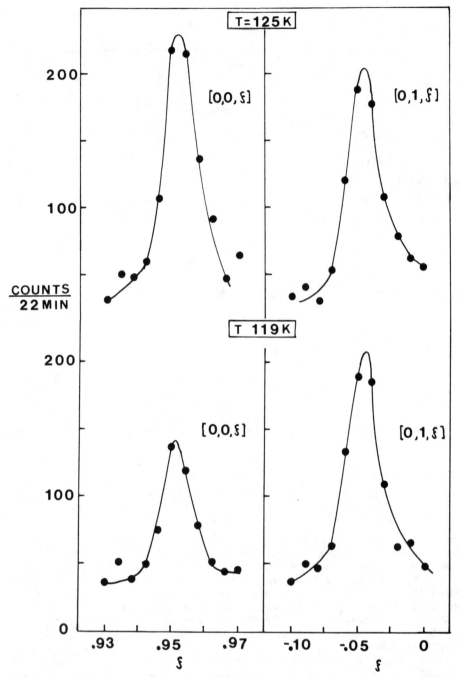

Fig. 8. Constant E (= 4 mev) scans near (0,0,1) and near (0,1,0) just
above the spin flip temperature and also just below T_{sf}.
(B.H. Grier, G. Shirane, S.A. Werner, ref. 11).

We have carried out experiments at higher energy transfers also, up to $\Delta E = 30$ mev. There are some perplexing features of this data. First of all, the spin wave peaks in the constant-ΔE scans are not centered above the incommensurate Bragg peaks, but are shift inward towards the (0,0,1) point. This means that the dispersion surfaces originating at these Bragg peaks are not simple conical surfaces, described by a single spin wave velocity parameter. Secondly, the intensity in the immediate region around (0,0,1) is considerably stronger than for the lower energy data. The overall trend of a wide assembly of this inelastic data is that as we move toward higher temperatures or toward higher excitation energies, the scattering cross section becomes dominated by the commensurate-diffuse mode.

Recently we have carried out experiments in the presence of a large dc magnetic field (up to 60 KOe). Previous diffraction experiments[12] have shown that the polarization of the spin density wave in the transverse state can easily be rotated in the x-y plane (normal to $\vec{Q}$) by applying a magnetic field $\vec{H}$ normal to $\vec{Q}$. The spin direction tends to allign perpendicular to both $\vec{Q}$ and $\vec{H}$, as is expected in any antiferromagnet where the dc perpendicular susceptibility $\chi_\perp$ is larger than the parallel susceptibility $\chi_\parallel$. Since $\Delta\chi$ ($= \chi_\perp - \chi_\parallel$) is quite small in Cr, and since a rather modest magnetic field of order 20KOe is sufficient to rotate all of the spins (in the best crystals), the magnetic energies per atom involved in this rotation process are very small. On the basis of these earlier experiments, one would therefore expect low energy transverse (to $\vec{Q}$) fluctuations of the polarization of the SDW. We had thought initially that the commensurate-diffuse scattering might be related to these modes. However, if this were the case, this scattering should be substantially affected by a large magnetic field. We observe no measurable effects of a field on either the commensurate mode or on the spin waves. We find this very surprising, since the field restrains the polarization direction of the static magnetization wave along, say the x-axis; whereas in zero field the spin direction is free to assume any direction in the x-y plane.

We have recently carried out careful experiments with very high energy resolution (down to 20 μV) to look for an energy gap in the spin wave spectrum. We observe well-defined spin waves down to less than 100 μV. This observation also appears surprising, if we think of Cr in terms spin wave theory of conventional, localized moment antiferromagnets in which a gap in the excitation spectrum is related to the exchange interaction and the anisotropy. We know that there is a four-fold anisotropy in the x-y plane from magnetic torque experiments and a very large anisotropy above T_{sf} forcing the polarization of the SDW to be transversely polarized. Since the spin wave velocity is so high, an effective exchange parameter J must also be large.

We show in Fig. 9 constant energy scans, at $\Delta E = 10$ mev, at various temperature in the paramagnetic phase up to temperatures corresponding to about 1.7 T_N. All of this data is similar in shape in momentum space, having a FWHM of $\Delta q \approx .07 \frac{2\pi}{a}$, while simply decreasing monotonically with increasing temperature. The intensity also

306

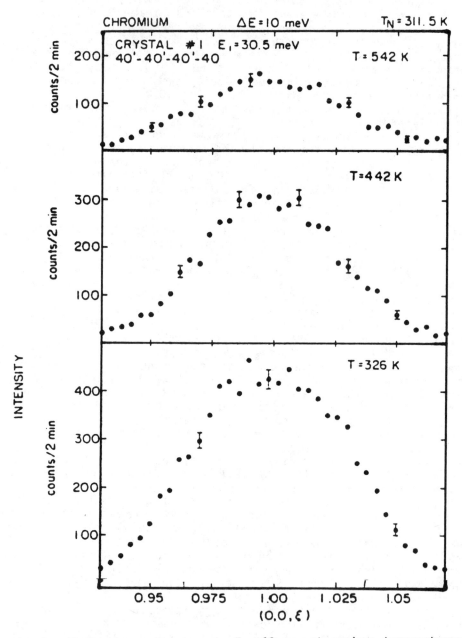

Fig. 9. Inelastic scattering at E = 10 mev at various temperatures
in the paramagnetic phase. (C.R. Fincher, G. Shirane, S.A.
Werner, ref. 10).

decreases monotonically with increasing energy transfer. From these observations we therefore must conclude that there are well-defined spin-spin correlations extending over 12 to 14 bcc unit cells.

CONCLUSIONS

Although additional experiments are required, especially at higher energy transfers, to understand this evolving picture of the magnetic excitation of chromium, we can already see similarities in these results with other itinerant magnetic systems. The diffuse, reasonably well-localized, scattering in the paramagnetic phase shown in Fig. 9 is qualitatively similar to observations in MnSi, which orders in a helically modulated ferromagnetic structure. This helical structure has a long period of order 150 Å with a $\vec{Q}$-vector in the <111> direction. Thus, the elastic magnetic scattering has a satellite structure, as does Cr. Precursory diffuse scattering qualitatively resembles some of the features of our Cr data. The scattering above T_c is not centered at the satellite positions in reciprocal space but rather around a commensurate Bragg point. As T approaches T_c, the scattering has an approximately spherical shape with a radius comparable to the wavevector distance of the satellites away from the central Bragg point - similar to the data of Fig. 9. The Moriya-Kawabata theory[12] of itinerant magnetism has been successful in understanding some of the results in MnSi, and therefore we might expect it to be helpful in understanding the results on chromium.

It is clear that there are several possible modes of excitations of the static incommensurate SDW in chromium:

1. Ordinarily spin waves which are precessional modes of the 3d atomic moments.
2. Amplitude modes in which the amplitude $\vec{M}_0$ of the SDW is allowed to fluctuate. These modes contribute to the longitudinal susceptibility $\chi_{zz}(\vec{q},\omega)$ which we apparently are seeing below T_{sf}.
3. Phason modes for which the phase of the SDW as a whole fluctuates, or for which the relative phase of the spin up and spin down electron density fluctuates. This relative phase fluctuation gives rise to dynamic charge density waves.

The relative importance of these three types of excitations in chromium is not known at the present time. Future progress will require a combination of additional experiments and new theoretical insights.

This work was carried out at the Brookhaven National Laboratory under contract DE-AC02-76CH00016 with the U.S. Department of Energy. One of us (SAW) would like to thank the neutron scattering group at BNL for their hospitality and to acknowledge the support of the NSF through grant NSF-PHY 7920979.

REFERENCES

1. A.W. Overhauser, Phys. Rev. 128, 1437 (1962).
2. W.M. Lomer, Proc. Phys. Soc. London 86, 489 (1962).
3. L.M. Corliss, J.M. Hastings, and R. Weiss, Phys. Rev. Lett. 3, 211 (1959).
4. S.A. Werner, A.S. Arrott, and H. Kendrick, Phys. Rev. 155, 528 (1967).
5. P.A. Fedders and P.C. Martin, Phys. Rev. 143, 245 (1966).
6. S.H. Liu, Phys. Rev. B2, 2664 (1970).
7. H. Sato and K. Maki, Int. J. Magnetism, 6, 183 (1974).
8. J. Als-Nielsen, J.D. Axe, and G. Shirane, J. Appl. Phys. 42, 1666 (1971); S.K. Sinha, S.H. Liu, L.D. Muhlestein, and N. Wakabayashi, Phys. Rev. Lett. 23, 311 (1969).
9. K. Mikke and J. Jankowska, J. Magn. and Magn. Mat. 14, 280 (1979).
10. C.R. Fincher, Jr., G. Shirane, and S.A. Werner, Phys. Rev. Lett. 43, 1441 (1979) and Phys. Rev. B (to be published Aug. 1981).
11. B.H. Grier, G. Shirane and S.A. Werner (paper in preparation).
12. S.A. Werner, A. Arrott, and M. Atoji, J. Appl. Phys. 39, 671 (1968).
13. T. Moriya and A. Kawabata, J. Phys. Soc. Japan 34, 639 (1973), ibid 35, 669 (1973).

MAGNETIC STRUCTURE OF NdFeO₃ AND PrFeO₃
INVESTIGATED WITH A HIGH RESOLUTION
NEUTRON TIME-OF-FLIGHT DIFFRACTOMETER

I. Sosnowska*, E. Steichele
Fakultät für Physik E 21, Technische Universität München
D - 8046 Garching

ABSTRACT

The magnetic moment direction of the Fe^{3+} ions in NdFeO₃ and
PrFeO₃ has been determined from high resolution neutron diffraction
data.

INTRODUCTION

The rare earth orthoferrites have orthorhombically distorted
perovskite structures with the space group D_{2h}^{16} - Pbnm. In these
systems the antiferromagnetic ordering of the Fe^{3+} ions is mostly
of G - type[1]. From the intensity ratio of the resolved doublet
{011}:{101} one can decide whether the moment is mainly parallel
to one of the orthorhombic axes (pure G_x, G_y, G_z type ordering).
In intermediate cases one needs to resolve additional lines, e.g.
{103}, {013}, {211}, {121} to find the complete orientation of the
spin direction[2].

In the light rare earth perovskites the distortion is quite
small (b/a = 1.0175 and 1.0240 for Pr - and Nd - ferrite respectively)[3] and, as only powder samples are available, one needs a
high resolution diffractometer. Using a double - axis - spectrometer with a pyrolytic graphite filter Pinto and Shaked[4] derived G_x
- type ordering for both NdFeO₃ and PrFeO₃. A better separation
of the interesting lines can be obtained with a Time - of - Flight
(TOF) diffractometer, which has the best resolution for small Q-
values, i.e. large interplanar spacings. Such a study was performed at Dubna[5] and from the splitted {101},{011} doublet deviations
from the pure G_x type antiferromagnetism were postulated for PrFeO₃
and NdFeO₃.

EXPERIMENT AND RESULTS

We performed a similar experiment at room temperature with
the high resolution TOF - diffractometer in Garching[6], which gives,
due to a 145 m long flight path and a backscattering geometry, a
resolution better than 10^{-3} in the region of the first magnetic
lines. In Fig.1 and Fig.2 we show diffraction patterns obtained
with neutrons of wavelength $\lambda \simeq 4.7$ Å, which lines couldn't be resolved in earlier experiments, and at $\lambda \simeq 9$ Å, the extreme end of

* Permanent address: Institute of Experimental Physics, Warsaw University, Warsaw, Poland.

ISSN:0094-243X/82/890309-03$3.00 Copyright 1982 American Institute of Physics

the primary spectrum, where the widely separated {101} and {011} reflexions can be seen.

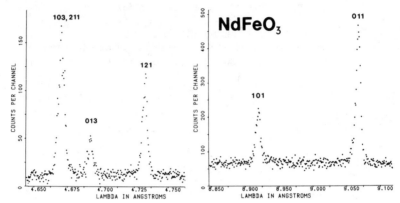

Fig.1. Neutron diffraction diagramm of NdFeO₃ at room temperature

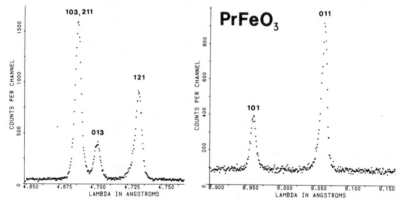

Fig.2. Neutron diffraction diagramm of PrFeO₃ at room temperature

For the intensity analysis we need the nuclear and magnetic contributions to the diffraction peaks, which were derived from a recent determination of the atomic positions and the magnetic moment of Fe^{3+} by a classical double - axis neutron diffraction measurement with profile analysis at Würenlingen[7].

We have calculated the R - factor for different assumptions of the angle ϕ between the a - axis and the moment component in the ab - plane and the angle α, the inclination of the moment to the ab - plane. From the results shown in Fig.3 and Fig.4 we conclude, that for PrFeO₃ the magnetic moment of the iron ion is parallel to the a - axis (G_x type), whereas in NdFeO₃ a deviation of $\phi = 15\pm 5°$ from the a - axis is realized. No significant inclination out of the ab - plane can be concluded from these datas. A complete high resolution structure analysis would help to improve our results.

311

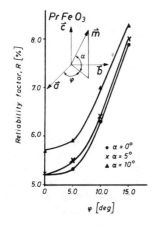

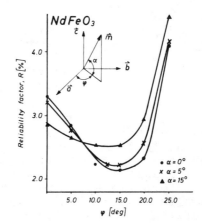

Fig.3. Reliability factor R for PrFeO₃ as a function of α and φ.

Fig.4. Reliability factor R for NdFeO₃ as a function of α and φ.

CONCLUSIONS

With the very high resolution one can determine the magnetic moment direction also in powder samples, when deviations from the main crystal directions occur.

One can realize the value of long neutron wavelengths in such an investigation of magnetic structures and therefore cold moderators at pulsed neutron sources could be very interesting from this point of view.

REFERENCES

1. W. C. Koehler, E. O. Wollan, and M. K. Wilkinson, Phys.Rev. 118, 58 (1960)
2. G. Shirane, Acta Cryst. 12, 282 (1959)
3. M. Marezio, J. P. Remeika, and P. D. Dernier, Acta Cryst. B26, 2008 (1970)
4. H. Pinto and H. Shaked, Solid State Comm., 10,663 (1972)
5. L. P. Kaun, B. Lippold, M. M. Lukina, W. Matz, B. N. Savenko, and K. Henning, Sov. Phys. Crystallogr. 21, 212 (1976)
6. E. Steichele, and P. Arnold, Phys. Lett. 44A, 165 (1973)
7. I. Sosnowska and P. Fischer, (1981) to be published

MAGNETIC STRUCTURE AND EXCITATIONS IN SINGLE-DOMAIN ANTIFERROMAGNETIC FCT γ-Mn-Cu AND γ-Mn-Fe ALLOYS

K. Mikke, J. Jankowska and E. Jaworska
Institute of Nuclear Research, 05-400 Świerk, Poland

ABSTRACT

Neutron diffraction studies of single-domain single crystals of fct γ-Mn-Cu and γ-Mn-Fe alloys have shown the presence of a small deviation of magnetic moment direction from c axis in both alloys and existence of additional perturbations in Mn-Cu. When compared with existing results for fcc γ-Mn alloys our data show no sensitivity of the SW dispersion relation to tetragonal distortion.

MAGNETIC STRUCTURE

The Mn-rich γ-Mn alloys were studied over the last 23 years. They were found to undergo a magnetic phase transformation accompanied by the martensitic fcc ↔ fct transition. The established magnetic structure is AF1, however an additional strong diffuse peak is observed around the (100) point. The aim of the present work was to obtain new information on the nature of this scattering and on the characteristics of spin waves in γ-Mn alloys by employing for the first time the single-domain single-crystal samples were performed by double and triple axis neutron spectrometer in Mn(10%Fe, 3%Cu) and Mn(18%Cu) alloys; in the last case for both quenched(homogeneous) and aged(chemically decomposed) samples. Line shapes and intensities were measured for various energy transfers and at several temperatures for the (001) and (100) points. Only (001) reflection was found in Mn-Fe, while in Mn-Cu diffuse peaks were also observed at (100) and (010).

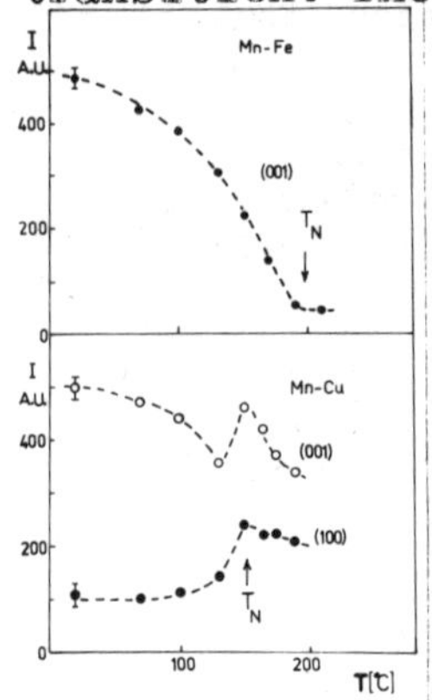

Fig. 1. I(T) for single-domain γ-Mn alloys

The temperature dependence of integrated intensities I(T) at (001) and (100) is shown in fig. 1 for Mn-Fe (upper part) and for aged Mn-Cu (lower part)(detailed discussion elsewhere [1]). Examples of the energy dependence of the intensity at the (001) point in Mn-Fe and at (001) and (100) in Mn-Cu shown in fig. 2 demonstrate a good separation of elastic from spin wave scattering in Mn-Fe but not in Mn-Cu. The data suggest that in Mn-Fe the spins deviate by a fixed angle from c axis and that this perturbation has infinite correlation range and lifetime. In Mn-Cu short range transverse correla-

ISSN:0094-243X/82/890312-03$3.00 Copyright 1982 American Institute of Physi

tions of magnetic moments for all three cubic directions
are required to explain the results. It is still an open
question whether these perturbations coexist with the
magnetic LRO or whether they appear only in those regions of the sample where the LRO disappears.

MAGNETIC EXCITATIONS

Magnetic excitations were measured around the (110),
(001) and (100) points over the energy transfer range
0÷34 meV at 300 K and in some cases also at 170 K and
80 K. Examples of constant momentum transfer scans are
shown in fig. 2. Resolution fit indicated in fig. 2 a

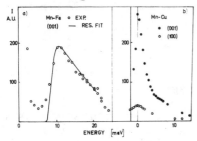

Fig. 2. Constant momentum
transfer scans for Mn-Fe
and quenched Mn-Cu

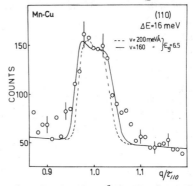

Fig.3. I(q) at 16 meV for Mn-Cu

by the solid line was obtained by convolving the sharp
SW dispersion relation for SW velocity v = 200 meVÅ and
SW energy gap E_g= 8 meV with the resolution function
of the spectrometer. An example of a constant energy
transfer scan I(q) is given in fig. 3 together with resolution fits for indicated parameters. From a set of
such data the following values were derived for v and E_g
For Mn-Fe at 300 K: v = 200±20 meVÅ and E_g= 8+1 meV,
for quenched Mn-Cu: v = 150±20 meVÅ at 300K and v = 160
±20 meVÅ at 80 K - at both temperatures the energy gap
could not be well determined. Our earlier values for the
polydomain Mn(25%Cu) samples were: for the aged sample
at 100 K: v = 250+30 meVÅ, E_g=8+2 meV and for the quenched sample at 100K: v = 140+30 meVÅ, E_g= 3+2 meV .These
data, excepting the quenched Mn-Cu alloy, fall within
the range of values found in fcc Mn-Fe(30÷70%Fe) alloys
where V = 160÷250 meVÅ and E_g= 6÷7 meV were reported [2].
Our data suggest the following conclusions:
1. Strong tetragonal deformation has no effect on the
isotropy of SW velocity. Within experimental errors
$\vec{v}\perp\hat{c}$ and $\vec{v}//\hat{c}$ do not differ. The absence of a pronounced
velocity anizotropy is puzzling in systems with a strong
tetragonal distortion 1-c/a = 0.02÷0.035.
2. There is no correlation between the tetragonal defor-

<u>mation and the parameters of the SW dispersion relation</u>.
The values of v and E_g for the Mn(10%Fe) with tetragonal
deformation 0.035 do not differ substantially from values obtained in cubic Mn-Fe alloys [2]. This fact is rather strange, since these alloys differ essentially in
the value of the magnetic moment per atom, which is
large in alloys with low Fe concentration and produces
much bigger disturbance of the band structure than in
alloys with the composition close to Mn(50%Fe). For cubic
Mn-Fe alloys, noncollinear MSDW magnetic structure was
suggested and one of the arguments for this was the isotropy of v. This argument looses most of its weight in
view of the fact that this isotropy is also observed in
tetragonal Mn-Fe alloy with an evidently collinear
structure.

3. <u>Pronounced inhomogeneities have no effect on SW</u>
<u>lifetime</u>. Annealing, giving rise to the chemical decom-
position, leads to an increase of both v and E_g in the
SW spectrum. This is consistent with the fact that annealing results also in an increase of Neel temperature
and of the magnetic moment per atom. However, there is
no indication that the appearance of two kinds of regions
one enriched and the other considerably depleted in Mn,
would disturb the spin waves. In particular, their lifetime is not visibly affected. This is a very surprising
fact considering that in the homogeneous Mn-Cu sample
the corresponding neutron groups are smeared out thus
indicating a serious reduction of the SW lifetime.

4. <u>There is still no good explanation of the origin of</u>
<u>the spin wave energy gap</u>. In principle, two kinds of in-
teractions might produce an energy gap in the SW spectrum: dipole-dipole interaction and spin-orbit interaction. The dipole-dipole interaction would imply the
magnetic structure with $\vec{\mu} \perp \vec{Q}_0$, while we observe $\vec{\mu} \parallel \vec{Q}_0$.
(Perhaps these interactions give origin to the trans-
verse polarization of "perturbations"?). The energy gap
results therefore from the spin-orbit interaction. It is
not clear, however, why this interaction should be so
much stronger in Mn than in other 3d metals where no indication for the energy gap in the SW spectrum was
found both in ferro- and in antiferromagnets.

REFERENCES

1. E. Z. Vintajkin et al. Sol. St. Comm. **37**, 295 (1981).
2. K. Tajima et al. J. Phys. Soc. Japan **41**, 1195 (1976).

MAGNETIC STRUCTURE OF RCo$_2$Ge$_2$ (R=Ho,Tb), A NEUTRON DIFFRACTION STUDY

H. Pinto[a], M. Melamud[a] and H. Shaked[a,b]

(a) Nuclear Research Center - Negev, Beer Sheva, Israel 84190
(b) Ben-Gurion University of the Negev, Beer Sheva, Israel 84120

ABSTRACT

A study of the magnetic structure of RCo$_2$Ge$_2$ (I4/mmm) with R=Tb, Ho was performed using neutron diffraction of powder samples. The 4.2 K diffraction patterns reveal a type I antiferromagnetic structure on the R sublattice. The magnetic moments lie along c, being 8.84β and 8.12β for Tb and Ho respectively. The possibility of order on the Co sublattice was also investigated, and the magnetic moment was found to be not larger then 0.12β (in TbCo$_2$Ge$_2$) and 1.7β (in HoCo$_2$Ge$_2$). The Neel temperatures are 31K and 8.5K respectively.

The family AB$_2$X$_2$ (A=lanthanide, actinide, B=transition metal, X= Ge,Si) crystallize in the BaAl$_4$ type structure[1], which belongs to the space group I4/mmm, with A at 2a, B at 4d and X at 4e. Some of the compounds in this family were studied by several techniques[1-8], including neutron diffraction.[9-12] The A sublattice is reported to be ordered, mostly as a type I antiferromagnet (HoCo$_2$Ge$_2$[9], NdFe$_2$Si$_2$[10], and NpCo$_2$Si$_2$[11]). There are also modulated (PrCo$_2$Ge$_2$[9,12]) and ferromagnetic (NpCu$_2$Si$_2$[11]) structures. YCo$_2$Ge$_2$ shows no magnetic ordering.[7,12]

For the present study powder samples of HoCo$_2$Ge$_2$ and TbCo$_2$Ge$_2$ were synthesized by arc melting under argon atmosphere. Neutron (λ= 2.45Å) diffraction patterns of the two compounds at 300K and 4.2K were taken (Fig. 1). The 300K patterns are in agreement with the rep-

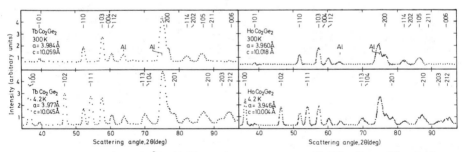

Fig. 1 Neutron (λ=2.45Å) diffraction patterns of RCo$_2$Ge$_2$. The indexing on each pattern is according to the cell constants as shown.

orted crystallographic structure[6], shown in Fig 2. The z parameter of the Ge site and the Debye-Waller constant were calculated (Table I) to give a best fit of the calculated to the observed integrated intensities. A random distribution of the Co and Ge ions, suggested by Ban et al[1], did not improve the fit.

The 4.2K patterns can be indexed according to the crystallographic cell constants (thermaly contracted, Fig 1), with additional re-

ISSN:0094-243X/82/890315-03$3.00 Copyright 1982 American Institute of Physics

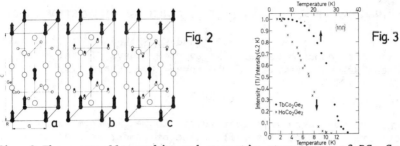

Fig. 2 The crystallographic and magnetic structures of RCo_2Ge_2. The magnetic space groups are: a. $I_p4/mm'm'$, b. $P4/mm'm'$, c. $I_p4m'm'$.

Fig. 3 Temperature dependence of the peak intensity of the {100} reflection in $HoCo_2Ge_2$ and $TbCo_2Ge_2$.

flections which are forbiden by I4/mmm, namely {hkℓ} with h+k+ℓ=2n+1, h or k ≠ 0 ; this is consistent with a magnetic structure were [½½½] is an antitranslation, and the magnetic axis is parallel to c. The four possible collinear configurations for an ordered Co sublattice, contribute to reflections with h+k of a single parity only. However, reflections with h+k of both parities are observed. Hence, ordering of the A sublattice is necessary to explain the observed pattern. Intensities were calculated for the magnetic structures shown in Fig. 2. The z parameter and magnetic moments were calculated to give a best fit of the calculated to the observed integrated intensities. For the magnetic form factors, f, absorption correction factors, $A_{hkℓ}$ and the Debye-Waller temperature factor, W, we used the approximation $U=\exp[-V(\sin\theta/\lambda)^2]$, where V=C, D and B for U=f, $A_{hkℓ}$ and W respectively. Since these parameters are highly correlated we kept constant the values of C, D and B (see Table I). A best fit to a structure with ferromagnetic order on the Co sublattice (Fig 2**b**) results in a zero moment on the Co for both compounds, being equivalent to Fig. 2a (Table I, columns 4 and 6). The result for a structure with antiferromagnetic Co sublattice (Fig 2**c**) is given in Table I columns 5,7.

The temperature dependence of the peak intensities of the {100}, {102} and {111} reflections in $HoCo_2Ge_2$ (Ho^{3+} in $^5G_8$) fit well the Brillouin curve with S=2, whereas in $TbCo_2Ge_2$ (Tb^{3+} in $^7F_6$) the data fall above the corresponding Brillouin curve (S=3). The result for {100} is shown in Fig. 3. The transition temperatures as deduced from these measurements are 8.5(5) K and 31(1) K for Ho and Tb compounds respectively, in agreement with the susceptibility results.[7]

A type I antiferromagnetic ordering was found in the present study for the A (Ho,Tb) sublattice. This type of order, with the same or different k vectors, was previously found in other AB_2X_2 compounds [9-12]. The R-R interaction in the RCo_2Ge_2 (R=lanthanide) series is of the RKKY type.[13] With this interaction, ordering temperature is expected to follow the deGennes function $c(g-1)^2J(J+1)$, which was shown to be the case in this series.[7] Different k vectors for compounds with different R, is characteristic of this type of interaction (e.g. the pure R metals). The question of possible ordering of the Co sublattice was considered in the present study, but remains unsolved. From

Table I Structural and magnetic parameters in RCo_2Ge_2 (R=Ho,Tb).

R=	300K		4.2K			
	Ho	Tb	Ho		Tb	
$\mu(R),\beta$	-	-	8.28(17)	8.12(15)	8.84(13)	8.84(12)
$C(R),Å^2$*	-	-	3.83	3.83	4.8	4.8
$\mu(Co),\beta$	-	-	NO ORDER	1.69(24)	NO ORDER	0.12(55)
$C(Co),Å^2$*	-	-	-	8.3	-	8.3
$D,Å^2$*+	-1.55	-1.2	-1.55	-1.55	-1.2	-1.2
$B,Å^2$	0.0(4)	0.3(4)	0.0*	0.0*	0.0*	0.0*
z(Ge)	0.374(1)	0.374(1)	0.372(2)	0.372(1)	0.370(1)	0.370(1)
R(%)†	5.1	6.2	6.6	5.5	3.8	3.8

* Kept constant during the refinement.
+ Corresponds to Σa of 1.40 cm^{-1} and 1.17 cm^{-1} in Ho and Tb resp.
† R(%)=100 $\{[(Iobs-Icalc)/\sigma]^2/\Sigma[Iobs/\sigma]^2\}^{\frac{1}{2}}$

group theory considerations, ordering of the R sublattice only belongs to a single irreducible representation (Γ_2^+) of $I4/mmm$ at k=(002π/c), whereas, ordering of R and Co sublattices (Fig. 2) does not belong to a single irreducible representation of I4/mmm. Hence, the former is consistent with a second order phase transition, whereas the latter is not. In the present analysis we have considered the two most symmetrical magnetic structures where both R and Co order (Fig. 2). We found that the analysis is not sensitive enough to ordering of the Co sublattice, because of the high correlation between the Co moment and other parameters. Good fits were obtained with and without ordering of the Co sublattice. The latter is consistent with the result that YCo_2Ge_2 shows no magnetic ordering.[7,12] The result for $TbCo_2Ge_2$ gives a magnetic moment which is practicaly zero, but appreciably different from zero for Co in $HoCo_2Ge_2$ (Fig 2c).

The authors are indebted to Dr. J. Gal for suggesting the problem, to Mr. S. Fredo for preparation of the samples and to Mr. H. Ettedgui for his technical aid.

1. Z. Ban and M. Sikirica, Acta Cryst. 18, 594 (1965).
2. W. Rieger and E. Parthe, Mon. fur Chem. 100, 444 (1969).
3. L. Omejec and Z. Ban, Z. fur Anorg. Allg. Chem. 380, 111 (1971).
4. I. Mayer and J. Cohen, J. Less Comm. Met. 29, 25 (1972).
5. I. Mayer, J. Cohen and I. Felner, J. Less. Comm. Met. 30,181(1972).
6. W. M. McCall, K. S. V. L. Narasimhan and R. A. Butera, J. Appl. Cryst. 6, 301 (1973).
7. W. M. McCall et al, J. Appl. Phys, 44, 4724 (1973).
8. S. K. Malik, S. G. Sankar, V. V. S. Rao and R. Obermyer, J. Chem. Phys. 37, 585 (1976).
9. A. Szytula, J. Leciejewicz and H. Binczycka, Phys. Stat. Sol 58, 67 (1980).
10. H. Pinto and H. Shaked, Phys. Rev. 37, 3261 (1973).
11. C. H. deNovion, J. Gal and J. L. Buevoz, JMMM 21, 85 (1980).
12. H. Pinto, M. Melamud and E. Gurevitz, Acta Cryst. A35, 533(1979).
13. C. Kittel, Quantum Theory of Solids (John Wiley, 1966), p. 366.

A POWDER NEUTRON DIFFRACTION STUDY OF ANTIFERROMAGNETISM IN $Er_6Mn_{23}D_{23}$

C. Crowder, B. Kebe, and W. J. James
Department of Chemistry and Graduate Center for Materials Research,
University of Missouri-Rolla, Rolla, MO 65401

W. Yelon
Missouri University Research Reactor, UMC, Columbia, MO 65201

ABSTRACT

The compound $Er_6Mn_{23}D_{23}$ has been prepared and a powder neutron diffraction study has been done. The compound is paramagnetic at 300°K and antiferromagnetic at 150°K. The crystallographic space group is I4/mmm with deuterium atoms occupying a, 1, m, n, and o sites. Since Er_6Mn_{23} belongs to the Fm3m space group, deuteration is responsible for a reduction in symmetry. At 150°K the Er and Mn moments in the deuteride are collinear with the c-axis and the Shubnikov magnetic space group is $P_I4/mm'm'$. This is also different from the non-deuterated Er_6Mn_{23} which has its easy axis along the cubic <111> direction (the tetragonal <101> direction). In the deuteride, the 'e' site Er atoms have moments of 2.5 μ_B while those on 'h' sites have moments of 1.2 μ_B. The Mn moments are approximately 3 μ_B for the 'b' and 'c' sites and approximately 1 μ_B for the two 'n' sites. The Mn atoms on the 'f' sites cannot have ordered moments due to the symmetry of the magnetic space group and, indeed, none is found. Magnetic susceptibility measurements are also reported.

INTRODUCTION

The addition of hydrogen to R_6Mn_{23} compounds dramatically alters their magnetic properties. Th_6Mn_{23} is a Pauli paramagnet whereas

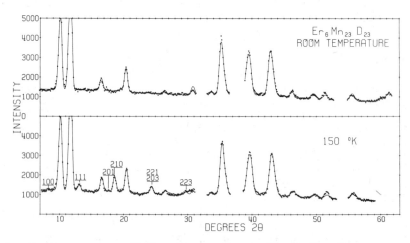

Fig. 1. Powder neutron diffraction patterns for $Er_6Mn_{23}D_{23}$.

ISSN:0094-243X/82/890318-03$3.00 Copyright 1982 American Institute of Physi

Th$_6$Mn$_{23}$H$_x$ shows long-range magnetic ordering at room temperature.[1] Y$_6$Mn$_{23}$ is ferrimagnetic[2] but Y$_6$Mn$_{23}$D$_{23}$ exhibits antiferromagnetism below ∿160°K[3,4]. Er$_6$Mn$_{23}$ is also ferrimagnetic[5] and new evidence reported in this paper shows Er$_6$Mn$_{23}$D$_{23}$ is antiferromagnetic at 150°K. New evidence also indicates the crystallographic symmetry of this deuteride is best described as body-centered tetragonal.

EXPERIMENTAL

Preparation of Er$_6$Mn$_{23}$D$_{23}$ was the same as given for Y$_6$Mn$_{23}$D$_{23}$[4]. Powder neutron diffraction patterns were obtained at 300 and 150°K and refined using a modified Rietveld profile technique[6,7]. Scattering lengths and form factors were obtained from Bacon[8]. Data and resulting profiles at both temperatures are shown in Fig. 1. Magnetic susceptibility measurements were obtained by the extraction method at temperatures ranging from 2–300°K and at fields up to 150 kOe.

RESULTS AND DISCUSSION

The pattern at 300°K was refined using both face-centered cubic (Fm3m) and body-centered tetragonal (I4/mmm) symmetry. Using cubic symmetry it is found that the 'a' site and one 'f' site are fully occupied with deuterium. A 'j' site is partially occupied with 10 deuterium atoms going into 24 locations. The weighted R-factor for this profile is 5.39% with an expected R-factor of 2.55% to give χ = 2.11. Using tetragonal symmetry, the 'j' site becomes an 'l' site and two 'm' sites. The refinement shows the 'l' site contains 3 deuterium atoms in 8 possible locations, one 'm' site is fully occupied with deuterium, while the other 'm' site is empty. This site occupation means the three-fold axis of the cube no longer exists, even though all metal atoms still occupy their cubic positions. The weighted R-factor for this profile is 4.18%, the expected R-factor is 2.55%. Thus χ = 1.64 which is a significant improvement over the cubic profile. A summary of the tetragonal structure at 300°K along with that for Y$_6$Mn$_{23}$D$_{23}$[4] is given in Table I.

The new reflections in the data at 150°K can be indexed as primitive tetragonal. This indicates a tetragonal P$_I$ magnetic Bravais lattice. No 00ℓ reflections are present so it is reasonable to assume

Table I. Refined Structures of Er$_6$Mn$_{23}$D$_{23}$ and Y$_6$Mn$_{23}$D$_{23}$

Site	position	parameter	Er$_6$Mn$_{23}$D$_{23}$*	Y$_6$Mn$_{23}$D$_{23}$*	Site	position	parameter	Er$_6$Mn$_{23}$D$_{23}$*	Y$_6$Mn$_{23}$D$_{23}$*
R 4e	0,0,z	N	2	2	D 2a	0,0,0	N	1	1
		z	0.205(2)	0.209(1)	D16l	x,y,0	N	2	2
R 8h	x,x,0	N	4	4			x	0.41(1)	0.42(1)
		x	0.205(2)	0.209(1)			y	0.16(1)	0.17(1)
Mn2b	0,0,½	N	1	1	D16m	x,x,z	N	8	8
Mn4c	½,½,0	N	2	2			x	0.367(3)	0.366(2)
Mn8f	¼,¼,¼	N	4	4			z	0.154(3)	0.152(3)
Mn16n$_1$	x,0,z	N	8	8	D16n	x,0,z	N	8	6
		x	0.352(2)	0.360(2)			x	0.207(3)	0.205(3)
		z	0.176(1)	0.180(1)			z	0.098(3)	0.092(3)
Mn16n$_2$	x,0,z	N	8	8	D32o	x,y,z	N	4	6
		x	0.252(2)	0.251(2)			x	0.228(5)	0.225(3)
		z	0.376(1)	0.375(1)			y	0.119(3)	0.113(3)
							z	0.168(3)	0.170(3)

Moments on metal atoms*	R(e)	R(h)	Mn(b)	Mn(c)	Mn(f)	Mn(n$_1$)	Mn(n$_2$)
Er$_6$Mn$_{23}$D$_{23}$ at 150°K	-2.5(3)	1.2(3)	-2.7(3)	3.4(2)	0	0.8(3)	0.9(2)
Y$_6$Mn$_{23}$D$_{23}$ at 80°K	0	0	-3.8(3)	3.4(2)	0	1.3(1)	1.3(1)

*Estimated errors in last expressed digits are given in parentheses.

320

a collinear structure with moments parallel to the c-axis. This means only eight Shubnikov space groups are possible. All eight were tried as possible models using the Rietveld profile technique. Seven groups resulted in magnetic R-factors ranging from 60-90%. Only the $P_I4/mm'm'$ magnetic space group gave a reasonable magnetic R-factor of 23%. Although this appears high compared to the nuclear R-factor, it is reasonable considering the small size of the magnetic reflections. The $P_I4/mm'm'$ space group is antiferromagnetic since atoms in body-centered related positions have equal moments in opposite directions. The 'f' site may not have an ordered moment along

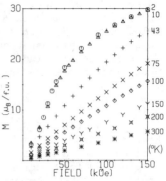

Fig. 2. Magnetization of $Er_6Mn_{23}D_{23}$.

the c-axis due to the symmetry in the space group. The magnitudes of the Er and Mn moments are given in Table I.

Plots of magnetization vs. field give linear results with near zero intercepts for temperatures above 75°K (see Fig. 2). This also supports a collinear antiferromagnetic structure. At lower temperatures, the magnetization curves are no longer linear. This phenomenon is not yet understood and was not observed for $Y_6Mn_{23}D_{23}$[4].

SUMMARY

$Er_6Mn_{23}D_{23}$ has a body-centered tetragonal crystallographic structure similar to that of $Y_6Mn_{23}D_{23}$. At 150°K, the compound is antiferromagnetic and belongs to the $P_I4/mm'm'$ space group. Magnetic susceptibility measurements support the powder neutron diffraction data. Below 75°K the magnetic structure shows evidence of change, the exact nature of which is not yet understood.

The authors would like to thank the U.S. Department of the Army for support under grant no. DAAG-29-80-C-0084.

REFERENCES

1. S. K. Malik, T. Takeshita, and W. E. Wallace, Solid State Comm. 23, 599 (1977).
2. K. Hardman and W. J. James, The Rare Earths in Modern Science and Technology, Vol. I, G. J. McCarthy, J. J. Rhyne, and H. E. Silber (eds.), (Plenum Press, New York, 1977), p. 408.
3. K. Hardman, J. J. Rhyne, H. K. Smith, and W. E. Wallace, to be published in The Rare Earths in Modern Science and Technology, Vol. III, (Plenum Press, New York, 1982).
4. C. Crowder, B. Kebe, W. J. James, and W. Yelon, ibid.
5. B. Kebe, C. Crowder, W. J. James, J. Deportes, R. Lemaire, and W. Yelon, ibid.
6. H. M. Rietveld, J. Appl. Cryst. 2, 65 (1969).
7. Modified Rietveld program received from E. Prince at NBS.
8. G. E. Bacon, Neutron Diffraction, (Clarendon Press, Oxford, England, 1975), pp. 38-41.

NEW NEUTRON DIFFRACTION RESULTS ON MAGNETIC PROPERTIES OF THE CUBIC RARE EARTH COMPOUNDS HoP AND PrX$_2$ (X = Ru, Rh, Ir, Pt)

P. Fischer and W. Hälg
Institut für Reaktortechnik, ETH Zürich
CH5303 Würenlingen, Switzerland

E. Kaldis
Laboratorium für Festkörperphysik, ETH
CH8093 Zürich, Switzerland

F. J. A. M. Greidanus
Kamerlingh Onnes Laboratorium der Rijksuniversiteit
Leiden, The Netherlands

K. H. J. Buschow
Philips Research Laboratories, Eindhoven, The Netherlands

ABSTRACT

Neutron diffraction studies performed on polycrystalline, NaCl type HoP in external magnetic fields yield ⟨100⟩ as easy directions of magnetization in the ferromagnetic state. The magnetic ordering of the MgCu$_2$ type Laves phase systems PrX$_2$ (X = Ru, Rh, Ir, Pt) was investigated on powdered samples by means of neutron diffraction. Simple ferromagnetic structures were observed. The determined Curie temperatures confirm bulk measurements, and the values of the ordered magnetic moments indicate crystal field effects.

INTRODUCTION

Both the rare earth compound HoP[1,2] and several PrX$_2$ systems[3,4] show anomalous magnetic properties. HoP undergoes at $0.88T_c$ a phase transition from a flopside structure to ferromagnetism[5]. Bulk physical properties of PrRh$_2$ and PrIr$_2$ display at low temperatures broad second maxima which might be due to magnetic phase transitions or caused by quadrupolar effects similar to PrMg$_2$[4]. In order to distinguish between these possibilities neutron diffraction investigations were performed on PrX$_2$ (X = Ru, Rh, Ir, Pt), in connexion with crystal field studies ([5]). Quadrupole (biquadratic) interactions are important for the magnetic phase transitions in HoP[6,7]. For theoretical considerations the easy directions of magnetization were determined in the ferromagnetic state, which previously were assumed to be ⟨110⟩.

EXPERIMENTAL

In the temperature range from 2 to 293 K powder samples were investigated on two-axis diffractometers at reactor Saphir, Würenlingen (neutron wavelength 2.3 or 1.1 Å). Al and for HoP V containers of cylindrical shape or plate geometry (PrRh$_2$ and PrIr$_2$) were used. Except for PrRu$_2$ (slightly contaminated by Ru) the samples proved to be single MgCu$_2$ type Laves phase systems. Similar to external magnetic fields $\underline{H}$ along the scattering vector $\underline{Q}$ were applied on a stoichiometric HoP powder sample with lattice constant a = 5.627 Å, using an electromagnet. Thus the easy directions of magnetization may be determined also in cubic systems[1], without problems due to extinction.

RESULTS AND DISCUSSION

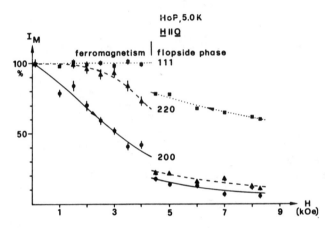

Fig. 1. Magnetic field dependence of normalized magnetic neutron intensities of polycrystalline HoP at 5.0 K.

At H = 0 the ferromagnetic state of HoP is stable in the temperature range from T_t = 4.75 K to T_C = 5.6 K[2]. At H = 8.3 kOe ($\underline{H}$ parallel to $\underline{Q}$) T_t increases to 5.3 K. The largest decrease of the 200 Bragg intensity shown in Fig. 1 proves that ⟨100⟩ are the easy directions of magnetization in ferromagnetic HoP, as at low temperatures[1]

Table I Neutron diffraction results for PrX$_2$ compounds (N), compared to other measurements. (* at 4.2 K)

X	$a^{(8)}$ (Å)	$T_C^{(3)}$ (K)	T_C^N	μ_o^N	$\mu_o^{CF(5)}$ (μ_B)	$\mu_{4.2\,K}^{M(3)}$	$\mu_o^{HF(3)}$
Ru	7.623	33.9(5)	35.3(4)	2.6(1)	2.9	1.7	2.29
Rh	7.575	7.9(5)		2.1(1)*	1.4/2.6	1.2	2.32
Ir	7.621	11.2(5)	10.8(3)	2.5(3)	3.1	2.1	2.44
Pt	7.711	7.7(5)	7.8(3)	0.9(2)	1.1	1.5	2.32

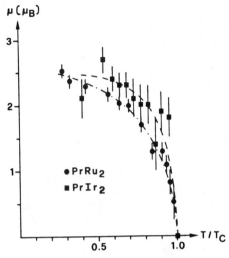

Fig. 2. Temperature dependence of ordered magnetic moments μ of PrRu$_2$ and PrIr$_2$ (Bragg peak 111).

and as expected from crystal fields. Similarly at 5.3 K zero magnetic intensity is first attained for reflection 200 at H ≈ 6 kOe.

Neutron scattering results for PrX$_2$ compounds are summarized in Tab. 1 and Fig. 2. No deviations from simple ferromagnetic ordering were detected. The determined Curie temperatures T_C confirm bulk measurements[3]. Due to crystal field effects the ordered magnetic moments appear to be essentially reduced below the free ion value gJ of 3.2 μ_B (Pr^{3+}, $^3H_4$). They agree reasonably with values calculated from crystal field parameters ([5], T_C[3] <100>). Presumably the bulk anomalies[3] are caused by quadrupolar effects similar to PrMg$_2$[4].

REFERENCES

1. H. R. Child et al., Phys. Rev. **131**, 922 (1963).
2. P. Fischer et al., J. Magn. Magn. Mater. **14**, 301 (1979).
3. J. C. M. van Dongen et al., J. Magn. Magn. Mater. **15-18**, 1245 and 1231 (1980).
4. A. Loidl et al., J. Appl. Phys. **52**, 1433 (1981).
5. F. J. A. M. Greidanus et al., to be publ. in Proc. 4th Int. Conf. on Crystal Field and Structural Effects in f-Electron Systems, Wrocław (1981).
6. A. Furrer and E. Kaldis, Crystalline Electric Field and Structural Effects in f-Electron Systems (Plenum, N. Y., 1980) p. 497
7. P. M. Levy et al., to be publ. in Phys. Rev. B (1982).
8. A. Iandelli and A. Palenzona, Handbook on the Physics and Chemistry of Rare Earths **2** (North-Holland, Amsterdam, 1979) p. 1.

THE STRUCTURAL AND MAGNETIC PROPERTIES OF MnAs$_{.88}$P$_{.12}$
STUDIED BY NEUTRON DIFFRACTION AND
MAGNETIZATION MEASUREMENTS

A.F. Andresen, H. Fjellvåg, A. Skjeltorp
Institute for Energy Technology, 2007 Kjeller, Norway

K. Bärner
Physikalisches Institut der Universität Göttingen,
3400 Göttingen, West-Germany

ABSTRACT

On cooling MnAs$_{.88}$P$_{.12}$ undergoes at 450K a transition from
NiAs to MnP-type structure. Magnetization measurements indicate
a Néel temperature at T_N = 220K followed by another transition at
$T_1 \approx$ 200K to a metamagnetic state. At even lower temperatures the
metamagnetic region ends in another transition at $T_K \approx$ 70K. Neutron diffraction measurements reveal a spiral spin structure with
propagation vector $\vec{\tau}$ = 0.082 · 2π · $\vec{a}$*. below T_N. On cooling the
spiral structure disappears at T_1 but reappears at T_K with approximately the same propagation vector. The disappearance of the helical structure in the intermediate temperature range can be related
to changes in the relative values of the exchange interaction
brought about by changes in the interatomic distances.

INTRODUCTION

MnAs takes the orthorhombic MnP-type structure only in the
small temperature range 306 - 393K. This structure can be stabilized to lower temperatures either by applying a pressure[1], by substituting with other transition metals (V,Cr,Fe or Co)[2], or by substituting a small amount of the anions with P[3]. In all these cases a
lattice contraction is effected and a transition from a high to a
low spin state is observed.

In many compounds with the MnP-type structure a spiral spin
structure is observed. This is not the case for MnAs, probably
because the MnP-type structure is not retained to low enough temperatures. However, by substituting with other transition metals spiral
spin structures appear[2]. This has also been observed for Mn$_{1-x}$P$_x$
(x = 0.075)[4]. We report here on susceptibility, magnetization and
neutron diffraction measurements carried out on a sample with x=0.12.

SUSCEPTIBILITY AND MAGNETIZATION MEASUREMENTS

Above 450K the inverse susceptibility follows a Curie-Weiss
law (fig. 1) leading to a moment value of 4.55+ 0.20 μ_B. In the
temperature interval 450 - 335K the susceptibility shows and anomalous behaviour which is connected with an observed lattice contraction
and transition to a low spin state. Below 335K the inverse suscepti-

ISSN:0094-243X/82/890324-03$3.00 Copyright 1982 American Institute of Physi

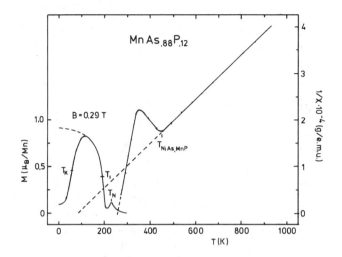

Fig. 1. Susceptibility and magnetization measurements on MnAs$_{.88}$P$_{.12}$

bility again follows a Curie-Weiss law with the smaller moment
value 2.20 μ_B.

Magnetization measurements show a small peak at 220K indicative
of a Néel point. This is followed by a metamagnetic region in
which a moment value of 0.8 μ_B/Mn is obtained at 0.29 T. The metamagnetic region stretches from 220K to 70K.

NEUTRON DIFFRACTION MEASUREMENTS

Neutron diffraction measurements show the presence of a spiral
spin structure below 220K. This is of the double spiral type pro-
pagating along the a-axis with propagation vector $\vec{\tau} = 0.082 \cdot 2\pi \cdot \vec{a}*$.
It is different from the spiral found in MnP[5] which propagates along
the c-axis, but similar to the spirals found in transition metal
substituted MnAs[2].

The temperature dependence of the 000$^{\pm}$ satellite reflection
shows the spiral structure to disappear around 70K but to reappear
with approximately the same propagation vector around 200K (fig.2).

MONTE CARLO CALCULATIONS

We assume that the disappearance of the spiral structure in
the intermediate temperature interval 70 - 200K is connected with
changes in the relative values of the magnetic exchange interactions
brought about by changes in the interatomic distances. Monte Carlo
calculations based on four near neighbour isotropic exchange interactions have been carried out to explore the effects of changing
the ratios between the interactions. The results show that the
onset of order is in particular sensitive to the ratio of the inter-

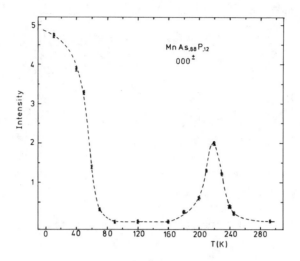

Fig. 2. Temperature dependence of
000$^{\pm}$ satellite reflection

actions between the second and fourth nearest neighbors. However,
to obtain definite results more information is needed about the
magnitudes of the interactions.

Tajima et al.[6] have found from spin wave dispersion relations
in MnP that the second neighbour interaction $J_{1\bar{2}}$ is almost temperature independent and equal to 10 meV. The fourth nearest neighbour
interaction J_{12} is small, negative, and temperature dependent. The
spiral structure becomes stable when $J_{1\bar{2}} < - 4J_{12}$.

It is interesting to note that in $MnAs_{.88}P_{.12}$ it is the
second nearest neighbour distance which changes most with temperature decreasing from 3.132 Å at 293K to 2.970 Å at 120K and then
increasing again. It is therefore possible that $J_{1\bar{2}}$ increases
above the stability limit for the spiral structure in the temperature range 70 - 200K.

REFERENCES

1. J.B. Goodenough and J.A. Kafalas, Phys.Rev. 157 (1967) 389.
2. K. Selte, A. Kjekshus, A.F. Andresen abd A. Zieba, J.Phys.
   Chem. Solids 38 (1977) 719·
3. N. Menyuk, J.A. Kafalas, K. Dwight and J.B. Goodenough, Phys.
   Rev. 177 (1969) 942.
4. S. Haneda, N. Kazama, Y. Yamaguchi and H. Watanabe, J. Phys.
   Soc. Japan 42 (1977) 31.
5. J.B. Forsyth, S.J. Pickart and P.J. Brown, Proc. Phys. Soc.
   88 (1966) 333.
6. K. Tajima, Y. Ishikawa and H. Obara, J. Mag.Mag.Mat. 15 (1980)
   373.

STRUCTURE AND MAGNETIC PROPERTIES OF TbB4

C. M. McCarthy and C. W. Tompson
Department of Physics and Research Reactor

F. K. Ross
Research Reactor and Department of Chemistry
University of Missouri, Columbia, Mo. 65211

Z. Fisk
Institute for Pure and Applied Physical Sciences
University of California-San Diego, La Jolla, Ca. 92093

ABSTRACT

Neutron powder diffraction analysis and single-crystal studies of TbB4 indicate that this compound is isostructural with ErB4 and DyB4 but has a different magnetic structure. The magnetic moments of the Tb atoms appear to lie in mixed antiferromagnetic domains of approximately [301] and [031] orientation.

INTRODUCTION

In a systematic study of rare-earth borides,[1] poly-crystalline samples of TbB4 (97% [11]B enriched) were studied by powder neutron diffraction and Rietveld refinement techniques. The compound is iso-structural with ErB4 and DyB4 at all temperatures investigated (78K, 28K and 4.2K) but the magnetic structure differs from the c-axis antiferromagnetic ordering reported[2] for the latter compounds. The powder diffraction data (Fig. 1) suggest a magnetic model containing antiferromagnetic ordering in (hk0) with mixed domains required to preserve tetragonal symmetry (Fig. 2). This assignment is based on

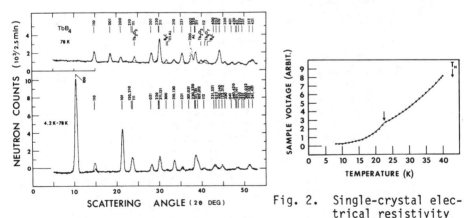

Fig. 1.   TbB4 powder diffraction data.

Fig. 2.   Single-crystal electrical resistivity
data for TbB4.

328

the large intensity of the 100 reflection compared to the 011; the
00ℓ reflections with ℓ odd are extinguished by virtue of the z = 0
coordinate for all Tb atoms. Rietveld refinement of the 4.2K data
with this structure gave poorer results than obtained for the other
rare-earth borides in the study. This was partly attributed to obvious sample contamination by terbium oxides and by B_4C (Fig. 1), but
the Rietveld residual for the magnetic structure (10%) was dominated
by the fit of peaks free from overlap with the contaminants. In
addition, electrical resistivity measurements made on TbB_4 single-
crystals show a discontinuity at 23K (Fig. 2), but comparison of the
28K and 4.2K diffraction data did not reveal any differences in the
crystal or magnetic structures. After obtaining single crystals of
TbB_4, we initiated further studies using 3-dimensional diffraction
data to define the magnetic structure of this material better.

EXPERIMENTAL

A well-formed polyhedron of TbB_4 (97% [11]B) showing 28 developed
faces was used and was approximated as a sphere of 1.05 mm diameter
for absorption and extinction corrections. Data were collected at
298K and 8K; after standing at 298K for several weeks, further data
were collected at 78K and 28K. Averaging of the symmetry-equivalent
reflections shows good internal agreement for all data except the 28K
data (Table I). The latter data contain more internal variation but
do not conclusively depart from 4/m mm symmetry. All data were
collected on the 2XE instrument (at MURR) which is equipped with a
40 cm i.d. Huber circle and a Displex cryostat.

TABLE I. Single-crystal data for TbB4

Temperature	Form	No. of Data	Stability	Data Averaging R(F)	$R_W(F)$	No. of Data for least-squares
298K	hkℓ	367	±0.9%	0.023	0.021	183
78K	±hkℓ	710	±1.8%	0.021	0.019	194
28K	±hkℓ	713	±1.6%	0.034	0.042	197
8K	hkℓ	362	±0.6%	0.019	0.017	191

ANALYSIS

The data were modeled using standard crystallographic least-
squares methods with locally-introduced modifications. The 78K data
were used to obtain reference parameters for atom positions, anisotropic thermal motion and anisotropic extinction. These parameters
were then used as constraints for the data at other temperatures;
only an overall scale factor and one isotropic thermal parameter, adjusted to compensate for changes in thermal motion, were used in the
fitting of this model. For magnetic structures, the least-squares
symmetry operations were replaced by 95 parameter constraints. This
permitted the generation of a magnetic cell having the same size as
but different symmetry form the crystal cell. Magnetic atoms were
entered as two contributions with different symmetry constraints but
with identical positional and thermal parameters. The projection of
magnetic domains onto the scattering planes was programmed as a fixed

calculation which was modified for reassignment of the magnetic vectors. Results of these models are tabulated in Table II.

Temperature	80K	298K	298K	28K	28K	8K	8K	8K
Model	(1)	(2)	(1)	(2)	(3)	(2)	(3)	(4)
R(F)	.0296	.0341	.0318	.123	.060	.199	.099	.090
Rw(F)	.0290	.0365	.0392	.255	.088	.262	.167	.125

TABLE II. Results of least-squares model refinement

(1) Standard crystallographic model. Variables are scale, atom coordinates and anisotropic temperature factors and anisotropic extinction.
(2) Constrained model. 80K parameters, scale and overall temperature factor.
(3) Constrained model (2) + magnetic atoms with variable occupancy and (100-010) projection.
(4) Constrained model (3) with (301-031) projection.

The extrapolation of the 78K parameters to 298K with a scale and a single parameter for thermal compensation gave good results. This model was applied to the 28K and the 8K data and magnetic scattering was introduced to minimize the residuals. Although this model can be altered to allow for a reduction in the symmetry of the crystallographic cell (by redefining some of the 95 constraints mentioned above), no evidence was found to justify a modification in the crystal structure. The nuclear cell is centrosymmetric and the magnetic cell has only B component contributions to the structure factors; this provides convenient identification of model effects and the fit of special reflections (h0ℓ with h even is purely nuclear while h odd receives only magnetic contributions). Comparisons of magnetic contributions referenced to various scattering directions gave best results for domains in the (h0ℓ) plane (and thus an equal mixture of (0kℓ) domains) with a minimum residual occurring approximately along [301].

CONCLUSIONS

The residual discrepancies in the fit of the 8K data appear to be predominantly in the magnetic model. The absence of any superlattice reflections constrains all magnetic models to coincide with the nuclear cell. Although the best fit occurs with an added ℓ-axis tilt of 20-30°, the R-factor obtained (9%) is higher than expected for the internal consistency of the 8K data. Modified Fourier mapping techniques are being employed to investigate this model further. In addition, the inconsistencies observed at 28K are probably due to unequal domain formation or annealing. These will be studied in future experiments.

REFERENCES

1. C. M. McCarthy, Ph.D. thesis, University of Missouri-Columbia (1981).
2. W. Schafer and G. Will, J. Chem. Phys. 64, 1994 (1976).

330

NEUTRON DIFFRACTION STUDY
OF THE WEAK ANTIFERROMAGNETISM IN ORTHOFERRITES

V.P.Plakhty, Yu.P.Chernenkov, M.N.Bedrizova
Leningrad Nuclear Physics Institute,
Gatchina, Leningrad distr. 188350 USSR

J.Schweizer
Institut Laue Langevin and DN/RFG Centre d'Etudes
Nucleares, 38042, Grenoble, France

ABSTRACT

Neutron diffraction study of the weak antiferromagnetism (hidden canting) in $YFeO_3$, $YbFeO_3$ and $ErFeO_3$ is carried out. The experimental data are in good agreement with the theory.

The antisymmetric exchange interaction

$$H_{as} = \sum_{ij} \vec{D}_{ij} \left[\vec{S}_i \times \vec{S}_j \right] \qquad (1)$$

can lead to weak ferromagnetism (WFM) or to weak antiferromagnetism (WAFM) – overt or hidden canting of antiferromagnetic sublattices. Both are allowed by the symmetry in orthoferrites $RFeO_3$ (R – rare-earth or yttrium). The WFM in these materials was studied by a number of authors of whom Treves et al[1] had obtained probably the most complete experimental data. As to the WAFM there were no experiments, except our neutron diffraction work where it was observed for the first time[2], and NMR study[3]. Both were performed on $YFeO_3$. In this paper we present the results of neutron diffraction study of the WAFM in yttrium, ytterbium and erbium orthoferrites in comparison with the theory[4,5].

The orthoferrite unit cell contains four Fe^{3+} ions as shown in Fig.1a. Magnetic ordering of the iron sublattice is described by the well known bases vectors $\vec{F}$, $\vec{G}$, $\vec{A}$, $\vec{C}$ of irreducible representations Γ_m of the space group Pbnm. The main antiferromagnetic structure is always $\vec{G}$, the WAFM component may be $\vec{A}$ or $\vec{C}$, and $\vec{F}$ is the

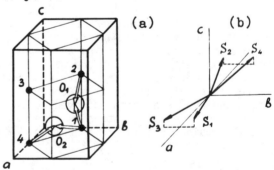

Fig.1. a– The orthoferrite unit cell. b– $\Gamma_4 (G_x A_y F_z)$ spin configuration.

WFM moment. Just below the Neel temperature all the orthoferrites have the same configuration $\Gamma_4(G_xA_yF_z)$ which is shown in Fig.1b. As the temperature decreases the reorientation may occur, for example, to $\Gamma_2(F_xC_yG_z)$ as in the case of erbium orthoferrite.

Reflections attributed to the WAFM components were expected to be four orders of magnitude weaker than nuclear ones or those due to the main antiferromagnetic structure. It means that they should be completely pure to be measurable. Fortunately the extinction law varies for different spin modes, moreover some of the WAFM reflections does not superimpose on nuclear ones. Thus it was not too time-consuming to measure such weak reflections having a single crystal of proper size (~4mm), but the special attention should be paid to avoid any false effect such as multiple scattering or contamination of higher orders in the beam. The neutrons wave length was 2.48 Å (reflection from (111)Ge), and higher orders were suppressed to the levels 10^{-4}-10^{-5} by the filter made of oriented graphite plates. The measurements with all the plates and without a part of them allowed to make correction for the remains of higher orders. As the multiple Bragg reflections can be hundreds times stronger than WAFM ones even their wings are dangerous. That is why before the measurement of any WAFM peak the proper crystal orientation was calculated where for the resolution used should be no traces of multiple scattering. Afterall that was verified by the gaussian form of the ω-scan profile. In the case of $YFeO_3$ polarization analysis was used to show that these reflections are actually magnetic. The details one can find in the ref.2. Here we briefly mention that neutrons were polarized with a neutron guide, the sample

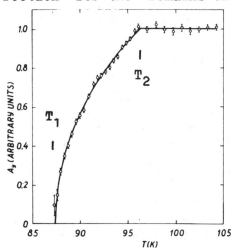

Fig.2. The temperature dependence of A_y component.

was placed in the white beam, and the analyzer (Heussler alloy) was used as a monochromator as well. With this set it was possible to do the polarization analysis on the medium flux reactor WWR-M. As a result of these experiments the value of A_y/G_x at room temperature was found to be $1.93(18) \cdot 10^{-2}$ for $YFeO_3$ and $1.67(6) \cdot 10^{-2}$

for YbFeO$_3$. Besides the temperature dependence of A$_y$ at
the spin reorientation in ErFeO$_3$ was obtained (Fig.2).
As it was shown by Moskvin and Sinitsyn[4] for the
superexchange interaction

$$\vec{D}_{ij} = \mathscr{D}_{ij}\left[\vec{r}_{io} \times \vec{r}_{jo}\right] \qquad (2)$$

where $\vec{r}_{io}$, $\vec{r}_{jo}$ connect the nearest magnetic ions i,j
with the intermediate anion, and $\mathscr{D}_{ij}$ depends on the bond
length and angle. Neglecting the small bond differences
two nonequivalent Dzialoshinski vectors D$_{12}$ and D$_{14}$
(Fig.1a) can be expressed through the only parameter $\mathscr{D}$
and known structure parameters. Leaving in the free energy[5] the terms corresponding to the iron-iron exchange
to the second order of antisymmetric terms one obtains

$$\Phi = 3J\vec{F}^2 - 3J\vec{G}^2 + J\vec{A}^2 - J\vec{C}^2 + 2D^x_{12}(G_yC_z - G_zC_y)$$
$$+ (2D^y_{12} + 4D^y_{14})(G_zF_x - G_xF_z) + 4D^z_{14}(G_xA_y - G_yA_x). \qquad (3)$$

Minimization of (3) for the Γ_4(G$_x$A$_y$F$_z$) leads to

$$F_z/G_x = (D^y_{12} + 2D^y_{14})/3J \ , \quad A_y/G_x = -2D^z_{14}/J \ . \qquad (4)$$

Using the experimental values of F$_z$/G$_x$[1] one can determine $\mathscr{D}$/J and then A$_y$/G$_x$ which occur to be 1.95·10^{-2} for
both orthoferrites in agreement with our experiment.
According to Yamaguchi[5] the reorientation of the
iron spins is caused by the anisotropic exchange interactions between iron and rare-earth which is polarized
in the field of the iron sublattice. Assuming that in
the spin-reorientation region the iron sublattice is saturated and the paramagnetic sucseptibility of the rare-
earth is proportional to inverse temperature[1] one can
obtain from (36d) of ref.5

$$A_y = -2D^z_{14}G_xT_1T_2\left[(1/T_1^2 - 1/T^2)/(T_2^2 - T_1^2)\right]^{1/2}/J \qquad (5)$$

The best fit of this function for T$_1$=87.4(1), T$_2$=96.4(1)
is shown in Fig.2 by the full line between T$_1$ and T$_2$.
The agreement with the experiment is very good (χ^2=1.5).

REFERENCES

1. D.Treves, L. Appl. Phys. **36**, 1093 (1965).
2. V.P.Plakhty, Ju.P.Chernenkov, J.Schweizer, M.N.Bedri-
zova, Proc. USSR National Conf. on Phys. Mag. Phenom.
p.396, Kharkov (1979); LNPI preprint N°597 (1980).
3. H.Lütgemeir, H.G.Bohn, M.Brejczewska, J. Mag. Mag.
Mat. **21**, 289 (1980).
4. A.S.Moskvin, E.V.Sinitsyn, FTT **17**, 2495 (1975).
5. T.Yamaguchi, J. Phys. Chem. Solids **35**, 479 (1974).

SPIN DENSITIES IN NON CENTRO SYMMETRIC STRUCTURES

J.X. Boucherle[x], B. Gillon[+o], J. Schweizer[x+]

x DN/RFG, C.E.A. 85X, 38041 Grenoble Cedex
+ Institut Laue-Langevin, 156X, 38042 Grenoble Cedex
° Centre de Mécanique Ondulatoire Appliquée, rue du Maroc, 75019 Paris

ABSTRACT

Spin densities can be calculated from magnetic structure factors F_M measured by the polarized neutron diffraction technique. In the case of non centro-symmetric crystals F_M cannot be deduced directly from the measurements. To overcome this difficulty we propose a treatment where the parameters of a multipole representation of the spin density are determined directly from the observations. This method has been applied to a nitroxide free radical : tanol $C_9H_{18}NO_2$.

Neutron diffraction can provide magnetic structure factors F_M which are the Fourier components of the spin density. Polarized neutron diffraction enables the measurements to be made with high precision. The flipping ratio R of the intensities diffracted on a Bragg peak for the two neutron spin states, is measured :

$$R = I^+/I^- = |F_N + F_M|^2 / |F_N - F_M|^2 \qquad (1)$$

In the case of centro-symmetric crystals, for which F_N and F_M, the nuclear and magnetic structure factors, are real, the ratio $\gamma = F_M/F_N$ can be obtained by solving the quadratic equation :

$$R = (1+\gamma)^2/(1-\gamma)^2 \qquad (2)$$

The choice between the two solutions is in general straightforward. If the nuclear structure is well known so that the nuclear structure factor F_N can be calculated, one gets the magnetic structure factor :

$$F_M = \gamma F_N \qquad (3)$$

Several corrections must be applied because of imperfect experimental conditions, extinction, and proton polarization. These are taken into account by modification of equation 2 but the general principle of the determination of F_M is unchanged.

In the case of non centro-symmetric crystals the previous method cannot be used. Both nuclear and magnetic structure factors are complex quantities : $F_N = F_N' + iF_N''$, $F_M = F_M' + iF_M''$ and the flipping ratio becomes

$$R = \frac{(F_N' + F_M')^2 + (F_N'' + F_M'')^2}{(F_N' - F_M')^2 + (F_N'' - F_M'')^2}$$

ISSN:0094-243X/82/890333-03$3.00 Copyright 1982 American Institute of Physics

From a knowledge of the crystal structure F_N' and F_N'' can be calculated but it is not possible to define the quantity γ. Measurement of R yields a relation between the two unknown quantities F_M' and F_M'' but does not enable to determine them.

To overcome this difficulty we propose to apply a model which provides an analytical description of the spin density. We have chosen an expansion of the density around the nuclei at rest. Such a model, first used to represent charge densities [1] has already proved to be very well adapted in the case of spin densities in centro-symmetric crystals [2]. The expansion consists of a superposition of aspherical atomic densities, each described by a series expansion in real spherical harmonic functions :

$$s(\vec{r}) = \sum_{atoms} \sum_{\ell=0}^{\infty} R_\ell(r) \sum_{m=-\ell}^{\ell} P_{\ell m} Y_{\ell m}(\hat{r}) \tag{5}$$

the $P_{\ell m}$ are population coefficients. The radial functions $R_\ell(r)$ are of Slater type :

$$R_\ell(r) = \frac{\zeta^{n+3}}{(n+r)!} r^n e^{-\zeta r} \tag{6}$$

The magnetic structure factors corresponding to eqs (1) and (2) become

$$F_M(\vec{H}) = \sum_{atoms} e^{2\pi i \vec{H}.\vec{r}} \left[\sum_{\ell=0}^{\infty} \phi_\ell(H) \sum_{m=-\ell}^{\ell} P_{\ell m} Y_{\ell m}(\hat{H}) \right] e^{-w} \tag{7}$$

with $\phi_\ell(H) = 4\pi i^\ell \int_0^\infty R_\ell(r) j_\ell(2\pi Hr) r^2 dr \tag{8}$

To establish the spin density map, the set of parameters $(\zeta, P_{\ell m})$ which characterizes the spin density and fits the experimental data must be determined. To do this, given a set of parameters, F_M' and F_M'' are calculated and, taking into account the usual corrections, a value R_{cal} for the flipping ratio of each measured reflection, is obtained from equations (7), (4). By comparing R_{cal} with the measured flipping ratios, it is possible to refine the parameters $(\zeta, P_{\ell m})$ minimizing the quantity

$$\sum_{reflections} w_i |R_{cal_i} - R_{obs_i}|$$

This method has been applied to tanol $C_9H_{18}NO_2$ (2,2,6,6 tetramethyl 4-piperidinol-1-oxyl), one of the smallest stable nitroxide radicals (non centrosymmetric space group Cm, z = 2) with a = 7.003Å, b = 13.686 Å, c = 5.695 Å, β = 118.88° at 10 K. To allow analysis of the polarized neutron experiment, the crystal structure (positions and thermal parameters of all atoms including hydrogen atoms) has been refined from neutron diffraction data at 10 K.

The nitroxide free radicals are characterized by one unpaired 2p electron, localized on the N-O group. Because of negative exchange interactions between molecules [3], the magnetic susceptibility does not increase below 10 K. A polarized neutron experiment has been performed at 4.2 K and with an applied magnetic field of 46.5 KOe. The flipping ratios have been determined for 173 independent reflections.

The multipole expansion has been restricted to the multipoles allowed by local symmetry : $\vec{x}$ // $\vec{CC}$, $\vec{y}$ // $\vec{NO}$. Only the significant populations have been refined : monopole, dipole y, quadrupole z^2 and the Slater exponent ζ for the two centres N and O. The best refinement yields an agreement factor χ^2 equal to 2. with the parameters given in table 1. The values of the monopole population correspond to a total moment of 0.15 μ_B for one molecule, in a very good agreement with a susceptibility of 250 emu/mole [3] at 4.2 K in a 46 KOe field

	N	O
$\zeta(\overset{\circ}{A}{}^{-1})$	4.4 (2)	5.8(5)
mono (μ_B)	0.089(8)	0.059(7)
y (μ_B)	-0.015(7)	-0.004(5)
z^2 (μ_B)	0.045(4)	0.030(6)

Table 1. Parameters of the multipole expansion

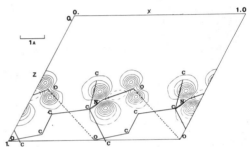

Fig. 1. Projection of the spin density along $\vec{b}$. Contours are from -.01 to 0.13 $\mu_B/\overset{\circ}{A}{}^2$ by step of 0.02 $\mu_B/\overset{\circ}{A}{}^2$.

The spin density map (Fig. 1) has been obtained from this representation. The spin density on the nitrogen atom is larger than that on the oxygen one in a proportion 0.6($\pm$.1)/0.4 ($\mp$.1). A previous experiment on tanol suberate, another nitroxide radical, which is centro-symmetric [2], has shown that the spin density is almost equally shared between N and O. However, in tanol, the distribution of the unpaired electron on the NO bond may be modified by the hydrogen bonding which exists between oxygen and hydroxyl proton group OH of the nearest neighbor molecule [4].

REFERENCES

1. N.K. Hansen and P. Coppens, Acta Cryst. A34 (1978) 909.
2. P.J. Brown, A. Capiomont, B. Gillon and J. Schweizer, J.M.M.M. 14 (1979) 289.
3. J. Yamauchi, T. Fujito, E. Ando, M. Nishigushi, Y. Deguchi, J. Phys. Soc. Jap. 25, 1558 (1968).
4. J. Lajzerowicz-Bonneteau, Acta Cryst. (1968) B24, 196.

THE SPIN DENSITY DISTRIBUTION IN CrCl₃ and CrBr₃

P.J. Brown and K.R.A. Ziebeck
Institut Max von Laue - Paul Langevin, Grenoble, France

P. Radhakrishna
Laboratoire Léon Brillouin, C.E.N. Saclay, Gif sur Yvette, France

ABSTRACT

The magnetisation distribution in the layered ionic compounds $CrCl_3$ and $CrBr_3$ has been studied using polarised neutron diffraction. The results show that in both compounds $\sim$ 20 % of the magnetic moment is not located in 3d- like orbitals centred on the chromium ions. This reduction of the 3d moment sets a lower limit ($A_{\pi}^2 > .04$) on the square of the covalent admixture parameter. The spatial distribution of the delocalised moment has been studied by Fourier techniques which indicate a significant moment density between chromium ions in the chromium layers.

INTRODUCTION

Magnetisation measurements on chromium tri-bromide and chromium tri-chloride [1] show that at 4.2 K fields of some 0.3 T are sufficient to produce essentially parallel alignment of all the chromium moments. In $CrCl_3$ this aligned phase is metamagnetic but $CrBr_3$ is truly ferromagnetic [2]. Both $CrCl_3$ and $CrBr_3$ have the rhombohedral $FeCl_3$ structure at low temperatures [3,4]. In this structure the chromium ions are octahedrally cordinated by their halide ligands and the octahedra share three of their twelve edges to form sheets perpendicular to the triad axis ; these sheets are then superposed to give the strongly layered structure. Magnetic interactions within the layers are ferromagnetic and pass through the shared halide ions. The very weak magnetic coupling between layers is antiferromagnetic for $CrCl_3$ and ferromagnetic for $CrBr_3$. The fact that the moments in both compounds can be ferromagnetically aligned in modest fields has made it possible to use the classical polarised neutron diffraction technique to study their magnetisation density and in particular the degree to which it is delocalised from the chromium ions by covalent interactions.

EXPERIMENTAL

Single crystals of both compounds were grown for us at the M.P.I. Stuttgart. Those used in the experiments were $\sim$ 5 x 5 x 1 mm cleaved from much larger boules. The crystals were mounted with an <01.0> axis parallel to the ω axis of the diffractometer. Two series of experiments were made, both at 4.2 K. In one using unpolarised neutrons and no applied field the integrated intensities of nuclear reflections in the zero first and second layers out to

$Sin\theta/\lambda = 0.78$ Å^{-1} were measured. In the other, using polarised neutrons and applied fields of 1.4 T the polarisation ratios of all reflections in the same layers were measured to a $Sin\theta/\lambda$ limit of 0.5 for $CrCl_3$ and 0.75 for $CrBr_3$.

RESULTS

The unpolarised neutron data were used to determine the structural parameters and the obverse/reverse twinning ratio in $CrCl_3$, the $CrBr_3$ crystal was essentially untwinned. The results of the structure refinement are given in Table I. The temperature parameters obtained for $CrBr_3$ are unrealistic because the large mosaic spread of the crystal made it impossible to integrate the reflections properly at the higher angles. Parameters obtained from the structure refinement were used to give nuclear structure factors with which the magnetic structure factors were obtained from the polarisation ratios. The experimental chromium form factors obtained from these structure factors were compared with a theoretical Cr^{3+} free ion 3d spherical form factor. In both compounds the experimental curve drops much more sharply at low angles but the outer parts of the curves have rather closely the same shape ; and for $CrBr_3$ where the measurements extend to $Sin\theta/\lambda = 0.75$ experimental and theoretical curves drop to zero at about the same angle.

Table I Structural Parameters in $CrCl_3$ and $CrBr_3$

	CrCl$_3$			CrBr$_3$	
Atom	Positional Parameters	Temperature Parameters	Atom	Positional Parameters	Temperature Parameters
Cr	z=0.3334(3)	B_{11}= 0.38(10)	Cr	z=0.3322(11)	B_{11}=2.6 (4)
		B = 0.65 (5)			B_{33}=2.2 (5)
Cl	x=0.0005(3)	B_{11}= 0.44 (5)	Br	x=0.0015(12)	B_{11}=0.97(17)
	y=0.3530(2)	B_{22}= 0.59 (5)		y=0.3494 (9)	B_{22}=0.87(19)
	z=0.9228(1)	B_{33}= 0.71 (4)		z=0.9216 (2)	B_{33}=2.20(14)
		B_{23}=-0.07 (3)			B_{23}=0.11(16)
		B_{31}= 0.00 (3)			B_{31}=0.07(14)
		B_{12}= 0.28 (4)			B_{12}=0.31(14)

DISCUSSION

The difference in shape between the experimental and theoretical free ion form factors is not unexpected since covalency in these compounds should be significant. For a ferromagnetically aligned material, the delocalisation associated with transferred spin should lead to a narrower form factor. To quantify the effect the

338

magnitude of the 3d-like part of the scattering has been estimated by fitting the moment value of a Cr^{3+} spherical 3d form factor to those data with $\sin\theta/\lambda > .25$. Reasonable fits were obtained with $\mu 3d = 2.46(4)$ for $CrBr_3$ and $2.24(6)$ for $CrCl_3$.

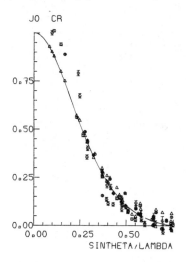

JO CR

SINTHETA/LAMBDA

Figure 1

The form factor for h0l reflections in $CrBr_3$. Circles mark observed values and triangles those corresponding to Cr^{3+} $^4\Gamma_2$.

Magnetisation and resonance measurements on chromic salts suggest that the 3d electrons in $CrCl_3$ and $CrBr_3$ should be described by the orbital singlet $^4\Gamma_2$ obtained as ground state from a ^4F term under an octahedral field. The scattering from this state is not isotropic and the calculated anisotropy of its form factor reproduces rather well the anisotropy of the observations as may be seen from Figure 1.

In a simple model including covalent effects, the magnetisation density can be split into a 3d part, a ligand part and an overlap part, this latter being negative for antibonding orbitals. Hence

$$\rho(r) = N(\rho_{3d}(r) + A^2\rho_{ligand}(r) - A\rho_{overlap}(r))$$

N is a normalising constant obtained by integrating the density over a unit cell so that $N = (1 + 3A^2 - 3AS)^{-1}$ where S is the value of the overlap integral. Since the observed 3d moment is reduced in $CrCl_3$ and $CrBr_3$, N must be less than unity and hence A > S a lower limit of $A^2 > 0.04$ is given if S is very small. A preliminary investigation of the delocalised moment, has been made using Fourier techniques. These give some evidence for magnetic density transferred towards ligand sites, but the most significant feature of the maps is an accumulation of density between near neighbour chromium ions within the chromium layer. This density demonstrates the importance of the shared edges of the octahedron of halide ions in the description of covalent interactions in these compounds.

REFERENCES

1.   H. BIZETTE and C. TERRIER, J. Phys. Radium 23 486 (1962)
2.   J.F. DILLON, J. Phys. Soc Japan 19 1662 (1964)
3.   B. MOROSIN and A. NARATH, J. Chem. Phys. 40 1958 (1964)
4.   H. BRAEKKEN Kgl Norske Videnskab Selskab. Forhandl 5 42 (1932)

ELECTRON THEORY OF THE MAGNETIC MOMENT STRUCTURE OF α-Fe, ε-Co, Ni FROM NEUTRON DIFFRACTION EXPERIMENT

S. H. Yü
Department of Physics, Jilin University
Changchun, Jilin, China

ABSTRACT

A direct method of analysis of valence electron distribution from their crystal structures, are applied to α-Fe, ε-Co, Ni based on the fine structure of atomic valence of type A hybridization of states of Fe, Co, Ni. Their valence electron distributions and theoretical magnetic moments of 3d are obtained. Using known spectroscopic splitting factor g and experimental total magnetic moment, the magnetic moment of the orbital m^{orb} and of the 4s electron m^{4S}, are calculated, whose negative sign is independent of the choice of known values of g. The results agree with the neutron diffraction experimental data to the first order of approximation. The theoretical bond lengths also agree with the values directly calculated from the crystal structures. Based on the valence electron distributions and the negative sign of m^{4S} deduced theoretically, the interpretation of the negative magnetic moment distributions as lattice electrons appears reasonable.

INTRODUCTION

Analyses of the valence electron distributions of α-Fe, ε-Co and Ni are made using a general method to be presented at the Ottawa congress for the analysis of valence electron distribution directly from crystal structure. This has been proved to be correct by the analyses of γ-, γ'-, δ-Fe, $Fe_I SiFe_I^2$, $Fe^C NFe_3^t$, FeP, FeO and FeAl. The present analysis provides a different aspect but at the same time equally severe test. This is because, although the metals are simpler but all get space distribution of negative magnetic moment beside the space distribution of positive magnetic moment in certain directions.

GENERAL THEORY OF ANALYSIS OF THE MAGNETIC MOMENT SPACE DISTRIBUTION

1. Theory of discontinuous hybridization of states

Suppose a metal atom in solid has an orbital ψ_α superimposed of two basic states ψ_h and ψ_t in the form $\psi_\alpha = a_h \psi_h + a_t \psi_t$. Then from quantum mechanical consideration it is found that for a fractional covalent bond of metal as first proposed by Pauling, the ratio $a_h / a_t \equiv k$ is in general discontinuous deduced from the stationary condition $(\partial \psi / \partial \alpha)_{\theta=0} = 0$ where θ is measured from the covalent bond direction passing through the two atoms u and v making the bond. Suppose ψ_h and ψ_t states to have 4s, 4p, 3d valence electrons and magnetic electron with numbers l, m, n, m_h and l', m', n', m_t respectively. Then we shall obtain:

$$k_a = \tau l' + m' + n' / \tau l + m + n \sqrt{l' + m' + n' / l + m + n} \cdot l \pm \sqrt{3}m \pm \sqrt{5}n / l' \pm \sqrt{3}m' \pm \sqrt{5}n' , \infty , 0 \qquad (1)$$

The interesting point is introduction of a new type of valence electron of s or s^2 denoted by ϕ or ϕ. It is called by lattice electron which is distributed in the interstitial spaces enclosed by more than three positive metallic ions. The distribution may or may not uniform. The negative magnetic moment distributions obtained by neutron diffraction experiments[2] for α-Fe, ϵ-Co, Ni are Typical examples. The other two are in NiAl from Cooper's experimental and in Cu and Ag each from their 6 bands of outer electron equidensity curves[3]. The parameter τ or τ' '=0 or 1 depending on s or s^2 to be lattice or covalent electrons. The relative composition of the h and t states are $C_{h\alpha}=a_h^2=k_\alpha^2/(k_\alpha^2+1)$, $C_{t\alpha}=a_t^2=1/(k_\alpha^2+1)$. Then we have the following parameters to be calculated for the α-th hybrid level k_α.

$n_{c\alpha}$ =covalent electron number= $(\tau l +m+n, C_{h\alpha}+(\tau'l'+m'+n')C_{t\sigma}$

$n_{l\alpha}$ =lattice electron number $=(1-\tau)l C_{h\alpha} +(1-\tau')l'C_{t\alpha}$

m_{3d}^α =magnetic moment of 3d magnetic electrons= $m_h C_{h\alpha} +m_t C_{t\alpha}$ (2)

$R_\alpha(1)$ =single bond radius= $R_h(1)C_{h\alpha}+R_t(1)C_{t\alpha}$

where R(1) is introduced by Pauling in the equation:

$$D_{uv}(n_\alpha)=R_u(1)+R_v(1)-\beta\log_\epsilon n_\alpha \quad (3)$$

where n_α is the fractional covalent bond electron pair number. $\beta=0.600$ in our cases.

2. Bond Length Difference (BLD) Method of Analysis of Valence Electron Distributions $n_\alpha(\alpha=A,B,\ldots N)$ and $n_{l\alpha}$ for pure Metals.

Suppose we have N types of $D_{uu}(n_\alpha)$ each has I_α equivalent bonds. If the metal atom u has $n_{c\alpha}^u$ covalent electron which is distributed among the bonds $D_{uu}(n_\alpha)$. Then we have

$n_{c\alpha}^u=I_A n_A +I_B n_B+I_C n_C\ldots+I_N n_N=n_A(I_A+I_B\gamma_B+\ldots+I_N\gamma_N)$

where $\gamma_\alpha=n_\alpha/n_A$, let Σ Iγ $=I_A+I_B\gamma_B+\ldots+I_N\gamma_N$,   (4)

Then $n_A=n_{c\alpha}^u/\Sigma$Iγ, $n_\alpha=n_A\gamma_\alpha$ $(\alpha=B,C,\ldots N)$ (5)

From (3) for pure metals we obtain

$$\log[\{D_{uu}(n_A)-D_{uu}(n_\alpha)]/\beta \equiv \Delta_{A\alpha} \quad (6)$$

Now $D_{uu}(n_\alpha)$ can be calculated from observed crystal structure:

$$
\begin{aligned}
bbc \quad &\alpha-Fe: \quad I_A=8, \quad D_{uu}(n_A)=\sqrt{3}/2\ a, \quad I_B=6, \quad D_{uu}(n_B)=a \\
fcc \quad &Ni: \quad I_A=12, \quad D_{uu}(n_A)=a/\sqrt{2}, \quad I_B=6, \quad D_{uu}(n_B)=a \\
hcp \quad &\epsilon-Co: \quad I_A=6, \quad D_{uu}(n_A)=[\tfrac{1}{3}+\tfrac{1}{4}(\tfrac{c}{a})^2]^{1/2}a, \quad I_D=2, \quad D_{uu}(n_D)=c \\
&I_B=6, \quad D_{uu}(n_B)=a, \quad I_E=6, \quad D_{uu}(n_E)=\sqrt{3}\ a \\
&I_C=6, \quad D_{uu}(n_C)=[\tfrac{4}{3}+\tfrac{1}{4}(\tfrac{c}{a})^2]^{1/2}a,
\end{aligned}
\quad (7)
$$

Hence $\gamma_\alpha(\alpha=B,C,\ldots N)$ can be obtained directly from observed lattice spacings a, or a and c for the three metals, so is ΣIγ as well. Then if $n_{c\alpha}^u$ is selected properly and substitute into (5) together with known ΣIγ, the valence electron distribution $n_\alpha(\alpha=A,B,\ldots N)$ is determined. To check the correctness of n_α we have to calculate $\bar{D}_{uu}(n_\alpha)$ from

$$\bar{D}_{uu}(n_\alpha)=2R_{u\alpha}(1)-\beta\log n_\alpha \quad (8)$$

and compare $\bar{D}_{uu}(n_\alpha)$ with $D_{uu}(n_\alpha)$ calculated from (7) to see whether $\Delta D=\bar{D}_{uu}(n_\alpha)-D_{uu}(n_\alpha)$ is small enough. $R_{u\alpha}(1)$ is known with $n_{c\alpha}^u$ at the same time from known $C_{h\alpha}$ and $C_{t\alpha}$ as shown in (2).

3. Fine Structure of Atomic Valence of Type A Hybridization of Fe, Co and Ni.

Great number of structure analyses leads to type A hybridization of Fe, Co, and Ni to be:

$$\begin{array}{llll}
 & 3d & 4s & 4p
\end{array}$$

Fe: h state: $\uparrow\uparrow\uparrow$ • • $\oplus$ $\underline{\bullet}$ o o $\left.\right\}$ $\ell, m, n, \tau = 2, 1, 2, 0$ $\left.\right\}$
    t state: $\|$ • • • • $\underline{\bullet}$ $\underline{\bullet}$ o o $\left.\right\}$ $\ell', m', n', \tau' = 1, 1, 4, 1$ $\left.\right\}$

Co: h state: $\uparrow\uparrow\uparrow$ • • $\oplus$ $\underline{\bullet}$ $\underline{\bullet}$ o $\left.\right\}$ $\ell, m, n, \tau = 2, 2, 2, 0$ $\left.\right\}$ (9)
    t state: $\|$ • • • • $\underline{\bullet}$ $\underline{\bullet}$ $\underline{\bullet}$ o $\left.\right\}$ $\ell', m', n', \tau' = 1, 2, 4, 1$ $\left.\right\}$

Ni: h state: $\|$ $\|$ • • o $\oplus$ $\underline{\bullet}$ $\underline{\bullet}$ $\underline{\bullet}$ $\left.\right\}$ $\ell, m, n, \tau = 1, 3, 2, 0$ $\left.\right\}$
    t state: $\|$ $\uparrow$ • • • $\underline{\bullet}$ $\underline{\bullet}$ $\underline{\bullet}$ $\underline{\bullet}$ $\left.\right\}$ $\ell', m', n', \tau' = 1, 3, 3, 1$ $\left.\right\}$

where ($\uparrow$), ($\|\equiv\uparrow\downarrow$), (•), ($\underline{\bullet}$) are magnetic, dumb pair, covalent, s-p equivalent d covalent electron, and (o), the empty orbit. The s-p equivalent d electron is 3d electron in origin but its electron cloud is so diffusely extended out that its effect to bond length is equivalent to a 4s or 4p electron. Substituting the values of $l, m, \ldots \tau'$ into (1) we obtain k_α and so obtain $C_{h\alpha}$, $c_{t\alpha}$, $n_{c\alpha}$, $n_{t\alpha}$, $R_\alpha^{(1)}$, m_σ^{3d} from (1) and (2). If α is arranged in ascending order of $C_{t\sigma}$ and arrange all the parameters in table form, we get type A hybridization tables of Fe, Co, Ni. Trial finds α-Fe, Ni are on the 8-th and 13-th hybrid level respectively, while ε-Co is near 11-th level with $C_{h\alpha}=0.6162$. The parameters of the three metals are:

	σ	$C_{h\alpha}$	$C_{t\alpha}$	$n_{c\alpha}$	$n_{t\alpha}$	$R_\alpha^{(1)}$ Å	$m_\sigma^{3d}(\mu_B)$	$m^T(\mu_b)$	g
α - Fe	8	.8015	.1985	3.5955	1.6030	1.1187	2.4045	2.216	2.070
Ni	13	.3138	.6862	6.3724	0.3138	1.156	0.6862	0.606	2.186
ε Co	~11	.6162	.3838	5.1514	1.2324	1.1409	1.849	1.716	2.171

m^T is the total atomic magnetic moment, g the spectroscopic factor. We have the orbital m^{orb} and 4s m^{4s} magnetic moment given

$$m_\alpha^{orb}=(\tfrac{g}{2}-1)m_\alpha^{3d}, \quad m_\alpha^{4s}=m^T-m^{3d}-m^{orb}=m^T-\tfrac{g}{2}m_\sigma^{3d} \quad (10)$$

Using these parameters in above table we calculate the values n_α, $\bar{D}_{uu}(n_\alpha)$, m_α^{orb}, m_σ^{+s} to compare with the values of $D_{uu}(n_\alpha)$, m^{3d}, m^{orb}, m^{4s} obtained from neutron diffraction experiments. A comparison of the theortical values with x-ray and neutron diffraction data are listed in table below.

α - Fe		Ni		ε - Co				
n_A	n_B	n_A	n_B	n_A	n_B	n_C	n_D	n_E
.3835	.0878	.5260	.0101	.4256	.4240	.0079	.0010	.0006

	α - Fe			Ni			ε - Co		
	Theo'l	Exp'l	$\Delta m, \Delta D$	Theo'l	Exp'l	$\Delta m, \Delta D$	Theo'l	Exp'l	$\Delta m, \Delta D$
$m^{3d}(\mu_B)$	2.4045	2.395	.0145	0.6862	0.656	.0302	1.809	1.86	0.013
m^{orb}	0.083	.085	.003	0.0638	0.055	.0088	0.158	0.13	0.028
m^{4s}	-0.249	-0.21	-0.039	-0.144	-0.105	.039	-0.293	-0.28	-0.013
$D(n_A)$ (Å)	2.4971	2.4824	.0047	2.4794	2.4916	-0.0122	2.5044	2.5045	-0.0001

It is seen that all Δm and ΔD are small enough that the agreement between theory and x-ray and neutron diffraction experiments appears satisfactory to the first order of approximation. From these results we can interpretate not only the negative but also the positive magnetic moment in terms of ℓ_g and t_{2g} d electrons much deeper.

REFERENCES

1. S. H. Yü, English translations of Chinese articles in Sci. Report of solid and Molecular State Phys. and Chem. Spec. Bulletin, Jinan Univ. 2-1 2-2 (1981).

2. C. G. Shull and Y. Yamada, J. Phys. Soc. Jap. 17. B. III. 1. 1962; C. G. Shull and H. A. Mook, Phys. Rev. Letters, 16 184 (1966); R. M. Moon, Phys. Rev. 136 A195 (1964); H. A. Moon, Phys. Rev. 148 495 (1966).

3. K. S. Wong, Y. F. Chen, and C. M. Wong, English Translation of Chinese articles in Sci. Report of solid and Molecular Phys. and Chem., Spec. Bulletin, Jinan Univ. 2-3, 2-4, 2-5 (1981)

4. A. J. P. Meyer and G. Asch, J. Appl. Phys. 32 3303 (1962)

5. Decicco and Kitz, Phys. Rev. 162 456 (1967).

PHONONS IN V_2O_3 AND Ti_2O_3*

M. Yethiraj, S.A. Werner and W.B. Yelon
Physics Department and Research Reactor Facility
University of Missouri-Columbia, Columbia, MO 65211

J.M. Honig
Department of Chemistry
Purdue University, West Lafayette, IN 47907

ABSTRACT

A neutron scattering study of the acoustic phonons in V_2O_3 has been carried out in the high temperature metallic phase (above 160 K) and in the low temperature antiferromagnetic insulating phase. The dispersion relations for acoustic phonons propagating along selected high symmetry directions in Ti_2O_3 have also been measured. In V_2O_3, there appears to be a low-lying optic mode in the metallic phase which disappears below the metal insulator transition temperature.

INTRODUCTION

The sesquioxides of vanadium and titanium have been a topic of considerable theoretical and experimental interest since it was found that there occurs a transition from metal to antiferromagnetic insulator in V_2O_3, and from metal to semi-metal in Ti_2O_3 with decreasing temperature.[1] Various investigations have shown that pure V_2O_3 goes through a first order transition at 160 K with a change in crystal symmetry from trigonal to monoclinic. The resistivity changes by seven orders of magnitude at the transition.[2] Ti_2O_3 has no long range magnetic order at low temperature and is seen to go through a gradual metal-semimetal transition over a wide temperature range between 300 K and 600 K without any change in symmetry.[3] In Ti_2O_3, there seems to be a general consensus that the transition occurs due to a gradual shrinking of the band gap with increasing temperature, which eventually leads to a band overlap. The mechanism driving the transition in V_2O_3 is still largely unknown. In this investigation a neutron scattering study of the lattice vibrational properties of these oxides has been carried out.

EXPERIMENTAL WORK

The experiments were carried out at the Missouri University Research Reactor on the triple axis instrument (3XE). Constant-Q and constant E scans were used as appropriate. The single crystal of pure V_2O_3 was grown at Purdue by the triarc process in an atmosphere of gettered argon. This sample was used for all the phonon measurements and was about 0.5 cm³ in dimensions. The crystal was wrapped in aluminum foil and epoxied to an aluminum post and then mounted in an Air Products Displex refrigerator. Phonons at room temperature were measured along the hexagonally indexed [0,0,1] and [1,1,0] directions. In addition, phonons along a third direction in the a-c plane, namely

*Work supported by grant NSF DMR 7727247 at Missouri and NSF DMR 7723798 at Purdue.

344

[1,1,3], were also measured. One set of phonons that were of particular interest were studied as a function of temperature. In the low temperature monoclinic antiferromagntic insulator (AFI) phase, the Bragg peaks were observed to be triply twinned. This twinning caused it to be difficult for the entire dispersion curve in the AFI phase to be traced out.

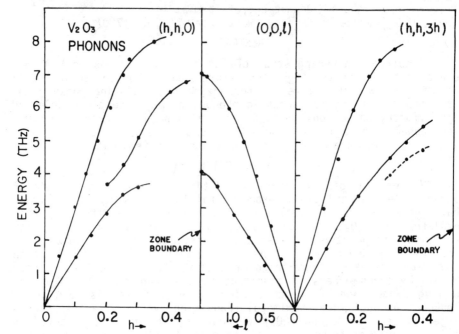

Fig. 1. Phonon dispersion relations in vanadium sesquioxide.

The disperson curves obtained for V_2O_3 in the metallic phase are shown in Fig. 1. Of particular interest is the low-lying transverse optic mode that was observed in the (h,h,0) direction. This branch was traced out towards the zone boundary and in towards the zone center as far as possible. There were, however, resolution problems close to the zone center as the contribution from the elastic Bragg intensity was too overwhelming for the phonon to be sharply resolved. It was observed that shoulders to the Bragg intensity were present.

On cooling the sample to the AFI phase, it was found that this optic mode was no longer observed at the low energy at which it was measured in the metallic phase. It was also found that the disappearance of this low lying optic mode was rather abrupt and coincided with the M-I transition temperature T_{MI}. The phonon was not seen a few degrees below T_{MI} and existed just about T_{MI}.

The Ti$_2$O$_3$ crystal was obtained from Prof. R.J. Sladek at Purdue University.[2] The crystal was prepared by a flux-growth technique and is about 1 cm^3 in size. For the room temperature measurements, the crystal epoxied onto an aluminum post, mounted on a goniometer and then affixed directly onto the sample table of the instrument. For the high temperature measurements, the crystal was mounted on a boron nitride post which had a groove cut in it that exactly fitted one edge of the sample. It was then held in position by a thin nickel wire that ran through the post and was secured over the top of the crystal. It was then placed in a furnace.

The room temperature acoustic phonon dispersion relations for Ti$_2$O$_3$[2] in the [1,1,0], [0,0,1] and [1,0,0] directions are shown in Fig. 2. We have also made measurements in the metallic phase at 625 K.

These measurements will be extended to include phonons propagating in other directions and to certain lower optic modes.

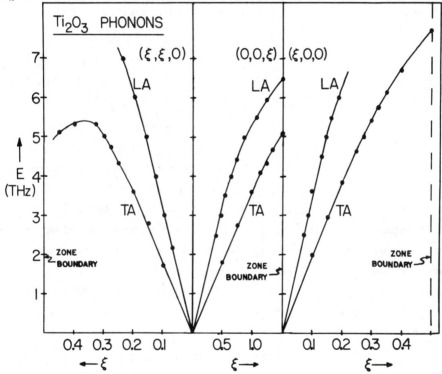

Fig. 2.   Phonon dispersion relations in titanium sesquioxide.

REFERENCES

1. F.J. Morin, Phys. Rev. Lett. 3, 34 (1959).
2. H. Kuwamoto, J.M. Honig and J. Appel, Phys. Rev. B 22, 2626 (1980).
3. C.N.R. Rao, R.E. Loehman and J.M. Honig, Phys. Rev. Lett. 27A, 271 (1968).

REFINEMENT OF CRYSTAL AND MAGNETIC STRUCTURE OF $NdFeO_3$ AND $PrFeO_3$

I.Sosnowska[*]

Fakultat fur Physik E 2I, Technische Universitat Munchen,
D - 8046 Garching.
P.Fischer
Institut fur Reaktortechnik ETH Zurich, CH-5303 Wurenlingen,
Switzerland.

ABSTRACT

The temperature dependence of structural parameters and magnetic ordering of $NdFeO_3$ and room - temperature data for $PrFeO_3$ are reported.

INTRODUCTION

The crystal structure of rare earth ferrites (space group Pbnm) was investigated with many techniques (cf. X - ray results of Marezio et al[1]). First neutron diffraction measurements were performed by Koehler et al[2] on several ferrites including $NdFeO_3$, yielding a G - type magnetic ordering of the Fe^{3+} ions for the latter compound. Based on the temperature dependence of the first two magnetic peaks evidence for a spin reorientation from G_x at room temperature to G_z at low temperatures in $NdFeO_3$ was found by Pinto and Shaked[3], in agreement with conclusion published by Mareschal and Sivardiere[4]. Recently Sosnowska and Steichele[5] have determined the magnetic moment direction in $NdFeO_3$ obtaining 15° deviation from the x - direction and purely G_x for $PrFeO_3$, using the structure parameters obtained in this paper.

EXPERIMENTAL AND DISCUSSION

In order to determine the temperature dependence of the nuclear and magnetic structures of $ReFeO_3$ (Re = Nd, Pr) neutron diffraction experiments were performed at reactor Saphir, Wurenlingen, using

---

[*] Permanent address: Institute of Experimental Physics, Warsaw Univ.,
Warsaw, Poland

ISSN:0094-243X/82/890346-03$3.00 Copyright 1982 American Institute of Physi

neutrons of wavelength λ = 2.333Å . In Fig. I we show the neutron diffraction pattern of $NdFeO_3$ at T = 8K with the Rietveld profile refinement data for the structure parameters shown in Tab. I. The results of Rietveld refinement are summarised in Tab. I. At room temperature the structure parameters are in good agreement with Marezio et al.[I] . For the determination of the ordered magnetic moment μ of Fe^{3+} the neutron magnetic form factor of Watson and Freeman[6] was used.

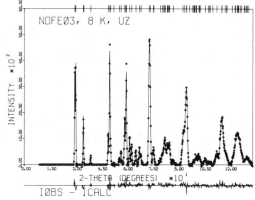

Fig. I. Neutron diffraction pattern of $NdFeO_3$ at T = 8K

        Fig. 2 gives clear evidence for the spin reorientation in $NdFeO_3$ in the temperature range of 70 to I70 K.

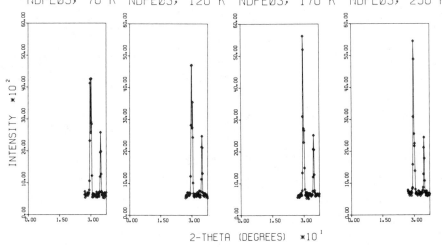

Fig. 2. Spin reorientation in $NdFeO_3$.

TABLE I

Parameters of crystal and magnetic structure of $ReFeO_3$.

Re	Nd			Pr	
	(I2I hkl)			(I3I hkl)	
T (K)	293	293[1]	8	293	293[1]
a (Å)	5.452(5)	5.453	5.442(5)	5.4828(5)	5.482
b "	5.59I(5)	5.584	5.587(5)	5.5778(5)	5.578
c "	7.767(7)	7.768	7.746(7)	0.7879(7)	7.786
B (Å²)	-0.09(8)		0.37(7)	0.92(5)	
x_{Re}	0.985(2)	0.9893I(4)	0.988(2)	0.984(2)	0.99097(4)
y_{Re}	0.0508(9)	0.0488I(5)	0.05I4(7)	0.044(I)	0.04367(5)
u_{Fe} (μ_B)	3.32(4)=μ_x		4.I3(4)=μ_z	3.73(4)=μ_x	
x_{OI}	0.089(2)	0.0876(7)	0.087(I)	0.084(I)	0.08I7(7)
y_{OI}	0.477(I)	0.4759(8)	0.476(I)	0.48I(I)	0.4788(9)
x_{O2}	0.705(I)	0.7052(5)	0.7023(9)	0.7065(7)	0.7075(5)
y_{O2}	0.2947(9)	0.2936(5)	0.2967(8)	0.2908(6)	0.29I9(5)
z_{O2}	0.0468(8)	0.0462(4)	0.464(7)	0.449(6)	0.437(5)
Rw_p	0.I2I		0.I04	0.09I	
R_n	0.049	0.028	0.04I	0.034	0.023
R_m	0.033		0.059	0.0I7	

References

I. M. Marezio, J.P. Remeika, P.D. Dernier, Acta Cryst. B26, 2008, (I970).

2. W. C. Koehler, E.O. Wollan, M. K. Wilkinson, Phys.Rev. II8, 58, (I960).

3. H. Pinto, H. Shaked, Solid State Comm. IO, 663, (I972).

4. J. Mareschal, J.Sivardiere, J.de Physique, 30, 967, (I969).

5 I.Sosnowska, E. Steichele, this conference.

6. R. E. Watson, A. J. Freeman, Acta Cryst. I4, 27, (I96I).

ONE YEAR'S EXPERIENCE WITH THE ORNL-NSF NATIONAL FACILITY
FOR SMALL-ANGLE NEUTRON SCATTERING*

W. C. Koehler, H. R. Child, R. W. Hendricks,[†]
J. S. Lin, and G. D. Wignall
Solid State Division and National Center for
Small-Angle Scattering Research
Oak Ridge National Laboratory
Oak Ridge, Tennessee 37830

ABSTRACT

At the time of this Conference, the ORNL-NSF National Facility
for Small-Angle Neutron Scattering will have been operating routinely for about one year. The Facility, part of a National Center
for Small-Angle Scattering Research, is a National User-Oriented
Facility created under an interagency agreement between the NSF and
the DOE, located at the High Flux Isotope Reactor at ORNL. It is
intended that this paper review some of the scientific highlights
of the past year.

INTRODUCTION

Small-angle neutron scattering (SANS) techniques have been applied to problems in biology, chemistry, materials science, polymer
science, and magnetism for more than 10 years. The explosion that
has taken place in the application of SANS to investigations of
structures on the scale of about 10 to 1000 Å is due in no small
part to the construction of long instruments at Jülich, at Saclay,
and at the Institut Laue-Langevin in Grenoble. The success of
these instruments is due in turn to the invention of position sensitive area detectors and to sophisticated data acquisition and
processing systems.

The applications of SANS to various disciplines have been
reviewed thoroughly; a representative but by no means complete
selection is listed below.[1-9] It is not our intention to contribute another such review paper to the literature of SANS; rather,
we shall survey the history of the first U.S. national user-
oriented facility for small-angle neutron scattering research.

The idea for establishing such a facility was conceived by Dr.
L. Nosanow during his term of office in the National Science
Foundation. Early in 1977, a proposal by two of us (WCK and RWH)

*Research sponsored jointly by the National Science Foundation Grant
No. DMR-77-244-49 and the Division of Materials Sciences, U. S.
Department of Energy under contract W-7405-eng-26 with Union Carbide
Corporation.

[†]Present Address: Technology for Energy Corporation, Knoxville,
Tennessee.

with the help of our colleagues in the neutron scattering program
was submitted to the NSF for the construction and operation of a
National SANS Facility at the Laboratory's High Flux Isotope
Reactor. The proposal was subsequently broadened to include part-
time use of several DOE constructed facilities, among them the 10-m
SAXS instrument designed by R. W. Hendricks.[10] The proposal was
accepted with the grant being effective on January 1, 1978. A
National Center for Small-Angle Scattering Research (NCSASR) was
thereby created under an interagency agreement between the NSF and
the DOE.

The Center is now well into its second three-year budget cycle
and the 30-m SANS instrument has been operating routinely for about
one year. It was formally dedicated in January of 1981.

INSTRUMENT CHARACTERISTICS

A small-angle scattering instrument is intended to permit
measurements of the intensities of neutrons scattered at small
values of the scattering vector K. For such measurements to be
meaningful one must have $\Delta K \ll K$. The various instruments in the
world that are used for SANS research have satisfied these requirements in different ways.

The ORNL-NSF instrument utilizes pin-hole geometry with colli-
mating slits of 0.5 - 3.0 cm separated by a distance of 8.3 m.
(See Fig. 1.) The detector is a position-sensitive proportional
counter of the Kopp-Borkowski design with an active area of 64 x 64
cm^2 and resolution element dimensions of 0.5 by 0.5 cm^2. It can be
positioned at any distance from 1.3 to 18.5 m from the specimen by
moving a motor-driven carrier along rails in the evacuated flight
path. The standard incident wavelength, provided by two banks of
six pyrolytic graphite crystals is 4.75 Å, although this can be
changed to 2.38 Å by substituting a tuned graphite filter for the
cold beryllium filter that is normally in place. The change-over
time is less than five minutes and the procedure permits experiments to be performed with increased flux over a wider range of
scattering angles.

The specimen chamber is designed to accommodate standard
samples, or if necessary, it can be fitted with specialized
ancillary equipment. At present this includes a cell holder for
liquids and solutions at various controlled elevated and reduced
temperatures, a Displex closed-cycle refrigerator, a sample changer
with eight positions, etc.

The data acquisition and processing system is based on a
ModComp II dedicated computer and it is adapted from a comprehen-
sive package used successfully for six years on the ORNL 10-m SAXS
camera. A user-oriented software package allows the user to carry
out preliminary data reduction and analysis at the same time that
data are being collected. Specifications for the instrument are
listed in Table I.

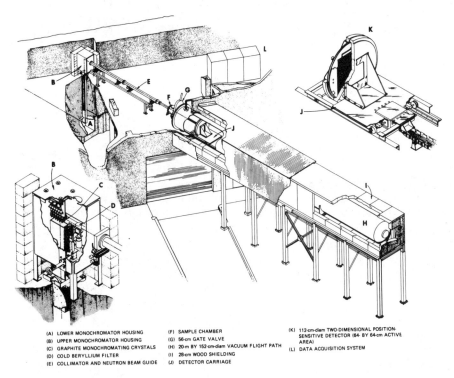

(A) LOWER MONOCHROMATOR HOUSING
(B) UPPER MONOCHROMATOR HOUSING
(C) GRAPHITE MONOCHROMATING CRYSTALS
(D) COLD BERYLLIUM FILTER
(E) COLLIMATOR AND NEUTRON BEAM GUIDE

(F) SAMPLE CHAMBER
(G) 56-cm GATE VALVE
(H) 20-m BY 152-cm-diam VACUUM FLIGHT PATH
(I) 28-cm WOOD SHIELDING
(J) DETECTOR CARRIAGE

(K) 112-cm-diam TWO-DIMENSIONAL POSITION-SENSITIVE DETECTOR (64- BY 64-cm ACTIVE AREA)
(L) DATA ACQUISITION SYSTEM

30-m SMALL-ANGLE NEUTRON SCATTERING FACILITY

Fig. 1. The 30-m SANS instrument

Table 1. The 30-m SANS instrument specifications

Beam tube: HB-4B (High Flux Isotope Reactor)
Monochromator: six pairs of pyrolytic graphite crystals
Incident wavelength: 4.75 Å or 2.38 Å
Wavelength resolution: $\Delta\lambda/\lambda$ = .06
Source-to-sample distance: 8.3 m
Beam size at specimen: 0.5-3.0 cm diam
Sample-to-detector distance: 1.3-18.5 m
K range: $3 \times 10^{-3} \leq K \leq 0.6$ Å^{-1}
Detector: 64 by 64 cm^2
Flux at specimen: 10^4 - 10^5 neutrons cm^2 s^{-1} depending
    on slit sizes and wavelength

Measurements of machine performance carried out prior to the commissioning of the instrument indicated that it should serve the needs of the majority of scientists in the polymer, materials, and

chemical sciences communities and that it should adequately meet
the requirements of many users in the biological sciences.
Experience with the instrument over the past year has confirmed
these expectations. Planned future developments will make the
facility useful to an even broader spectrum of scientists.

UTILIZATION OF THE INSTRUMENT

For low resolution experiments the differential scattering
cross section per atom can be expressed as

$$\frac{d\sigma}{d\Omega} = \frac{1}{N} \left| \int \rho_b(\vec{r}) \exp i\vec{K}\cdot\vec{r} \; d^3\vec{r} \right|^2 \tag{1}$$

where $\rho_b(\vec{r})$ is the scattering length density at position $\vec{r}$ and $\vec{K}$ is
the scattering vector $\vec{k}_0 - \vec{k}'$ with magnitude $4\pi \sin\theta/\lambda$ where $\vec{k}_0$ and
$\vec{k}'$ are the incident and scattered wave vectors and θ is one-half
the scattering angle. When K is small equation (1) can be written
as

$$\frac{d\sigma}{d\Omega} = \frac{1}{N} \left| \int \{\rho_b(\vec{r}) - \bar{\rho}(r)\} \exp i\vec{K}\cdot\vec{r} \; d^3\vec{r} \right|^2 \tag{2}$$

In this expression $\bar{\rho}$ is the average scattering length density. The
result given in (2) shows that for small-angle scattering experiments
it is the fluctuation in scattering length density that is important.

To proceed further one needs to specify the experimental system
more exactly. For example, for a system consisting of N_p identical
particles embedded in a matrix M and with uniform scattering length
densities ρ_p and ρ_M, the differential scattering cross section takes
the form

$$\frac{d\sigma}{d\Omega} = \frac{V_p^2 N_p}{N} (\rho_p - \rho_M)^2 \left| F_p(K) \right|^2 \tag{3}$$

where V_p is the particle volume, N is the total number of atoms

and $F_p(K) = \frac{1}{V} \int \exp i\vec{K}\cdot\vec{r} \; d^3\vec{r}$ is the single particle form factor.

This system is approximated in nature by two phase systems, voids or
precipitates, say, in a matrix. The single particle form factor can
be calculated for particles of various shapes.

The contrast shown in the expression (2) can be varied by substituting H with D and this has made possible tremendous progress in
understanding polymers in the bulk. Experiments have been carried
out at Saclay, Jülich, and the ILL on suitably tagged systems giving
rise to results of the type listed in Table II. Similar experiments
are now being done on the ORNL-NSF instrument. Indeed, of the 73
proposals so far submitted for the SANS instrument 41 are in the
area of polymer science. We give here a few selected results.

Table II. Typical information from SANS from polymers

Forward Scattering

$$I(K = 0)$$

M_W and A_2 in bulk via isotopic substitution. Test for compatibility in blends

Guinier Region

$$I(K) \cong 1 - \frac{K^2 R_g}{3}$$

R_G in bulk polymers, blends, concentrated solutions, networks, copolymers

Intermediate K Region

Chain statistics in bulk, mode of reentry in semi-crystalline polymers

Porod Region

$$I(K) \sim K^{-4}$$

Specific surface

Integrated Intensity

$$Q_0 = 4\pi \int_0^\infty K^2 I(K) dK$$
$$= C(1-c) \Delta \bar{\rho}^2$$

Partitioning of units between phases in copolymers

SANS experiments on amorphous polymers have confirmed the general predictions of Flory's Random Coil Model that the overall molecular size as measured by the radius of gyration in the bulk state is the same as in an ideal solution.[11] These experiments are currently being extended to higher values of K to test how far the local configuration in the melt is described by this model. Projects are in progress on a range of polymers including polyisobutylene (Professor P. J. Flory - Stanford) and polymethyl methacrylate (Dr. J. O'Reilly - Kodak Company). The first results on the latter polymer indicate that the scattering at intermediate K values, $0.1 < K < 0.4$ Å^{-1}, is markedly dependent on the chain tacticity and that the qualitative shape of the scattering curve as a function of tacticity is consistent with the Random Coil Model using rotational isomeric statistics. The model calculations will now be refined by Professor Flory's group using the appropriate tacticity parameters for the three polymer samples measured to check the degree of quantitative agreement of theory and experiment.

Until recently SANS experiments on polymers have been conventionally performed with small relative concentrations (<5%) of labeled polymer, but recent measurements on the 30-m SANS instrument and elsewhere, have confirmed that the same information may be obtained with concentrations of labeled molecules up to 50%.[12] This development means that experiments can be performed with much greater statistical accuracy, and has been particularly useful in the intermediate K-range measurements described above, where the scattered intensities are relatively low.

Several projects are in progress aimed at developing a fuller understanding of polymer-polymer interactions in multicomponent polymer blends (alloys). SANS studies have shown that polystyrene-polyphenylene oxide, marketed commercially as Noryl, forms a truly compatible blend with a statistical distribution of the two components.[13] Extensive studies are in progress on both compatible and incompatible (phase segregated) blends by Professor Stein's group (University of Massachusetts). Experiments are also in progress aimed at providing information on the domain organization, chain configuration within the domain, and nature and extent of the interface regions in phase separate diblock copolymers (C. V. Berney and R. Cohen, MIT).

Several projects are in progress or scheduled to investigate the response of individual molecules to a macroscopic deformation of a sample. The first results have been collected on extrudates of oriented polystyrene samples prepared by a new technique developed by Professors R. S. Porter and R. S. Stein (University of Massachusetts) to produce polymers of ultra high orientation and chain extension. The results show that the radii of gyration transform in the same manner as the external dimensions (affine deformation). This result is in marked contrast to studies made on drawn polystyrene in which the chain deformation has been shown to be strongly non-affine.[14] Further projects in hand will extend these studies to semi-crystalline polymers.

Proposals in the materials and chemistry area are the next most numerous, there being 29 of 73. Two groups are currently investigating micellar systems. These are significant in many practical areas, enhanced oil recovery, detergent action, catalysis, as well as in fundamental surface and solution chemistry. Studies are underway as well of the unfolding of model protein systems.

For spherical micelles, in the range of concentration where interparticle interaction occurs, the intensity is proportional to a product of two terms $I(K) \sim P(K)S(K)$. The first term is the single particle form factor and it is a monotonically decreasing function of K. The second term is the inter-particle form factor and, in the range of K covered by the SANS experiments, it is monotonically increasing. The net effect is an "interaction" peak in the small-angle scattering spectrum. In most cases this peak is well within the Guinier region so that the particle dimensions can be extracted from the K region beyond the peak.

Chen and Bendedouch (MIT) have studied the inter- and intra-particle structure factors of protein bovine serum albumin (BSA) and micelles formed by the ionic detergents sodium dodecyl sulfate (SDS) and lithium dodecyl sulfate (LiDS) in aqueous solutions of various ionic strengths at temperatures ranging from 20°C to 65°C. At low ionic strengths and moderate concentrations a dominant electrostatic interaction causes a dramatic sharpening of the inter-particle

structure factor in such a way that a prominent interaction peak in
the SANS pattern is observed. As the ionic strength is increased
(by adding salt) the structure factor is gradually washed out with
the result that the pattern is reduced to a normal intra-particle
form factor for spheres. This latter effect is shown in Fig. 2 for
LiDS at 50°C with and without additions of LiCl. The ultimate aim
of this series of experiments is to extract the inter-particle
structure factor from the data and study it as a function of concentration, ionic strength, and temperature. The experimental results
are to be compared with recent theoretical results of liquid structure factors for charged hard spheres in the presence of neutra-
lizing counter ions.

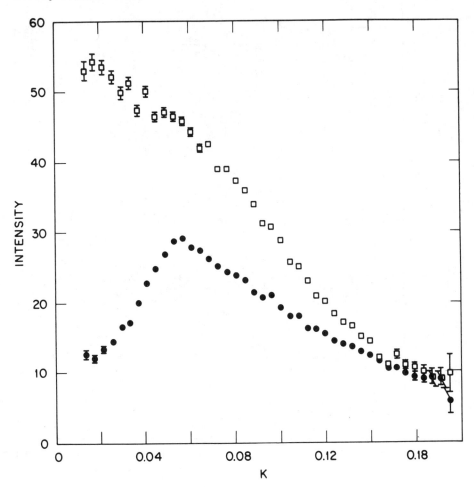

Fig. 2. I(K) for LiDS at 50°C with LiCl (upper curve) and without
salt (lower curve).

A group at Oak Ridge, Johnson, Triolo, Magid, and Ho, have investigated aqueous solutions of the surfactants sodium p (1-pentyl-heptyl) benzene sulfonate ($6\phi C_{12}SNa$) and sodium p (1-heptylnonyl) benzenesulfonate ($8\phi C_{16}SNa$).

Scattering curves for several $6\phi C_{12}SNa$ solutions are shown in Fig. 3. A prominent feature is the existence of an interaction peak

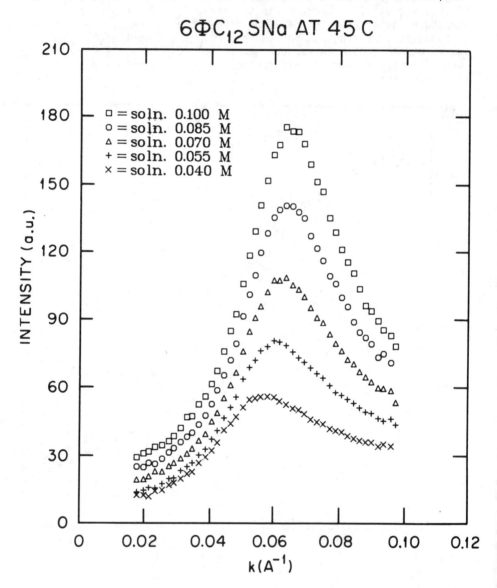

Fig. 3. SANS of sodium p(1-pentylheptyl) benzenesulfonate effect of concentrations (moles/k_gD_2O) 45°C.

at the higher surfactant concentrations due to intermicellar interactions. The Rg values, determined from the K region beyond the
peak range from 11.3 Å for the most dilute to 14.9 Å for the most
concentrated solutions. The main effect of the addition of salt in
this system is to increase I(0). Calculations of S(K) are being
made using the Mean Spherical Approximation. The effect of additions of 1-butanol to the solution was studied with the result that
little or no change in micelle size was observed.

In the metal physics area two related experiments on the formation of grain boundary cavities have been performed (J. Weertman
at Northwestern and Yoo and Ogle at Oak Ridge). Phase transitions
in Fe-Cr alloys (Schwartz, Northwestern) and in TiO_2-SiO_2 glasses
(Labarbe, Montpelier) were investigated.

A few experiments in the biological sciences have been carried
out. However, another paper in this symposium[15] deals exclusively
with applications of SANS to biological systems and we will not
discuss it further. Similarly, applications to phase transitions in
glasses are discussed by Wright.[16]

CLOSING REMARKS

In general, the last year's experience with the ORNL-NSF SANS
Facility has been satisfactory for the staff and for the user-
community. There remain a few instrumental problems to solve and
sources of background to track down and eliminate, but generally
even the weakest scatterers studied (J. Weertman's Cu samples) have
given useful data. For many specimens the turn around time is as
low as ten minutes but a more usual measuring period is a few hours.

We propose to replace some of the monochromators with better
quality graphite and it is our intention to install beam guide sections in the presample flight path. For experiments not requiring
high resolution the latter should result in a large increase in
intensity.

The software is being modified so as to be more convenient for
the user and automatic sample changers are under construction. A
hard-copy plotter, attached to the CR terminal, has been acquired.

For the time being the staff is able to keep up with the flow
of users. In the near future it will be necessary, if the users'
needs are to continue to be satisfied, to augment the staff with one
or more technicians.

REFERENCES

1. W. Schmatz, T. Springer, J. Schelten, and K. Ibel, J. Appl.
   Crystallogr. 7, 96 (1974).
2. J. Schelten and R. W. Hendricks, J. Appl. Crystallogr. 11, 297
   (1978).
3. H. B. Stuhrman and A. Miller, J. Appl Crystallogr. 11, 325
   (1978).
4. Julia S. Higgins and Richard S. Stein, J. Appl Crystallogr. 11,
   346 (1978).

358

5. V. Gerold and G. Kostorz, J. Appl. Crystallogr. 11, 376 (1978).
6. G. Kostorz, "Small-Angle Neutron Scattering and Its Applications to Materials Science" in Treatise on Materials Science and Technology, ed. by G. Kostorz, Academic Press, 1979.
7. J. S. Higgins, "Polymer Conformation and Dynamics" in Treatise on Materials Science and Technology, ed. by G. Kostorz, Academic Press, 1979.
8. G. Zaccai, "Application of Neutron Diffraction to Biological Problems" in Neutron Diffraction, ed. by H. Dachs, Springer-Verlag, 1978.
9. J. Schelten, NATO ASI Series B, ed. by S. H. Chen, B. Chu, and R. Nossal, Plenum Press, NY 1981.
10. R. W. Hendricks, J. Appl. Crystallogr. 11, 15 (1978).
11. J. S. Higgins and R. S. Stein, J. Appl. Crystallogr. 11, 346 (1978).
12. G. D. Wignall, R. W. Hendricks, W. C. Koehler, J. S. Lin, M. Wai, R. S. Stein, and N. Thomas, Polymer 22, 886 (1981).
13. G. D. Wignall, H. R. Child, and F. Li-Aravena, Polymer 21, 131 (1980).
14. C. Picot, R. Dupplessix, D. Decker, H. Benoit, J. P. Cotton, M. Daoud, B. Farnoux, G. Jannink, M. Nierlich, A. Devries, and P. Puncies, Macromolecules 10, 436 (1977).
15. D. L. Worcester, this volume.
16. A. F. Wright, this volume.

SMALL ANGLE SCATTERING FROM HETEROGENEITIES IN GLASSES

A. F. WRIGHT
Institut Laue-Langevin
Avenue des Martyrs, 156 X, 38042 Grenoble Cédex, France

ABSTRACT

Small angle X-ray and neutron scattering studies have recently
added considerably to the understanding of the formation and growth
of heterogeneities in glass. These extend from fluctuations through
glass-in-glass phase separation to crystallisation. By way of introduction, the parameters controlling their formation and growth are
discussed in terms of temperature and viscosity in order to understand the cumulative effects of thermal history, which are readily
observed in small angle scattering data. Particular emphasis is
given to SANS data of glasses during two-stage heat treatment schedu-
les used in the preparation of glass ceramics. Because of interparticle interference small angle scattering is particularly sensitive to
the study of nucleation.

INTRODUCTION

The separation of a glass into two immiscible vitreous phases,
and the crystallisation of a glass are phase transitions which are
controlled both by thermodynamic and kinetic forces. Crystallisation
is favoured by a large free energy difference between glass and
crystal, and hence by a large undercooling. At the same time the
high viscosity at large undercoolings hinders the necessary re-
arrangements, so that the nucleation and growth rates are affected
unequally by temperature. Crystallisation behaviour, and to a less
extent immiscibility, are therefore strongly dependent on
thermal history, even to the rate of cooling of the glass melt.
This must be borne in mind in the interpretation of experimental
results concerning precipitation in glasses, and particularly in the
interpretation of neutron small angle scattering experiments because
these have been shown to be particularly sensitive to thermal treatment which promotes homogeneous nucleation of the precipitating
phase [1]. Only recently have determined efforts been made to use
S.A.N.S. in the study of crystallisation phenomena in glasses and so
before discussing this work, the factors which contribute to the crystallisation of a glass, and the role played by viscosity in influen-
cing the nucleation and growth rates will be presented. The parameters which contribute to any particular SANS result on crystallisation in glasses will then be more easily recognised

STABILITY OF GLASSES

Whilst the glassy state is always metastable with respect to
the crystalline state at all temperatures below the liquidus (T_ℓ)
crystallisation does not occur in the absence of germs or nuclei
which precede the transformation. Crystallisation can start at

a surface imperfection or impurity (heterogeneous nucleation) or as
a result of fluctuations in the bulk of the material (homogeneous
nucleation) [2,3]. Fluctuations give rise to transitory appearance of
clusters of different sizes, compositions, and densities whose equilibrium abundance varies according to their excess free energy and
supersaturation. At high viscosities the equilibrum distribution is
slow to develop. According to the classical theory of nucleation [4]
clusters of the precipitating phase exceeding a certain size (critical radius r_c), pass a saddle point in free energy such that additional
growth is then energetically favourable. The critical size decreases
with increased supersaturation or undercooling. We can define the
homogeneous nucleation temperature T_h, below which the rate of
creation of nuclei by this process becomes detectable. At high supersaturation ratios, homogeneous nucleation is the dominant
process.

When the separating phase has a composition different to that
of the bulk glass, we can treat the system as that of a precipitation
from a supersaturated solution where the solubility is temperature
dependent [5]. Consider a solution of concentration C_0 (fig.1) being
slowly cooled from the melt. At T_ℓ the solution becomes saturated,
but in the absence of heterogeneous nucleation, crystallisation is
inhibited. Homogeneous nucleation begins at T_h where the critical
radius, which decreases as a function of the supersaturation, $r_c =$
$2\sigma v/(RT \ln \pi)$, approaches the cluster radius.
Here π is the supersaturation ratio C/C_s
    σ is the surface free energy
and v is the molar volume of the precipitating phase.
π_1 (fig.1) then corresponds to a critical supersaturation just sufficient to induce nucleation. Since in this example $T_h \gg T_g$, the
system closely resembles a supercooled liquid, and any nuclei formed
grow rapidly causing the supersaturation ratio to fall below π_1,
thus stopping the nucleation. We may describe this as uncontrolled
crystallisation which results in a very coarse microstructure.

It is possible to pass through the zone between T_ℓ and T_g
without causing crystallisation if the liquid is cooled in less time
that it takes for the cluster distribution to form. At $T \ll T_g$ the
viscosity is too high to permit any major reorganisation in reasonable time. Crystallisation can now be approached on increasing temperature from below T_g where the supersaturation and viscosity factors
are reversed. The critical radius near T_g is very much smaller than
near T_h because of the very much higher supersaturation π_2, but the
high viscosity strongly slows growth. At long times therefore a
high density nucleation of very small crystallites is favoured, and
the crystallisation can be controlled. Taking these parameters into
account, the growth and nucleation rates can be presented schematically as in fig.2 [6]. We can understand the maximum in the two curves
as due to two competing effects : (i) increase in the rates of
growth and nucleation as a result of supersaturation, and (ii)
reduction due to the increasing viscosity. Nevertheless the nucleation rate curve is displaced to lower temperatures because of the
temperature dependence of r_c. In the region of T_g there is therefore

a temperature zone where over relatively long timescales nucleation occurs, but is almost unaffected by the growth process.

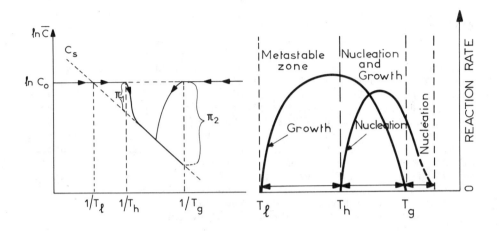

Fig. 1. Nucleation in a super-saturated glass. T_h is the homogeneous nucleation tempe-rature. Following a quench nucleation is obtained at supersaturation π_2.

Fig. 2. Scheme of nucleation and growth rates for an unstable glass.

We may distinguish 3 types of glass according to the position of T_h and the viscosity.
i) Glasses with a very low nucleation rate at all temperatures ($T_h \ll T_g$). These glasses are usually very viscous even at T_ℓ, and are very resistant to crystallisation. Commercial glasses of this type are unaffected by thermal cycling between T_g and T_ℓ. The stability is partly due to a complex composition far from any stoi-chiometric crystalline phase.
ii) Glasses for which T_h is slightly above T_g can be readily made and cooled normally below T_g without crystallisation. Crystallisation occurs at the first heat treatment above T_g. Glass ceramics are made from glasses of this kind, where the crystallisation is controllable. We can modify the composition of a stable glass by adding a few percent of a nucleating agent which will provoke crystallisation. For silicate glasses, titania is most widely used in the formulation of glass ceramics [7].
iii) Where the nucleation rate is very high in the temperature range, between T_ℓ and T_g, the only way to prepare a glass from the liquid is to cool extremely rapidly. Metallic glasses are formed in this way. These glasses may contain nuclei of crystals which have not

had time to grow.

THERMAL TREATMENT AND MICROSTRUCTURE

To obtain a fine controlled microstructure of small crystals as in a glass ceramic, a heat treatment is normally carried out in two stages. A first treatment of several hours near T_g to create the nuclei, followed by a second treatment at a higher temperature to grow the new phase. With multicomponent glasses, this second treatment may extend to very high temperatures in order to complete phase transformations in the crystallised part of the composite. The second treatment is easy to follow with modern physical methods such as X-ray diffraction, D.T.A. and electron microscopy [8], whereas the nucleation process is difficult to observe. When clusters are of the order of 10 Å it is hardly possible to define the state of order within the structure, and it is possible that an intermediate process of amorphous phase separation occurs in some cases. Such a species would only be identified where the limit of the method was sufficiently fine (a few Ängstroms). The nuclei which are formed in the first treatment must therefore be grown larger than the detection limit, but this in turn is likely to modify the result of the nucleation treatment.

In the case of a growth treatment at an elevated temperature, the initial value of r_c increases in response to the lower supersaturation and for particles of radius $r < r_c$ dissolution occurs. In addition the growth of particles of radius $r > r_c$ causes a further decrease in the mean supersaturation and a corresponding continued rise in r_c. The larger particles dominate the evolution of concentration such that smaller particles which fail to remain ahead of the critical radius are continuously eliminated from the system.

Growth at high viscosity is not influenced by the mean concentration, but by the local concentration around each particle. Thus slow rates of diffusion ensure that a depleted zone forms around each crystal [9,10] The saturation remains in equilibrium with the crystal radius, and the growth rate is determined by the transport of material across the depleted zone. This situation remains stable until overlap of depleted zones occurs, whereupon the larger crystals contribute unequally to the depletion and redissolution of the smaller particles begins. In the absence of redissolution, the particle size distribution depends only upon the unequal time of growth since nucleation. A narrow particle size distribution can be obtained by an extended heat treatment at the lowest temperature which allows nucleation but limits growth, followed by a growth treatment at a slowly increasing temperature.

The presence of depleted zones around growing particles modifies the probability of nucleation and later redissolution through its effect on critical radius to an extent that the original random spatial distribution of particles becomes partially ordered as the volume fraction of depleted zones becomes important. In particular, in the depleted zone of a particle, the probability of nucleation of a second particle of the same compound is greatly reduced, whereas the nucleation of a different compound may be enhanced. Also, the rate of dissolution of neighbouring particles depends upon the degree of penetration into the depleted zone by the neighbour. Consequently the equivalent of a short range order interaction potential develops as a function of the depleted zones, and ensures that neighbouring particles are well separated.

NEUTRON SMALL ANGLE SCATTERING EXPERIMENTS

Studies of crystallisation in glasses use the same techniques which have been developed for studies of alloys [11,12]. Where there exists a short range order between particles, and the particle size distribution is not too large, the small angle scattering curve is modified by an interparticle interference term which can produce a maximum in $I(q)$. We may roughly interpret the position of this maximum according to $q_m = 2\pi/\Lambda$ where Λ is the mean interparticle separation [1,13]. q_m is therefore very sensitive to the number density of precipitated particles. Where there is no long range order however, the curve of $I(q)$ at $q > q_m$ is not disturbed by the interparticle effects ($S(q) \rightarrow 1$) and the isolated particle function is reestablished. Within the limits of the Guinier approximation we can write $I(q) = I(o) \exp(-q^2 Rg^2/3)$ where Rg is the radius of giration of the particle and for spherical particles $Rg = (3/5)^{1/2} Rs$. It is therefore possible to obtain, at the same time, the mean particle separation and the size of the precipitate. However, given that the particle size itself has an effect on the peak position, it is desirable to fit the whole curve to a model incorporating the interaction potential, and so obtain the interparticle interference term $S(q)$ in the absence of particle size effects [13].

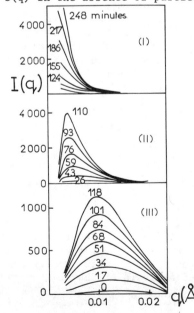

One of the earliest studies of crystallisation in glasses by SANS was of a magnesium aluminium silicate glass incorporating TiO_2 as nucleation agent [14]. A single stage heat treatment was used and the evolution of the growth process with time indicated an initial glass-in-glass phase separation preceding the crystallisation, but within the limited q-range available, no maximum was observed. A two-stage heat treatment on a very similar glass gave results shown in Fig.3 [1].

Fig. 3. SANS curves as a function of growth time at 835°C for 3 different nucleation treatments at T_g.
(i) No treatment (ii) 2 hours
(iii) 10 hours.

Each series of curves shows the growth of I(q) at 835°C following
a first stage treatment at Tg = 760°C for different periods. Without pretreatment (i) there is no interference peak, but after 2 hours
(ii) and then 10 hours (iii) the maximum develops and is displaced
to a higher value of q. Following first stage heat treatments for
various times and temperatures, by a constant growth treatment,
Fig.4 is obtained. A saturation effect in the displacement of q_m
for the higher temperatures studied [15], corresponds to the limit
of nucleation due to the high volume fraction of the depleted zones.
Note the effectiveness of the very long treatments at 720°C

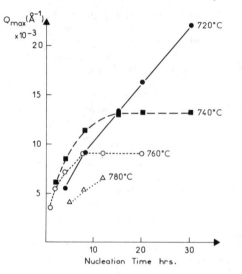

Fig. 4. Position of maximum in I(q)
as a function of nucleation treatment temperature and time.

(Tg – 40°C) where no
saturation effects
are observed. Here the
mean interparticle sepa-
ration is of the order
of 300 Å and the parti-
cles radius R ≈ 50 Å
when fully grown.
In this case, the com-
posite body is transpa-
rent. Interference
effects have been noted
also in glass-in-glass
phase separation by both
neutron and X-ray small
angle scattering. In
some cases the periodic
structure giving rise to
interference originates
from spinodal decomposi-
tion rather than the
depleted zone effect in
nucleation and growth.
Excellent treatments of
this mechanism of phase separation observed by small angle scattering
in glasses have been given by Zarzycki [16,17]. Coarsening effects
occur in the later stages of growth of both phase separation and
precipitation, resulting in an increase in the spatial correlation
wavelength or the mean interparticle separation. This is most
easily observed by following the evolution of the maximum in I(q)
as a function of time of growth. For long times Λ is observed to
vary as $t^{1/3}$ in the case of phase separation [17,18,19] and crystallisation [15]. The $t^{1/3}$ law suggests an Ostwald ripening process [20]
where the differences in solute concentration due to particle size
variations cause the larger particles to grow at the expense of the
smaller ones. Kinetic small angle scattering measurements are particularly useful in the study of this evolution of microstructure, which
is barely observable by other techniques. Indeed, in the crystallisation of a basalt glass Labarbe et al. [15] have shown that ripening can
be effectively controlled by heat treatments near Tg which modify the
particle size distribution. More detailed analysis of the scattering

curves based on a least squares fit to a hard sphere particle-interaction potential [21] are in course of study [22]. Although a hard sphere potential is a very over-simplified model for the depleted zone interaction, giving too much structure in $S(q)$ at $q > q_m$ (Fig.5) the initial results are encouraging. Interesting results have also been obtained for more complicated scattering systems in glasses. Larché et al. [23] have investigated the nucleation and growth of both chemical and magnetic heteogeneities by SANS in a strong magnetic field, taking advantage of the magnetic scattering of neutrons by the magnetized sample to produce an anisotropic scattering pattern. The magnetic and chemical cross sections were then separated and analysed in the usual way.

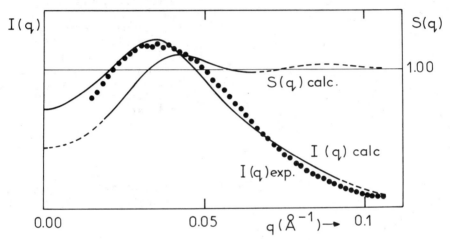

Fig. 5. Least squares fit of a hard sphere model to SANS curves of precipitation in a basalt glass.   Particle Diameter 69 Å
Sphere Diameter   140 Å
Vol. fraction   0.105

McMillan [24] has followed by electron microscopy the decomposition of Lithium disilate glass through nucleation and growth of phase separation, followed by crystallisation of the phase rich in Li_2O. At the intermediate stage crystalline and glassy precipitates coexist with different well separated size distributions. The evolution of these precipitates is easily observed by Guinier plots of the SANS curves (Fig.6) which show the growth of two linear regions of different slopes as a function of heat treatment. The complementarity of small angle scattering with electron microscopy is well demonstrated in this case as in many others [16,17,18]. Although the ability to make kinetic measurements with SANS is very advantageous, ambiguities may arise through the difficulty of identification of the inhomogeneity. In the study of an aqueous electrolyte glass at 140 K, Elarby et al. have combined SANS with neutron diffraction [26] to follow the change in short range atomic order of the glass occurring during nucleation of ice crystals. This approach to precipitation

366

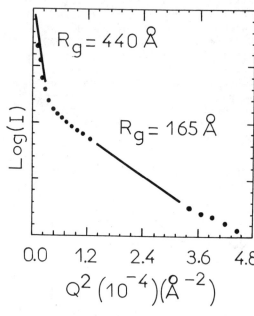

$R_g = 440 \text{ Å}$

$R_g = 165 \text{ Å}$

Log(I)

0.0   1.2   2.4   3.6   4.8

$Q^2 (10^{-4})(\text{Å}^{-2})$

will probably be very useful in the study of crystallisation in metallic glasses, where substantial short range order changes during heat treatment have been observed [27].

The author wishes to thank Pierre Labarbe, Peter McMillan, Brian Fender and Jerzy Zarzycki for their help and collaboration.

Fig. 6. Guinier Plot with 2 particle dimensions in a Lithium disilicate glass : Phase separation of silica spheres, Rg = 165 Å. Crystallisation of Lithium Disilicate, Rg = 440 Å.

REFERENCES

1. A.F. Wright, J. Talbot, B.E.F. Fender, Nature 277 (5965) 366 (1979).
2. A.G. Walton, "Nucleation" Ed. A.C. Zettlemoyer, 255-307, Marcell Dekker Inc. New York (1969).
   W.J. Dunning, Idem, 1-67.
3. H. Reiss, Faraday discussions of the Chemical Soc. N° 61 "Precipitation" 213-223 (1976).
4. D. Turnbull and R.E. Cech, J. Appl. Phys. 21 804-810 (1950)
5. M. Kahlweit, "Precipitation and Ageing", Physical Chemistry, An Advanced Treatise 10, Ed. Eyring, Hendersin, Jost. Academic Press (1970).
6. E.M. Levin, Phase Diagrams in Materials Science and Technology Vol. 6 - 171. Ed. A.M. Alper, Academic Press, 143-236 (1970).
   D.R. Ullman, J. Non Cryst. Solids 25, 43 (1977)
7. D.R. Stewart, Advances in Nucleation and Crystallisation in Glasses (American Ceramic Soc., Columbus, Ohio), p. 83 (1971).
8. A.G. Gregory and T.J. Veasey, J. Mat. Sci. 6, 1312 (1971)
9. R.D. Maurer, J. App. Phys. 33, 6,2137 (1962).
10. B. Lewis, "Crystal Growth", Ed. Pamplin B.R. Pergamon Press (1981).
11. G. Kostorz, Treatise on Materials Science and Technology, Vol. 15, 227, Academic Press (1979).
12. G. Lazlaz, P. Guyot, G. Kostorz, Journal de Phys. Colloque C7 suppl. N° 12,38,406 (1977).
13. G. Lazlaz, G. Kostorz, M. Roth, P. Guyot, R.J. Stewart, Phys. Stat. Sol. 41,577 (1977).
14. A.A. Loshmanov, V.N. Sigaev, R.YA. Khodakovskaya, N.M. Pavlushkiv and I.I. Yamzin, J. Appl. Cryst. (1974) 7, 207.

15. P. Labarbe, A.K. Bandyopadhyay, J. Zarzycki and A.F. Wright, J. Non Cryst. Sol. 43, 3, (1981).
16. J. Zarzycki, J. Apply. Cryst. 7, 200 (1974).
17. J. Zarzycki and F. Naudin, J. Non Cryst. Sol. 1.215 (1969).
18. J. Zarzycki and F. Naudin, Phys. Chem. Glass 8.11 (1967).
19. M. Roth and J. Zarzycki, J. Non Cryst. Sol. 16, 93 (1974).
20. S.C. Jain and A.E. Hughes, J. Mat. Sci 13 1611 (1978).
21. M.S. Wertheim, Phys. Rev. Letters, 10 321 (1963).
22. P. Labarbe, A.F. Wright and J. Zarzycki (to be published)
23. F. larche , M. Roth, J. Zarzycki, Conference Proceedings, "Structure of Non Crystalline Solids" Ed. Gaskell, P. Cambridge (1976).
24. P. Hing and P.W. McMillan, J. Mat. Sci. 8, 340-348 (1973).
25. P. McMillan and A.F. Wright (to be published)
26. A. Elarby, J.F. Jal, J. Dupuy, P. Chieux, A. Wright, Congrès S.F.P. Clermont-Ferrand, 1981. To appear in J. de Physique.
27. E. Balanzat, C. Mairy, J. Hilleret, J. de Physique C8. 41, 871 871 (1980).

THE ROLE OF NEUTRON SCATTERING IN MOLECULAR AND CELLULAR BIOLOGY

D.L. Worcester
Institut Laue-Langevin, Grenoble, France

ABSTRACT

Neutron scattering measurements of biological macromolecules
and materials have provided answers to numerous questions about
molecular assemblies and arrangements. Studies of ribosomes, viruses,
membranes, and other biological structures are reviewed, with
emphasis on the importance of both deuterium labelling and contrast variation with H_2O/D_2O exchange. Although many studies of
biological molecules have been made using contrast variation alone,
it is the deuterium labelling experiments that have provided the
most precise information and answers to major biological questions.
This is largely the result of the low resolution of scattering data
and the consequent rapid increase of information content that specific deuterium labelling provides. Procedures for specific deuterium
labelling 'in vivo'are described for recent work on myelin membranes
together with basic aspects of such labelling useful for future research.

INTRODUCTION

The majority of biological studies using neutron scattering
have been low resolution studies of biological macromolecules in
solution or in partially ordered arrangements such as membranes,
gels, fibres and intact tissue. Crystallography has had an important
role, particularly in elucidating hydrogen bonding in small biological molecules[1] and more recently in answering specific questions
about protein structures which had not been answered by X-ray diffraction studies[2-4]. But crystallographic studies of biological molecules using neutrons continue to be relatively few. Inelastic neutron scattering has been applied to biological molecules and materials, to determine dynamical properties, but not extensively, and
most work has been of an exploratory nature.[5-7] This paper reviews
basic aspects of the low resolution studies, which make neutron
scattering particularly useful and often unique in the information
obtained. These aspects are all variations on the theme of hydro-
gen/deuterium manipulation. In addition to the original manipulations
used in early neutron scattering studies, several new variations
have been used in recent work.
There are two fundamental categories of hydrogen/deuterium
manipulation in neutron scattering. They are deuterium exchange
and deuterium labelling. Deuterium exchange occurs primarily in the
aqueous solvent when water is changed from one isotopic composition
to another. This is the basis of the widely used contrast variation
technique originally developed for neutron scattering by Stuhrmann.[8,9]
Deuterium labelling refers to the introduction of covalently bound
deuterium to the biological molecules. This may be done 'in vivo',
or by chemical means 'in vitro', or by a combination of the two.

ISSN:0094-243X/82/890368-10$3.00 Copyright 1982 American Institute of Physi

CONTRAST VARIATION AND DEUTERIUM LABELLING

Contrast variation using H_2O/D_2O exchange was the first and is still the most widely used technique for studying biological molecules by neutron scattering. Its value is in studying macromolecules or macromolecular assemblies which contain components of different scattering density. The major components of biological materials are proteins, nucleic acids, carbohydrates and lipids. The scattering densities of these components are all different, but are between the scattering density of H_2O and that of D_2O. Determination of the relative spatial arrangements of such components is greatly facilitated by contrast variation. Radius of gyration measurements in different H_2O/D_2O mixtures are readily interpreted by making a Stuhrmann plot of R_g^2 vs. the reciprocal of the contrast, $\bar{\rho}$ (the difference between the mean scattering density of the particle, and that of the solvent).[10] The dependence of R_g^2 on $\bar{\rho}$ is of the form

$$R_g^2 = R_{go} + \frac{\alpha}{\bar{\rho}} + \frac{\beta}{\bar{\rho}^2}$$

R_{go} is the radius of gyration at infinite contrast, and is therefore determined only by the water excluding volume of the particle. Alpha reflects the distance of the different components from the particle centre of mass. Positive alpha indicates components of higher scattering density at the outside of the particle, whereas negative alpha indicates these components are inside. Beta reflects the separation of the centres of mass of components with different scattering densities, hence particle asymmetry. Beta must be negative for physical separations. In addition to analysis of radius of gyration scattering, the more extensive scattering curves are analyzed by contrast variation to establish the three basic scattering functions of a particle[8-10] : one for the volume of the particle (I_v), one for the fluctuation of scattering density about the mean value (I_F), and an interference term : $I(\bar{\rho})=I_F+\bar{\rho}I_I+\bar{\rho}^2I_v$. Contrast variation data can be reduced to these three basic scattering functions as a simple consequence of the <u>linearity</u> of the scattered wave amplitude with contrast. This linearity is also used in a slightly different way for contrast variation studies of partially ordered structures. Using these basic aspects of contrast variation analysis in solution scattering, studies have been made of lipoproteins,[11] the nucleosome,[12-16] several proteins of nonuniform scattering density such as ferritin,[10,17,18] protein-detergent complexes,[19,20] and many viruses.[21-27]

Contrast variation has also been used in structural studies of partially ordered biological molecules. In studies of membranes and model membrane systems, it has provided a means of dealing with the phase problem, particularly for centrosymmetric structures.[28] It also was used to locate water and exchangeable hydrogens in these structures.[29,30] Because neutron scattering amplitudes of the atoms in biological molecules are real numbers, the amplitude of the scattered wave is a linear function of the D_2O mole fraction X. This means of course that in the general case, the real and imaginary parts are separately linear, and in centrosymmetric structures (no imaginary part) the square root of the intensity is linear

with χ. This linearity is of great value in data processing for
deuterium labelling experiments and is discussed later in this
paper. The linearity provides directly the magnitude of the amplitude (structure factor) of all exchangeable hydrogens (F_D) and the
magnitude of the phase angle between this amplitude and the amplitude of the structure with no deuterium exchanged (F_H). Thus with
D_2O mole fraction χ :

$$|F_\chi|^2 = |F_H|^2 + \chi^2|F_D|^2 - 2\chi|F_D||F_H|\cos \alpha$$

where alpha is the phase angle between F_H and F_D. Note the similiari-
ty of this equation to that for the three basic scattering functions
of solution scattering. There is in fact no difference between them.
They both are direct mathematical consequences of the Fourier transformation between the scattered wave and the structure composed of
atoms with real scattering amplitudes. Thus, measurements in three
different H_2O/D_2O mixtures are sufficient to determine $|F_H|$, $|F_D|$
and $|\alpha|$, or in solution scattering, the three basic scattering
functions. For centrosymmetric structures, alpha is 0 or π, and
hence

$$F_\chi = F_H + \chi F_D.$$

Although the Fourier transform relation is essential to this
contrast variation analysis, the reason for its validity in neutron
scattering is not often explained, nor is attention drawn to limitations of its validity. Because neutrons behave as De Broglie
waves, the scattered wave satisfies the Schrödinger wave equation,
and for weak scattering is given by the approximate solution derived by Born (hence called the Born approximation) for which the
scattered wave is the Fourier transform of the scattering structure
(often called the potential).[31] If the scattering is not weak,
Fourier transformation no longer applies. This is the familiar case
of primary extinction in crystals and partially ordered structures.
What is not so familiar is that it can also be the case for macromolecules in solution. The mathematical requirement is that the
phase shift of the neutron wave be small traversing the particle
and depends on the size of the particle, the neutron wavelength,
and the contrast : $\lambda D\bar{\rho} \ll 1$. As an example the Born approximation
noticeably begins to break down in D_2O buffer (high $\bar{\rho}$) for a
molecule of about 1000 Å diameter and neutrons of 10 Å wavelength
or longer (specifically for proteins in D_2O, $\lambda D \ll 2 \times 10^5 Å^2$). This
usually is manifested by the filling in of minima in the scattering
curve, and by a non-linear Guinier plot.

The Fourier transform relation specifies the maximum information content of scattering data. This information content is
often rather limited because the data do not extend far in Q
(i.e. low resolution). Deuterium labelling provides a very valuable
way of increasing the information content. In many cases very detailed structural analysis can be made from scattering data of limited Q range. One of the best examples of this is the analysis of
ribosome structure made by Engelman and Moore.[32-35] The work

depends on the fact that the organism 'E. Coli' can be grown in high D2O medium to produce deuterated ribosomes, and that specified proteins from these ribosomes can be extracted and reconstituted with unlabelled ribosomal proteins and RNA to form intact, fully functional ribosomes. By analysing the scattering from ribosomes with different pairs of proteins labelled, to determine primarily the distances between these proteins, the structural arrangement of proteins in the ribosome has been considerably elucidated.

In an early study of model membranes, deuterium labelling was used to provide additional information on the structure of lecithin and cholesterol bilayers.[30] Several other studies of this type have since been made[36-40]. In such work, specific deuterium labelling is achieved by chemical means and the model membranes are constructed from mixtures of the components. Thus reconstitution is again an essential aspect. The sensitivity of the scattering to deuterium labelling is illustrated by the data in Fig.1 which was obtained from bilayers of dimyristoyl lecithin and cholesterol, in which only two deuterium atoms are present at different positions of one of the two hydrocarbon chains. The different diffraction intensities are due only to the deuterium being at different positions. The analysis of a series of such data with different deuterium labelling is best done using the linearity of structure factors with H_2O/D_2O mixtures. Thus, for fixed water content, each Bragg order (n) has structure factors given by

$$F_n(X) = F_{H,n} + XF_{D,n}$$

where F_H and F_D have the same meaning as before. For a sample with deuterium at position i, the structure factors are given by

$$F_{n,i}(X) = \alpha_i(F_{H,n} + \Delta_{n,i} + XF_{D,n})$$

$\Delta_{n,i}$ is the change of the structure factors due to the deuterium labelling, and α_i is a scale factor between labelled and unlabelled samples, since they are usually of different sizes. Thus, H_2O/D_2O exchange provides 'direct scaling', as used in the first labelling studies of lecithin-cholesterol bilayers.[30] Neutron diffraction measurements establish values for the left hand side of the equations, and provided that measurements are made with at least three H_2O/D_2O mixtures for each sample, a non-linear, variance weighted least squares fit to all the data provides the values and uncertainties for all the parameters on the right hand side of the equations. The values for $\Delta_{n,i}$ give the Fourier profiles for the deuterium labels of each sample. In the example of Fig. 1, this analysis provides structural information at atomic resolution from very limited diffraction data.

SPECIFIC DEUTERIUM LABELLING 'IN VIVO'

These examples of deuterium labelling have one very important limitation however. They all depend on reconstruction of the object

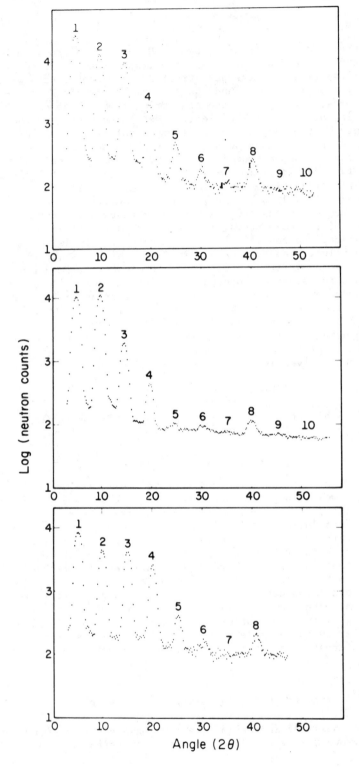

Fig. 1. Neutron diffraction from dimyristoyl lecithin and cholesterol bilayers with deuterium on carbon 2(top), 6(middle) and 12(bottom) of one hydrocarbon chain.

studied after deuterium labelling has been achieved in separate
components. Such reconstitution is often difficult work, usually
brings the biological integrity into question, and may not be
possible for the majority of biological structures that are of
interest to analyse. If the value of deuterium labelling is to
be widely utilized in biological studies by neutron scattering, then
specific deuterium labelling methods must be developed which are
not dependent on reconstitution. This may seem a very difficult
task, but in fact has been done fairly easily for a number of specific 'in vivo' labelling experiments with laboratory rats. These
experiments were originally for studies of myelin membranes, but
the procedures for specific labelling may be of value for many
studies of different types.

Specific 'in vivo' deuterium labelling is possible as a consequence of specific utilizations of foods by organisms and animals.
These may require detailed knowledge of metabolic pathways to ensure
that sufficient labelling occurs, and that the labelling occurs
almost exclusively where it is wanted rather than among many components of the structure studied. Recent work on 'in vivo' labelling
of choline groups in myelin membrane lipids is a good example.[41]

Choline ($HOCH_2CH_2N(CH_3)_3$) has long been considered by nutri-
tionists to be an essential metabolite. Higher animals can in fact
live without it, but their health is more or less impaired depending
on additional nutritional factors. The reasons for this were clarified by detailed biochemical studies. Disrupted production of the
lipid phosphatidylcholine was the main factor, but it was found that
direct utilization of dietary choline was not the only production
mechanism. Thus 'in vivo' labelling of choline-containing lipids
(phosphatidylcholine and sphingomyelin) can be achieved with diets
rich in deuterated choline, but to obtain high degrees of labelling,
attention must be given to all biochemical pathways by which choline
is produced and utilized by the animals. Biochemical work during the
last fifty years has clarified one such pathway: choline can be produced by stepwise methylation of phosphatidylethanolamine. In this
pathway, the methyl group transfer takes place from S-adenosylmethio-
nine. Thus, if sufficient methionine is present in the diet (0.5% for
rats), choline can be absent without harmful effects.[42] (This is not
true, however for birds, since the enzyme for the first methylation
of ethanolamine is lacking.) For deuterium labelling experiments on
rats, either the methionine methyl groups must also be labelled, or
the amount of methionine must be minimized. The latter is easily
achieved using casein as the protein source, for example. The S-aden-
osylmethionine which remains acts as a methyl donor, but converting
thereby to S-adenosylhomocysteine, it accepts a methyl group from
an oxidation product of choline (betaine) to become a methyl donor
again. Thus, with limited methionine available, nearly all methyl
groups come directly or indirectly from choline.

There is yet another pathway for choline synthesis however,
which involves the vitamins folic acid and B_{12}. In this pathway,
carbon from any of several small molecules (formaldehyde, formate,
serine, formiminoglycine) is attached to the nitrogen in the 5
position of tetrahydrofolic acid and is successively hydrogenated
to form a methyl group. This is transferred first to vitamin B_{12}

100% yield of deuterium labelled choline. This was a considerable improvement over previous methods which used bases that were not sterically hindered and hence had side reactions that required extensive purification of the choline produced.

Another specific 'in vivo' deuterium labelling study of myelin membranes has examined the distribution of cholesterol between the two halves of the membrane bilayer.[43] Cholesterol labelling represents a different type of problem than choline labelling because cholesterol is extensively synthesized 'de novo' from acetate in very many animal cells. Labelled cholesterol obtained from the diet is therefore diluted. Endogenous synthesis is inhibited however, by cholesterol itself, in a negative feedback mechanism. High specific labelling can therefore be achieved by feeding large amounts of labelled cholesterol in the diets (typically 1% by wt.). But although this was found to be true for plasma lipoprotein and erythrocyte membrane cholesterol, it was not true for cholesterol in the nervous system. Labelling of peripheral nerves of weanling rats was only 10% (determined by tritium tracer analysis) while labelling of the central nervous system was less than 1%.[43] This is apparently due to the very strong chemical protection provided to nervous systems (the blood-brain barrier). It may be possible to increase the amount of nervous system labelling by continued feeding past weaning since cholesterol synthesis is most active in nervous systems early in development, and the blood-brain barrier gradually weakens with age.

A procedure for efficient production of deuterium labelled cholesterol was important for the prolonged feeding experiments since substantial quantities were needed. Labelling with five deuterium atoms at the 2,3 and 4 positions (adjacent to the hydroxyl group) was achieved beginning with keto-enol exchange in Δ^4cholesten-3-one in a hot alkaline mixture of singly deuterated ethanol (CH_3CH_2OD) and D_2O followed by reduction of the enol acetate derivative with sodium borodeuteride. The crude product was recrystallised from acidified ethanol and provided 92% pure cholesterol, the remainder being the 3-alpha isomer, epicholesterol.

Specific deuterium labelling of proteins is also of great interest for diffraction studies of myelin and other membranes, and probably for other structures as well although at present, all experiments have been on membranes. In rats, extensive protein labelling can be achieved because of the essential amino acid requirements.[44] That is, the following amino acids cannot be produced by rats, and must come from the diet if protein synthesis is to occur: tryptophan, phenylalanine, lysine, threonine, methionine, leucine, isoleucine, valine, histidine and arginine. Tyrosine as well would only come from the diet since it is produced from phenylalanine. Thus, over half of the amino acids used for protein synthesis in rats must come from foods and therefore can be provided only in deuterated form. There is some variation between animals, but only plants have no external amino acid requirements.

Production of substantial quantities (tens of grams) of deuterated amino acids is problematic. For aromatic amino acids, it can be achieved by exchange with D_2O using Adam's catalyst or Rainey nickel. Such quantities have not been efficiently achieved for other amino acids however. Extraction of these amino acids from organisms grown in D_2O may be the best method.

Labelling of specific amino acids in 'Halobacterium halobium' has recently been done for neutron diffraction studies of the protein structure in purple membranes.[45] In such single-celled organisms there are often specific amino acid requirements, sometimes controlled by culture conditions, or by selection of particular mutants, so that high specific labelling occurs.

In protein labelling experiments, the difficulties of introducing deuterium into other compounds are great. Metabolic breakdown of amino acids leads to small molecules such as acetate which are used in synthesis of macromolecules (lipids in the case of acetate). Thus deuterium can be widely dispersed, depending on the amino acid used, but much is also lost to water. The pathways available in each case must be examined.

These examples do not exhaust the possibilities for specific 'in vivo' deuterium labelling. Many other examples are being or will be formulated. They will be very important if the most effective use of neutron scattering is to be made for studies of biological materials. Although reconstitution procedures have been very important in developing the field, their limitations must now be emphasized and more preference given to 'in vivo' deuterium labelling methods.

ACKNOWLEDGEMENT

Fellowship support from the Science Research Council of Great Britain, held at Queen Elizabeth College, London for much of the work on 'in vivo' deuterium labelling of myelin membranes described in this paper, is gratefully acknowledged.

REFERENCES

1. H. Fuess, in Modern Physics in Chemistry (E. Fluck and V.I. Goldanskii, eds.), Vol. 2, pp. 1-193, Academic Press, London. (1979).
2. A.A. Kossiakoff and S.A. Spencer. Nature 288, 414-415 (1980).
3. S.E.V. Phillips and B.P. Schoenborn. Nature 292, 81-82 (1981).
4. A. Wlowdawer and L. Sjölin. Proc. Natl. Acad. Sci. USA 78, 2853-2855 (1981).
5. V.D. Gupta, S. Trevino and H. Boutin. J. Chem. Phys. 48 3008-3015 (1968).
6. H.D. Middendorf and J.T. Randall. Phil. Trans. Roy. Soc. Lond. B290, 639-655 (1980).
7. J.T. Randall, H.D. Middendorf, H.L. Crespi and A.D. Taylor. Nature 276, 636)638 (1978).
8. H.B. Stuhrmann. J. Appl. Cryst. 7, 173_178 (1974).
9. H.B. Stuhrmann, in Brookhaven Symposia in Biology 27, p. IV-3 (1976).

10. K. Ibel and H.B. Stuhrmann. J. Mol. Biol. 93, 255-265 (1975).
11. H.B. Stuhrmann, A. Tardieu, L. Mateu, C. Sardet, V. Luzzati, L. Aggerbeck and A.M. Scanu. Proc. Nat. Acad. Sci. USA 72 2270-2273 (1975).
12. J.F. Pardon, D.L. Worcester, J.C. Wooley, K.E. Van Holde and B.M. Richards. Nucleic Acids Research 2, 2163-2176 (1975).
13. R.P. Hjelm, G.G. Kneale, P. Suau, J.P. Baldwin and E.M. Bradbury. Cell 10, 139-151 (1977).
14. J.F. Pardon, D.L. Worcester, J.C. Wooley, R.I. Cotter, D.M.J. Lilley and B.M. Richards. Nucleic Acids Research 4, 3199-3214 (1977).
15. J.F. Pardon, R.I. Cotter, D.M.J. Lilley, D.L. Worcester, A.M. Campbell, J.C. Wooley and B.M. Richards. Cold Spring Harbor Symposia on Quantitative Biology. Vol. XLII. pp.11-22 (1978).
16. B.M. Richards, J.F. Pardon, D.M.J. Lilley, R. Cotter and D.L. Worcester. Cell Biol. Int. Reports 1, 107-116 (1977).
17. H.B. Stuhrmann, J. Haas, K. Ibel, M.H.J. Koch and R.R. Crichton. J. Mol. Biol. 100, 399-413. (1976).
18. H.B. Stuhrmann and E. Duée. J. Appl. Cryst. 8, 538-542 (1975).
19. H.B. Osborne, C. Sardet, M. Michel-Villaz and M. Chabre. J. Mol. Biol. 123, 177-206 (1978).
20. D.S. Wise, A. Karlin and B.P. Schoenborn. Biophys. J. 28, 473-496 (1979).
21. B. Jacrot, P. Pfeiffer and J. Witz. Phil Trans Roy. Soc. London B276, 109-112 (1976).
22. B. Jacrot. Rep. Prog. Phys. 39, 911-953 (1976).
23. B. Jacrot, C. Chauvin and J. Witz. Nature 266, 417-421 (1977).
24. C. Chauvin, J. Witz and B. Jacrot. J. Mol. Biol. 124, 641-651 (1978).
25. S. Cusack, A. Miller, P.C.J. Krijgsman and J.E. Mellema. J. Mol. Biol. 145, 525-543 (1981).
26. J. Torbet. FEBS Letters 108, 61-65 (1979).
27. J. Torbet, D.M. Gray, C.W. Gray, D.M. Marvin and H. Siegrist. J. Mol. Biol. 146, 305-320 (1981).
28. D. L. Worcester. Brookhaven Symposia in Biology 27 p.III-37 (1976).
29. D.L. Worcester. In "Biological Membranes, Vol. 3" pp. 1-48. D. Chapman and D.F.H. Wallach, eds. Academic Press (1976).
30. D.L. Worcester and N.P. Franks. J. Mol. Biol. 100, 359-378 (1976).
31. L.I. Schiff. Quantum Mechanics. Third Edition. p. 324. McGraw Hill. (1968).
32. P.B. Moore, D.M. Engelman and B.P. Schoenborn. J. Mol. Biol. 91, 101-120 (1975).
33. P.B. Moore, J.A. Langer, B.P. Schoenborn and D.M. Engelman. J. Mol. Biol. 112, 199-234 (1977).
34. J.A. Langer, D.M. Engelman and P.B. Moore. J. Mol. Biol. 119, 463-485 (1978).
35. D.G. Schindler, J.A. Langer, D.M. Engelman and P.B. Moore. J. Mol. Biol. 134, 595-620 (1979).
36. G. Büldt, H.U. Gally, J. Seelig and G. Zaccai. J. Mol. Biol. 134, 673-692 (1979).
37. G. Zaccai, G. Büldt, A. Seelig and J. Seelig. J. Mol. Biol. 134, 693 (1979).

38. G. Büldt and J. Seelig. Biochemistry 19, 6170-6175 (1980).

39. G.I. King, W. Stoeckenius, H.L. Crespi and B.P. Schoenborn. J. Mol. Biol. 130, 395-405 (1979).

40. G.I. King, P.C. Mowery, W. Stoeckenius, H.L. Crespi and B.P. Schoenborn. Proc. Nat. Acad. Sci. USA 77, 4726-4730 (1980).

41. D.L. Worcester, S. Scott, K.R. Bruckdorfer, R. Skarjune and E. Oldfield. (manuscript in preparation.)

42. H.F. Tucker and H.C. Eckstein. J. Biol. Chem. 131 567 (1937).

43. S.C. Scott, K.R. Bruckdorfer and D.L. Worcester. Biochemical Society Transactions (1981).

44. W.C. Rose, M.J. Oesterling and M. Womack. J. Biol. Chem. 176, 753-762 (1948).

45. D.M. Engelman and G. Zaccai. Proc. natl. Acad. Sci. USA (1981).

RECENT NEUTRON SPIN ECHO EXPERIMENTS

F. Mezei

Institut Laue-Langevin, 156X, 38042 Grenoble Cédex, France

ABSTRACT

In the past few years Neutron Spin Echo (NSE) has been used in routine fashion for the high resolution study of both magnetic and non-magnetic inelastic scattering effects. Applications include the investigation of macromolecular motion in polymer and micelle solutions and biological matter; spin relaxation in paramagnets and spin glasses; critical fluctuations in magnetic and structural phase transitions, etc. In this paper the quasielastic scattering experiments are reviewed. It is shown that the NSE method opened up new fields in both lineshape and momentum dependent linewidth studies, partly by its very high resolution partly by its capability of providing badly needed model independent results in a number of cases.

INTRODUCTION

Neutron Spin Echo (NSE) is not only a method to provide high resolution in inelastic neutron scattering, but at the same time it represents a conceptually new approach. Quite naturally, this conceptual novelty leads to very particular features relevant to the practical utilisation of the method. Of course, both classical inelastic neutron scattering techniques and NSE give information about the same scattering function $S(\vec{q},\omega)$ of the sample, but this information is provided in different forms. Therefore, depending on the type of information sought for, the effectiveness of one or the other approach varies strongly from one concrete experiment to the other, and it is not only characterised by the sheer energy resolution capability. The basic features of NSE, which determine its experimental utilisation, are the following :

a) High resolution: 2 neV is routinely achieved in quasielastic scattering, and 1 μeV in the study of linewidth of elementary excitations.

b) No instrumental resolution broadening. Due to the Fourier transformation inherently involved in the method, the directly measured spectrum is the product of the sample scattering spectrum and the easily measurable instrumental response function, and not their convolution as for the classical methods. Therefore the data reduction does not imply the otherwise inevitable and often ambiguous integral operations like deconvolution or fitting to model functions.

c) Directly measured quantity is the real part of the time dependent correlation function, i.e. the intermediate scattering law $S(\vec{q},t)$ as a function of the time t (and averaged over the instrumental $\vec{q}$ resolution spot, of course).

d) Wide dynamic range : inelastic linewidths are studied over 4 orders of magnitude from 2 neV to 20 μeV on the same instru-

ISSN:0094-243X/82/890379-13$3.00 Copyright 1982 American Institute of Physics

ment, and, what is more important, the time variable, can span a
range as big as 1 : 1000 in a single scan without any geometrical
changes.
e) Inherent polarization analysis : unambiguous identification and
selective study of magnetic scattering effects, both with or without
a strong magnetic field on the sample.

Note that while points a) b) d) and e) represent very advantageous
features in all cases when they may be useful, point c) is a mixed
blessing. In some cases, it is what makes the experiment possible
at all, as illustrated by examples below, but it makes the use of NSE
impractical if a multitude of sharp structures separated by not more
than a few μeV is to be investigated.

For an introduction to the NSE concept and a description of the
method's basic features the reader is referred to previous papers [1,2].
A number of further contributions by different authors contained in
Ref.3 present various aspects and applications.

The NSE spectrometer used in the experiments described below
is the IN11 instrument at the Institut Laue-Langevin, which is shown
schematically in Fig.1, and has been described elsewhere [4]. Two
essential improvements have been made since. By the introduction
of "Fresnel spiral" coils [5] the homogeneity of the precession fields
could be improved substantially, which allows to reduce the instrumental
imperfections to the equivalent of less than 20 neV linewidth, or,
to increase the beam cross section at the detector. The latter
opened the way to the installation of a prototype large solid angle
polarization analyser of 85 x 95 mm cross section, consisting of a
600 mm long Soller arrangement of 24 super mirrors of dimension 100x.

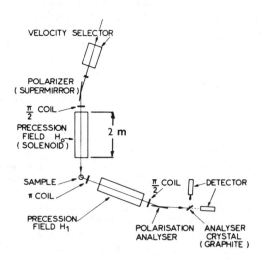

Fig. 1. Schematic lay-out of the NSE spectrometer IN11. The
graphite analyser before the detector is optional, it is only
used when good momentum resolution is required.

200 mm. At present the best luminosity of the instruments corresponds
to 3 counts/sec counting rate for 1% isotropic scattering at $\lambda \approx 5$ Å.
Thus magnetic cross sections have been identified with a sensitivity
of 2 mbarns sterad.

In what follows the two archetypes of quasielastic NSE experiments will be described through a large variety of examples. First
we will consider inelastic lineshape studies, where the application
of NSE opened up a new field, virtually inaccessible by conventional
methods. The second class of experiments consists of the study of
the momentum dependence of the quasielastic line width in which case
prime use was made of the high resolution provided by IN11 at small
scattering angles. Other types of NSE experiments, like the investigation of phonon lifetimes [6] and of magnetic crystals in high
fields up to 4 Tesla will not be discussed here.

LINESHAPE STUDIES

The two unique features b) and c) as described above, make NSE
a particularly adequate tool for the study of the functional form of
$S(\vec{q},\omega)$ concerning the ω variable. It happens that the different
forms which were actually observed in quasielastic scattering are
easier to distinguish in the $S(\vec{q},t)$ intermediate scattering functions (correlation function) which describes the response of the
system in real time. In what follows we will use the notation
$S\vec{q}(t) = S(q,t)/S(q,0)$ where the bar refers to the average over the
actual instrumental momentum resolution spot around the nominal
momentum value $\vec{q}$. Note, that in all of the examples given below the
role of the finite q resolution has been checked in detail and it
was found negligible in each case. The simplest physical situation
is described by the exponential decay, which corresponds to the familiar Lorentzian lineshape :

$$S_{\vec{q}}(t) = e^{-\Gamma t} \quad \text{i.e.} \quad S_{\vec{q}}(\omega) = \frac{1}{\pi} \frac{\Gamma}{\Gamma^2 + \omega^2}$$

In effect exponential relaxation is the one most frequently found in
nature, which just underlines the physical interest of systems
showing a different form of fluctuation spectrum.

In fig.2, sample $S_q(t)$ functions are shown which were measured
in two macromolecular solutions and which are related to the specific
diffusion mechanisms. In the semilog plot, the slope determines the
Lorentzian width Γ for an exponential line, as indicated. Non-expo-
nontial decay behaviour has been predicted theoretically for segmental diffusion of polymers in solution [9] and the measured curve shown
was the first experimental evidence [8]. An effective linewidth Γ^*
can be defined in this case as the initial slope of the plot (Fig.2).

Quasielastic lines with several components represent another
important need for precise lineshape analysis. The example shown
in Fig.3 is the $S_q(t)$ spectrum measured [10] for the critical fluctuations in paraterphenyl (PTE) at the position of the superlattice
reciprocal lattice point, 0.6 K above the critical temperature

382

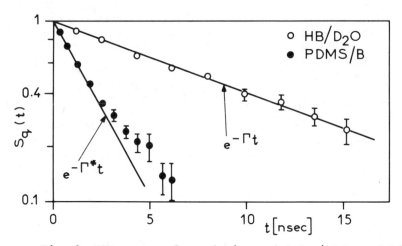

Fig. 2. NSE spectra for solutions of 0.1 g/ml hemoglobin
(HB) in heavy water [7] and 0.05 g/ml polydimethylsiloxane
(PDMS) in deuterated benzene [8], at $T = 18°$ C, $q = 0.152$ Å^{-1}
and $T = 72°$ C, $q = 0.096$ Å^{-1}, respectively. ($\Gamma = 60$ neV and
$\Gamma^* = 295$ neV)

(179.5 K) of the antiferrodistorsive phase transition. The initial
fast decay, corresponding to inelastic scattering with more than
20 μeV energy change (i.e. too big to be looked at in detail by NSE),
are due to phonons. The contributions B and C (cf. Fig.3) represent
the critical fluctuations and they show marked critical slowing down.
The very existence of two components, and in particular that of the
"central peak" C is a surprising discovery, which was not apparent
in the previous classical type neutron backscattering experiment [11].
This was just due to the inevitable ambiguity of classical deconvolution or model fitting procedures. Indeed, a later reanalysis
of the same data brought out the existence of the central peak
too [12]. Note that the spectrum in Fig.3 only implies that the
component C is much narrower than B (γ< 0.03 μeV compared with
Γ = 1.1 μeV) but does not exclude that it is strictly elastic (γ=0).
Higher temperatures spectra suggest that $\gamma\neq 0$, although γ<<Γ always
holds. No particular attention was focussed to this point in the
test experiment done until now.
Another antiferrodistorsive phase transition which is known to
show a central peak above the phase transition temperature has also
been studied [13] by NSE (actually in 1977). Fig.4 shows two $S_q(t)$
spectra for $SrTiO_3$ at the $q = 1/2$ (1,1,3) superlattice position.
It is seen that besides the phonons (initial drop) there is only a
single contribution which was found to be elastic within error for
all temperatures investigated viz the 70% confidence level upper
limits for γ were found to be 15, 20, 35, 85 and 100 neV at
T_c + 0.5, 1.6, 4.0, 8.0 and 15.0 K respectively. The sample used

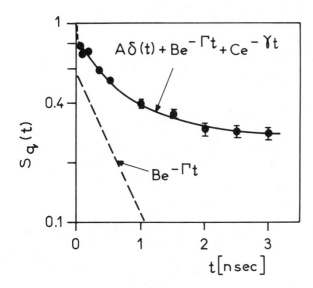

Fig. 3. NSE spectrum measured at 0.6 K above the temperature of structural phase transition in PTE [10] showing the
critical fluctuation. The fitted curve corresponds to
A = 0.15, B = 0.58, C = 0.27, Γ = 1.11 µeV and γ = 0.

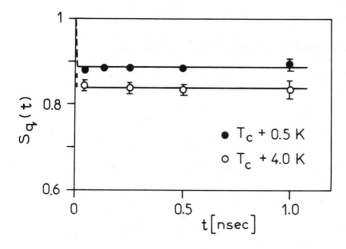

Fig. 4. NSE spectra measured above the temperature of
structural phase transition in $SrTiO_3$ (ref. 13) on the
"central peak" (T_c = 99.5 K).

was the same Sanders crystal with $T_c = 99.5$ K studied by the
Brookhaven group,[14] who have kindly lent it to us.
    The most striking example of non-exponential relaxation which
has been studied by NSE is the spin dynamics in spin glasses. The
identification of magnetic scattering effects by the use of the
paramagnetic neutron spin echo (PNSE) technique [2] and the large
dynamic range it provides were fundamental in these experiments.
The sample spectra shown in Fig.5 illustrate the very particular
"freezing" behaviour of the spin-spin correlation function which is
characterised by the very slow decay of the Sq(t) function over
several orders of magnitude in time. This can be represented by a
spectrum of relaxation times $1/\Gamma$ covering several decades from 10^{-12}
to 10^{-1} sec or beyond. In the case of the Cu(Mn) alloys the neutron
scattering data seem to correspond to the q=0 limit, and therefore
a comparison to AC susceptibility results is of great direct
interest [16]. As shown in the figure, the two sets of data merge
nicely together into a coherent pattern. The striking similarity
in the time and temperature dependence of the spin correlation
function for the variety of spin glasses investigated suggests the
universality of the observed behaviour. The present NSE results
offer an explanation to all of the other experiments on spin glass
dynamics, (ESR, μSR, classical neutron scattering which themselves did
not provide quite sufficient information for a reconstruction of
the full dynamic picture). In particular the data in Fig.5. imply
the existence of the characteristic cusp in the temperature dependence of the susceptibility upto frequencies as high as 1 GHz. The
cusp temperature is frequency dependent and there it is not accompanied by any sudden change of the correlation function S(q,t).
None of the existing theories can predict S(q,t) in as much detail
as it is known from this experiment, but different theories
predict one or another qualitative feature.

MOMENTUM DEPENDENT LINEWIDTHS STUDIES

    Quasielastic neutron scattering from objects performing simple
random walk type diffusive motion is described by the well known
scattering law :

$$S(q,\omega) = \frac{1}{\pi} S(q) \frac{\Gamma^2}{\Gamma^2 + \omega^2}$$

or

$$S_q(t) = S(q,t)/S(q,o) = e^{-\Gamma t}$$

where $\Gamma = Dq^2$, with D being the diffusion constant. The same simple
behaviour applies for spin systems with a weak, spin conserving
(e.g. exchange) interaction. An example of this simple behaviour
has been found for hemoglobin solutions in D_2O at various concentrations [7] (Fig.6). It is seen that the strong interactions between
solute molecules reduce substantially the diffusion constant for
high concentrations, but, surprisingly enough, they do not affect
the simple diffusion dynamics : the $\Gamma(q)$ function remains a straight

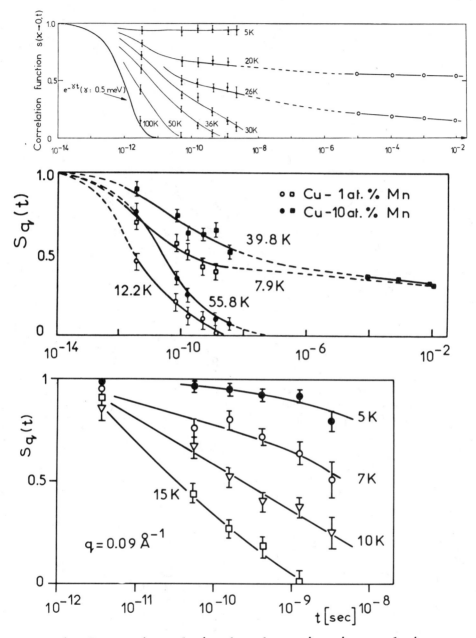

Fig. 5. Experimental time dependent spin-spin correlation functions for various spin glass alloys: Cu - 5% Mn (top, ref. 15), Cu - 1% Mn and Cu - 10% Mn (middle, ref. 17) and La$_{.7}$Er$_{.3}$Al$_2$ (bottom, ref. 18). The data points for times shorter than 10^{-8} sec were directly measured by NSE, the others were calculated [16] from AC susceptibility data. The lines are guides to the eye.

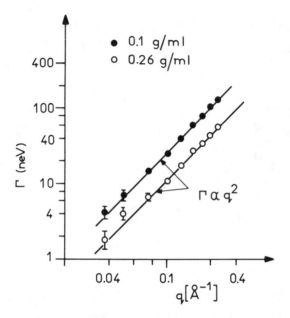

Fig. 6. Momentum dependence of the quasielastic scattering
linewidth in hemoglobin/heavy water solutions at different
concentrations [7] (T = 18° C).

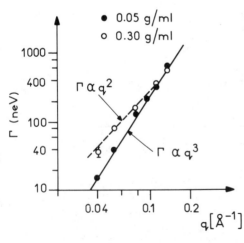

Fig. 7. Momentum dependence
of the quasielastic scattering linewidth in PDMS/deu-
terated benzene solutions of
different concentratins [8]
(T = 72° C).

line of slope 2 in the log-log plot.

The most interesting applications of quasielastic neutron scattering in Γ vs. q studies, however, are just those concerned with deviations from the simple Dq^2 law. NSE proved to be an ideal tool for this type of experiments on polymer [8,19,20] and macroionic [21] solutions, because it provides uniquely high energy resolution and it can be easily used for small scattering angles. In effect since 1978 this field became a very active area in inelastic neutron scattering.

Deviations from the $\Gamma = Dq^2$ law can be due to various mechanisms. For example, scaling theories for the motion of segments of polymer chains in solutions have predicted non-lorentzian quasielastic lineshape, (cf. Fig.2) and the effective linewidth Γ is expected to be proportional to q^3 beyond a critical momentum value q^*, which latter corresponds for entangled system to the inverse screening length and increases with solvent concentration [9]. The NSE results [8] in Fig.3 show clear evidence for this behaviour in PDMS solutions in deuterated benzene at 72 $^\circ$C. The non-trivial behaviour of Γ vs. q is ultimately related to a strong interaction mechanism between solute molecules, which interaction in the case of segmental motion in polymers is just due to the very existence of long chains.

In macroionic solutions the Coulomb forces are responsible for the strong solute-solute interaction leading to another anomalous shape of the $\Gamma(q)$ function, as illustrated by the results [21] shown in Fig.8. The relative depression of Γ with respect to the infinite dilution limit D_oq^2 appears around q values corresponding to the maximum in the interparticle correlation function, which is 0.11 $\overset{\circ}{A}{}^{-1}$ for the case shown. Such q values are not accessible for the standard light scattering techniques for diffusion studies, and NSE provides for the first time the necessary high energy resolution in neutron scattering.

The last two examples to be presented are related to spin dynamics in paramagnetic phase. The use of the IN11 spectrometer was indispensable in these experiments because of the need for high energy resolution, small momentum transfer and/or unambiguous identification of the magnetic scattering effects above a high nuclear scattering background.

The first of these examples is the determination of the anisotropy of spin diffusion in quasi-one-dimensional magnetic crystal TMMC at room temperature. At this temperature the static anisotropy is negligible and the spin fluctuations are fast, as it was found by NMR. Spin conserving exchange interactions give rise to random jump type diffusion of the spins from one site to the other, and for tetragonal crystal symmetry this leads to the following $\Gamma(\vec{q})$ function :

$$\Gamma = D_{||} q_{||}^2 + D_\perp q_\perp^2 + \Gamma_o$$

where $||$ and $\perp$ refers to the c-axis, and Γ_o represents the spin-lattice (spin non-conserving) relaxation. Indeed, as shown in Fig.9, a strong anisotropy of Γ has been found in the NSE experiment [22], which not only provided strong evidence for the predomi-

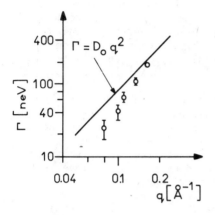

Fig. 8. Momentum dependence
of the quasielastic scattering
linewidth in 0.8 M sodium
dodecyl sulphate / D_2O solution [21] (T = 313 K).

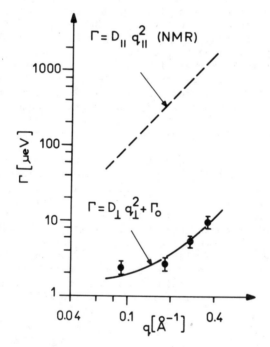

Fig. 9. Anisotropic momentum dependence of the paramegnetic
quasielastic scattering linewidth in $(CH_3)_4NMnCl_3$ (TMMC)
single crystal at room temparature [22].

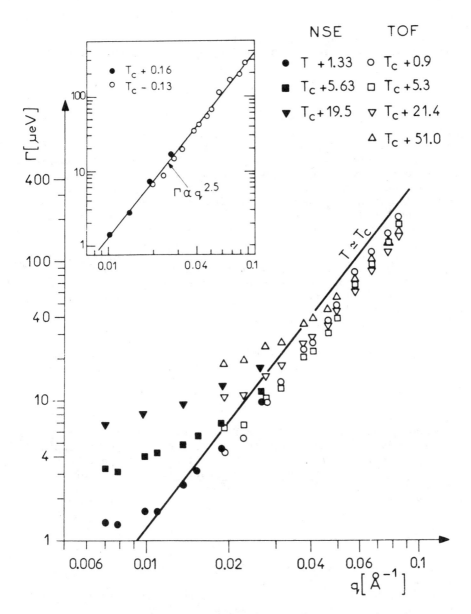

Fig. 10. NSE (full symbols) and high resolution (TOF,open symbols) results on the momentum dependence of the quasi-elastic scattering linewidth of crytical fluctuations in the paramagnetic phase of iron at various temperatures above the Curies point. ($T_c \sim 1040$ K). The insert shows the results obtained at $T \sim T_c$. The straight lines in both the main figure and the insert correspond to $\Gamma = Aq^{5/2}$ with A = 128 meV $\mathring{A}^{5/2}$

nance of the exchange interaction, but it was the only way of determining $D_\perp$ and Γ_0.

Finally, Fig.10 shows the $\Gamma(q)$ curves for paramagnetic critical fluctuations obtained by NSE and by the high resolution time-of-flight spectrometer IN5 at various temperatures above the ferromagnetic Curie point in iron [23]. This problem has been extensively studied by neutron scattering [24]. Guided by dynamical scaling theories, the data evaluation in previous experiments was made under the assumption that the $\Gamma = Dq^2$ law holds in the hydrodynamic regime (i.e. for $q \ll q_c(T)$, where the inverse coherence length q_c is known from static measurements), and D was determined as a function of the temperature. By the extraordinary progress of time-of-flight spectroscopy (IN5) and the introduction of NSE, it became possible to obtain the model independent data in Fig.10. They show strinkingly that the above assumption was wrong ($q_c = 0.08$, 0.025, 0.068 and 0.13 Å^{-1} for $T = T_c + 1$, 5, 20 and 50 K, respectively), and dynamical scaling does not apply. Instead, the $\Gamma(q)$ function seems to converge to finite, temperature dependent, values as $q \to 0$. This would imply that spin non-conserving interactions (e.g. spin-orbit coupling) play a fundamental role in the phase transition. On the other hand, the often considered dipolar interactions [25] seem to have no serious effect, since they should be of increasing importance on approaching T_c while the experimental data converge to the predicted $\Gamma \propto q^{5/2}$ scaling behaviour at $T = T_c$. In sum it is clear that the Curie point in iron does not follow dynamical scaling laws, but it remains to be seen if the apparent violation of dynamical scaling can be resolved by invoking a competing interaction.

CONCLUSION

We have seen that by its unusual features NSE proved to complement powerfully classical neutron scattering and other microscopic methods like NMR, μSR, inelastic light scattering. In particular, it allowed to obtain model independent results in a number of cases, which, I think, is the ultimate goal of experimental work. It is little surprise that these results sometimes shed fundamentally new light on old problems.

REFERENCES

1. F. Mezei, Z. Physik 255, 146 (1972).
2. F. Mezei, in Ref. 3, pp 3-26.
3. Neutron Spin Echo, edited by F. Mezei, Lecture Notes in Physics Series, Vol. 128 (Springer Verlag, Berlin 1980).
4. P.A. Dagleish, J.B. Hayter and F. Mezei, in Ref.3, pp. 66-71.
5. F. Mezei, idem, pp. 183-187.
6. F. Mezei, idem, pp. 113-121, Phys. Rev. Letters, 44, 1601 (1980).
7. Y. Alpert and F. Mezei, to be published.
8. D. Richter, J.B. Hayter, F. Mezei and B. Ewen, Phys. Rev. Letters 41, 1484 (1978).

9. P.G. de Gennes, Macromolecules, 9, 587 and 594 (1976).
10. H. Cailleau, A. Heidemann, F. Mezei and C.M.E. Zeyen, to be published.
11. H. Cailleau, A. Heidemann and C.M.E. Zeyen, J. Phys. C 12, L411 (1979).
12. A. Heidemann, W.S. Howells and G. Jenkin, in Ref.3, pp. 132-135.
13. F. Mezei and J.B. Hayter, unpublished.
14. S.M. Shapiro, J.D. Axe, G. Shirane and T. Riste, Phys. Rev. B6, 4332 (1972).
15. F. Mezei and A.P. Murani, J. of Magn. and Magn. Mat. 14, 211 (1979).
16. F. Mezei, Proc. of 1980 Annual Meeting of Cond. Matter Division of European Physical Society (Plenum Press, N.Y.) in press.
17. A.P. Murani, F. Mezei and J.L. Tholence, Proc. Int. Conf. on Low Temp. Phys. LT16 (Los Angeles, 1981) in press.
18. F. Mezei and A.P. Murani, to be published.
19. D. Richter, B. Ewen and J.B. Hayter, Phys. Rev. Letters, 45, 2121 (1980); D. Richter, A. Baumgärtner, K. Binder, B. Ewen and J.B. Hayter, idem 47, 109 (1981).
20. J.S. Higgins, L.K. Nicholson and J.B. Hayter, Polymer, 22, 163 (1981) and Macromolecules, 14, 836 (1981).
21. J.B. Hayter and J. Penfold, in Ref. 3, pp. 80-86 and J.C.S. Faraday I, in press.
22. K. Holczer, F. Mezei and J.P. Boucher, to be published.
23. F. Mezei, to be published.
24. See e.g. the review article by G. Parette, Ann. Phys. 7, 299 and 313 (1972), and references therein.
25. J. Villain, J. Phys. (Paris) Colloque 32, C1 (1971); S.V. Maleev, Zh. Eksp. Teor. Fiz. 66, 1809 (1974) and 69, 1398 (1975) [Sov. Phys - JETP 39, 889 (1974) and 42, 713 (1976)].

The Decomposition of a Ni-12.5 at.% Si Alloy

J. E. Epperson†, D. F. R. Mildner††, and H. Chen†††

† Materials Science Division, Argonne National Laboratory, Argonne, IL 60439
†† Research Reactor Facility, University of Missouri, Columbia, MO 65211
†††Department of Metallurgy and Mining Engineering, University of Illinois at Urbana-Champaign, Urbana, IL 61801

ABSTRACT

A survey, using electrical resistivity measurements, shows that for a Ni-12.5at.% Si alloy the decomposition behavior is dramatically different depending on whether the alloy decomposition takes place above or below about 440°C, and these differences are also manifest in neutron small angle scattering experiments. At 551°C, a broad diffuse maximum in the small angle scattering profile has been observed, and this is attributed to a non-random spatial distribution of Ni_3Si precipitate particles. Below 440°C, the decomposition occurs by a continuous process.

Introduction

Many of the desirable high temperature mechanical properties of a class of Ni-base alloys are dependent on the presence of ordered q' precipitate particles. These precipitates are produced by allowing a supersaturated solid solution to decompose partially. Conditions under which alloy decomposition occurs can greatly influence the size, shape, number density, volume fraction and distribution of second phase particles. Hence, there is considerable practical interest in understanding alloy decomposition processes in detail in order that this knowledge can be utilized to modify beneficially the decomposition microstructure. The purpose of the present contribution is to report results from an investigation on the decomposition of a Ni-12.5 at.% Si alloy being carried out to determine, first of all, the mode of phase separation operating in the different temperature ranges.

Observations and Discussion

An electrical resistivity survey investigation has been made of the decomposition behavior of a Ni-12.5 at.% Si alloy following rapid quenching from 1050°C, and the results are summarized in Fig. 1. After 15 second isochronal annealing of the quenched alloy in the range from 300° to 650°C, one finds a rather well defined peak in the electrical resistivity change (relative to the "as-quenched" value) at about 440°C as shown in Fig. 1a. That this peak at 440°C corresponds to a change in the mode of decomposition is suggested by the isothermal recovery at

ISSN:0094-243X/82/890392-03$3.00 Copyright 1982 American Institute of Physics

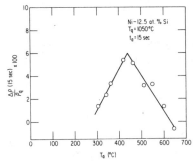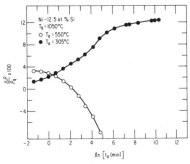

Fig. 1.   Electrical resistance survey of the decomposition
          behavior of a Ni-12.5 at.% Si alloy: a (left), 15 second
          isochronal annealing and b (right) isothermal annealing
          at 305°C and 550°C.

305° and at 550°C shown in Fig. 1b. At 550°C, a monotonic
decrease in resistivity is superimposed on a rapid rise of 3.28
percent, the latter occurring within 15 seconds. In contrast,
however, after 522 hrs at 305°C the resistance continues to
increase and is some 12.4 percent above the quenched value.

If one cycles the annealing temperaure for short times (15
sec) within the high temperature branch of Fig. 1a, a strong
tendency toward reversibility of the electrical resistivity is
observed. Some hysteresis is noted which is thought to be
associated with the interfacial energy of small Ni_3Si particles.
If, however, the cycling experiment is performed within the low
temperature branch, no suggestion of reversibility is observed;
rather, the resistance increases irregardless of whether the
annealing temperature is raised or lowered. As a consequence of
these cycling experiments, it is concluded that 440°C indeed
represents an instability temperature for this alloy.

Neutron small angle scattering results from polycrystalline
bulk samples are also dramatically different for the high and low
temperature branches of Fig. 1a. After 22 hours at 551°C, an
intense small angle scattering characterized by a well defined
maximum is produced as shown in Fig. 2a. This corresponds to
particles of Ni_3Si with an average Guinier radius of 67 Å and
a mean interparticle spacing of 251 Å. For a sample annealed 70
hrs at 305°C, the small angle scattering is weak, but
significantly above background and above the scattering for the as
quenched sample. It is worth noting that this anneal did not
produce a maximum in the q-range available at the University of
Missouri instrument as shown in Fig. 2b. These data correspond to
a mean Guinier radius of approximately 13 Å.

Conclusions

Based on thermodynamic considerations (e.g., de Fontaine[1]),
it is to be expected that if a supersaturated alloy is held at a

394

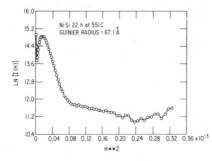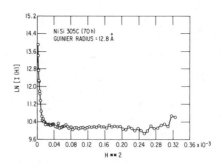

Fig. 2.   Neutron small angle scattering from polycrystalline
bulk Ni-12.5 at.% Si (a) (left) 22 hrs at 551°C and
(b) (right) 70 hrs. at 305°C.

temperature only slightly below the incoherent solubility limit,
decomposition should occur by more or less conventional nucleation
and growth of second phase particles. Within this temperature
interval the alloy is metastable, and a nucleus of critical size
must exist before the system can lower its free energy by
separating into two phases. However, if the annealing takes place
at sufficiently low temperature, a different mode of decomposition
is to be anticipated. It is predicted (Cahn[2]) that continuous
transformation takes place within the region of the phase diagram
bounded by the locus of points for which

$$f'' + 2\eta^2 Y = 0, \qquad (1)$$

where f'' is the second derivative with respect to composition of
the Helmholtz free energy and $\eta^2 Y$ is a coherent strain energy
term. Within this latter region of the phase diagram, it is not
necessary to produce nuclei of a critical size, hence the designa-
tion continuous transformation; one merely requires thermal energy
to drive the diffusion process.

    The experimental observations are consistent with the concept
outlined above. That is, above 440°C Ni-12.5 at.% Si decomposes
by nucleation and growth of discrete particles of Ni_3Si. Elastic
interactions between the second phase particles cause them to
assume a non-random spatial distribution (Khachaturyan[3]). Below
440°C, decomposition takes place by a continuous process.

*Work supported by the U.S. Department of Energy

References

1. D. de Fontaine, Treatise on Solid State Chemistry, vol. 5,
   edited by N. B. Hannay, Plenum Press, N.Y. pp. 129-178 (1975).
2. J. W. Cahn, Trans. AIME 242, pp. 166-180 (1968).
3. A. G. Khachaturyan, private communication, July 1979.

THE NATIONAL BUREAU OF STANDARDS
SMALL-ANGLE NEUTRON SCATTERING SPECTROMETER

C. J. Glinka
National Bureau of Standards, Washington, DC 20234

ABSTRACT

A new facility for small-angle neutron scattering is near
completion at the NBS Research Reactor. The instrument utilizes a
65 x 65 cm^2 position-sensitive detector, variable incident wavelength, and a novel converging beam collimation system. The instrument and its capabilities are discussed along with measurements
indicative of its performance.

INTRODUCTION

The application of small-angle neutron scattering (SANS)
techniques to the study of structural and magnetic inhomogeneities
in the 10 to 10,000 Å range has increased dramatically in the last
ten years. This growth has been due in part to the development of
specialized instrumentation for SANS, in particular, the development
of large two-dimensional position-sensitive detectors. Interest at
NBS in SANS for research in polymers, metallurgy, biology, critical
scattering and for nondestructive testing has led to the construction of a dedicated instrument for SANS at the NBS Reactor which
utilizes a large area detector. This facility is now nearly completed with full operation expected in the fall of 1981.

DESCRIPTION OF THE INSTRUMENT

A drawing of the layout of the new SANS instrument at NBS is
shown in Fig. 1. The neutron source is a direct thermal beam from
the reactor which passes through a liquid-nitrogen cooled filter,
consisting of 25 cm of beryllium and 15 cm of single-crystal bismuth, to remove core gamma rays and fast neutrons from the beam.
The beam is then further refined by a rotating, helical-channel
velocity selector. By tuning the frequency of the velocity selector, the mean wavelength of the beam can be varied from 4 to 10 Å
with a constant $\Delta\lambda/\lambda$ (FWHM) of about 0.25. Following the velocity
selector, the beam enters a 4.5 m long evacuated flight tube to the
sample position. The beam is collimated either by irises at both
ends of the flight tube or by a novel high-resolution collimation
system, [1] shown schematically in Fig. 2, for use with larger samples
($\sim$1.5 x 1.5 cm^2). A series of twelve masks, with nine apertures
each, are aligned along the flight tube so as to produce nine
independent beams which converge to a point at the center of the
detector. In this way, the intensity on the sample (n/sec) is
increased ninefold over what would be obtained with a single pair of
irises with the same angular divergence.

ISSN:0094-243X/82/890395-03$3.00   1982 American Institute of Physics

NBS SANS SPECTROMETER

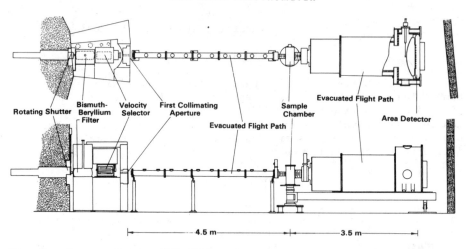

Fig. 1. Layout of the NBS SANS spectrometer.

Focussing Collimation System

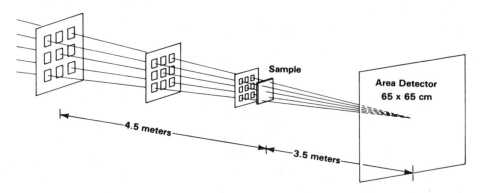

Fig. 2. Schematic of the converging beam collimation used in the
NBS SANS facility.

The evacuated post-sample flight path consists of three
shielded cylindrical sections which are mounted on a frame that can
be rotated about the sample position to reach larger scattering
angles. The section which houses the detector rolls along rails in
the frame to allow the sample-to-detector distance to be changed
from 3.5 m to 2.0 m. Built into the detector shield is an exter-
nally adjustable positioning device for locating a beam stop in
front of the detector.

Two area detectors of the Borkowski-Kopp type [2] have been purchased [3] for the instrument. The primary detector has an active area of 65 x 65 cm^2 with 5 mm spatial resolution. The smaller backup detector has a 25 x 25 cm^2 area and 4 mm resolution. A dedicated minicomputer processes the signals from the detector and provides real-time imaging of the data. Completed data sets are then transferred to a larger computer for reduction and analysis as well as display on an interactive color graphics terminal.

PERFORMANCE

Absolute measurements of the flux at the entrance to the SANS flight path indicate that the flux at the sample position, for an incident wavelength of 6 Å and reactor power of 10 MW, is 1 x 10^4 n/cm^2-sec under the highest resolution collimation and increases to 2 x 10^5 n/cm^2-sec as the collimation is relaxed. These numbers will double when the reactor power is increased to 20 MW in 1982. The converging beam collimation system has been tested and works as expected. The 25 x 25 cm^2 detector has also performed as expected; in particular, its pixel versus position relationship is linear over the entire active area. The larger detector has been received and will be installed in August 1981. When this detector is calibrated, the instrument will become fully operational. The measurement capabilities of the instrument are summarized in the following table.

Table I Characteristics of the NBS SANS facility

Wavelength:	variable from 4 to 10 Å, $(\Delta\lambda/\lambda) = 0.25$
Collimation:	single pair of irises or nine channel converging beam collimation
Minimum Q:	0.003 Å^{-1} at λ = 6 Å, 0.002 Å^{-1} at 9 Å
Q range:	0.002 Å^{-1} to 0.5 Å^{-1}
Sample size:	0.4 to 2.5 cm diameter
Flux at sample:	10^4 to 4 x 10^5 n/cm^2-sec depending on slit sizes and wavelength (at 10 MW)
Detector:	65 x 65 cm^2 position-sensitive counter with 5 x 5 mm^2 resolution

REFERENCES

1. A. C. Nunes, J. Appl. Cryst. 11, 460 (1978).
2. C. J. Borkowski and M. K. Kopp, Rev. Sci. Instrum. 46, 951 (1975).
3. Area detectors are the products of the Technology for Energy Corp., Knoxville, TN.

AIP Conference Proceedings

		L.C. Number	ISBN
No.1	Feedback and Dynamic Control of Plasmas	70-141596	0-88318-100-2
No.2	Particles and Fields - 1971 (Rochester)	71-184662	0-88318-101-0
No.3	Thermal Expansion - 1971 (Corning)	72-76970	0-88318-102-9
No.4	Superconductivity in d-and f-Band Metals (Rochester, 1971)	74-18879	0-88318-103-7
No.5	Magnetism and Magnetic Materials - 1971 (2 parts) (Chicago)	59-2468	0-88318-104-5
No.6	Particle Physics (Irvine, 1971)	72-81239	0-88318-105-3
No.7	Exploring the History of Nuclear Physics	72-81883	0-88318-106-1
No.8	Experimental Meson Spectroscopy - 1972	72-88226	0-88318-107-X
No.9	Cyclotrons - 1972 (Vancouver)	72-92798	0-88318-108-8
No.10	Magnetism and Magnetic Materials - 1972	72-623469	0-88318-109-6
No.11	Transport Phenomena - 1973 (Brown University Conference)	73-80682	0-88318-110-X
No.12	Experiments on High Energy Particle Collisions - 1973 (Vanderbilt Conference)	73-81705	0-88318-111-8
No.13	π-π Scattering - 1973 (Tallahassee Conference)	73-81704	0-88318-112-6
No.14	Particles and Fields - 1973 (APS/DPF Berkeley)	73-91923	0-88318-113-4
No.15	High Energy Collisions - 1973 (Stony Brook)	73-92324	0-88318-114-2
No.16	Causality and Physical Theories (Wayne State University, 1973)	73-93420	0-88318-115-0
No.17	Thermal Expansion - 1973 (lake of the Ozarks)	73-94415	0-88318-116-9
No.18	Magnetism and Magnetic Materials - 1973 (2 parts) (Boston)	59-2468	0-88318-117-7
No.19	Physics and the Energy Problem - 1974 (APS Chicago)	73-94416	0-88318-118-5
No.20	Tetrahedrally Bonded Amorphous Semiconductors (Yorktown Heights, 1974)	74-80145	0-88318-119-3
No.21	Experimental Meson Spectroscopy - 1974 (Boston)	74-82628	0-88318-120-7
No.22	Neutrinos - 1974 (Philadelphia)	74-82413	0-88318-121-5
No.23	Particles and Fields - 1974 (APS/DPF Williamsburg)	74-27575	0-88318-122-3
No.24	Magnetism and Magnetic Materials - 1974 (20th Annual Conference, San Francisco)	75-2647	0-88318-123-1
No.25	Efficient Use of Energy (The APS Studies on the Technical Aspects of the More Efficient Use of Energy)	75-18227	0-88318-124-X

AIP Conference Proceedings